T0391270

Fundamentals of Electronic Materials and Devices

A Gentle Introduction to the Quantum-Classical World

Other World Scientific Titles by the Author

Nanoelectronics: A Molecular View
ISBN: 978-981-3144-49-1
ISBN: 978-981-3146-22-8 (pbk)

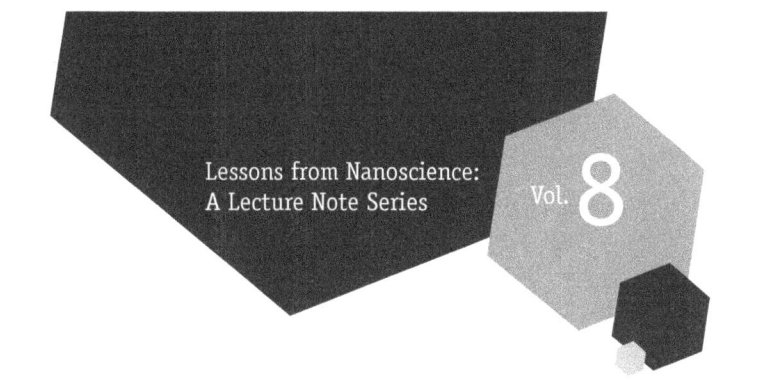

Lessons from Nanoscience:
A Lecture Note Series

Vol. 8

Fundamentals of Electronic Materials and Devices

A Gentle Introduction to the Quantum-Classical World

Avik Ghosh

University of Virginia, USA

World Scientific

NEW JERSEY · LONDON · SINGAPORE · BEIJING · SHANGHAI · HONG KONG · TAIPEI · CHENNAI · TOKYO

Published by

World Scientific Publishing Co. Pte. Ltd.

5 Toh Tuck Link, Singapore 596224

USA office: 27 Warren Street, Suite 401-402, Hackensack, NJ 07601

UK office: 57 Shelton Street, Covent Garden, London WC2H 9HE

Library of Congress Control Number: 2022951010

British Library Cataloguing-in-Publication Data
A catalogue record for this book is available from the British Library.

Lessons from Nanoscience: a Lecture Notes Series — Vol. 8
FUNDAMENTALS OF ELECTRONIC MATERIALS AND DEVICES
A Gentle Introduction to the Quantum-Classical World

ISBN 978-981-126-595-2 (hardcover)
ISBN 978-981-126-657-7 (paperback)
ISBN 978-981-126-596-9 (ebook for institutions)
ISBN 978-981-126-597-6 (ebook for individuals)

For any available supplementary material, please visit
https://www.worldscientific.com/worldscibooks/10.1142/13131#t=suppl

Desk Editor: Joseph Sebastian Ang

Typeset by Stallion Press
Email: enquiries@stallionpress.com

To my mentors — Camille Bouche, John Wilkins, Supriyo Datta and Mark Lundstrom

Preface

In the autonomous community of Castile-León, Spain, just a stone's throw-away from Disney's inspiration for the Sleeping Beauty castle, stands the 17-km aqueduct of Segovia, silhouetted against sparkling snow-covered mountains. It is best savored from across the cobbled streets with a glass of fruity Duero wine and a slice of Ponche Segoviano — a fluffy delicate sponge cake laced with cream and marzipan. This architectural marvel consisted of 20000 granite blocks held together without a single clamp, cement or mortar. And yet the Romans built this 1500 years before Newton came along, armed with just a rudimentary working knowledge of simple machines, architecture and commonly available materials of the day. In contrast, today's high-end bridge-building technology is an enormous enterprise, involving sophisticated CAD design, 3D printing, acoustic imaging, AI, LIDAR sensing, GPS tracking, advanced materials like fiber-reinforced composites, self-healing concretes and thermoplastics. When we push technology to its absolute limits, a simple working knowledge or a convenient rule-of-thumb often proves inadequate. Instead, an in-depth understanding of core physical principles, both macroscopic and microscopic, top-down vs bottom-up, becomes essential.

We find ourselves today at a similar crossroads in semiconductor device technology, where a simple working knowledge of solid state electronics may no longer suffice. The technology has arguably been pushed to its limits, with a few billion transistors on a chip switching at a few billion times per second. Faced with the prohibitive energy consumed in computing and the corresponding slowdown of device scaling and the global decentralization of chip manufacturing, the semiconductor industry has been looking beyond silicon CMOS devices, Von Neumann architecture and Boolean logic by exploring alternate platforms — 2-D nano-materials, spintronics, analog processing and quantum engineering. And software driven hardware (accelerators and ASICs), instead of the other way around. The $53 billion CHIPS for America Act was just passed in 2022 to address the hardware challenges facing the US semiconductor industry. Accordingly, as textbook materials rapidly grow obsolete, we need a back-to-basics exposure to the old and the new, crossing multiple departmental boundaries while maintaining an overall simplicity — no doubt a daunting task.

I have attempted to accomplish that task in this book, combining top-down classical device physics with bottom-up quantum transport in a single venue to provide the basis for such a scientific exploration. Having finished a companion reference book ("Nanoelectronics — a Molecular View" or as I refer to in this book, NEMV, World Scientific 2016), where I lay out the mathematical foundations in detail, I decided to focus here on a textbook that I consider to be essential, easy reading for beginning undergraduate and practicing graduate students, for physicists unfamiliar with device engineering and engineers unversed in quantum physics. With just a modest pre-requisite of freshman maths, the book works quickly through key concepts in quantum physics, combining Matlab exercises and original homeworks, to cover a wide range of topics from chemical bonding to Hofstader butterflies, domain walls to Skyrmions, solar cells to photodiodes, FinFETs to Majorana fermions. This required reframing cherished physical principles in a different light — old wine in a new bottle. For instance, an exercise on a quantum tunneling transistor brings home a new concept for the engineer, and a new context for the physicist.

The contents of the book have been employed as core material for two university courses I teach back-to-back — a graduate course in solid state devices and a course in nano-electronics for senior undergraduates and graduate students. These courses have been taken for credit by both practicing academics and part-time industry interns and have earned strong reviews and the occasional accolade. Together with the homeworks, they can easily be adopted for comparable curricula across multiple learning institutions. The book and courses are unique for combining classical and quantum device physics, starting at opposite ends — Newton vs Schrödinger, towards the same technological goal, namely emerging hardware for the future of computing. The success of these courses over the years has earned this author University of Virginia's highest teaching award.

Keyed online lectures from these courses, a solution manual available to instructors, and a deeper dive through a companion reference book (NEMV, mentioned above) bring this knowledge to a contemporary, research-ready level that touches upon the current state-of-the-art. In their online form and with supplementary homework problems, I aim to make this ultimately an evolving document, accommodating additional insights and newer topics (quantum entanglement, topology, analog computing) as our knowledge in these domains progressively solidify.

I have many people to thank for this labor of love — a journey adopted primarily because I felt the need to document my understanding somewhere and also provide a convenient starting point for newbies entering the field. Mi familia — a constant fixture throughout my career — providing much needed inspiration. My mentors, who I tried to capture in a light hearted cartoon, provided unadulterated insights into the field right from its frontlines. My students and post-docs, whose patient research over days and nights has helped distil my understanding to the point where I could boil their work down into compact and illustrative homework problems. If indeed in failure lies success, I have many failed research proposals to thank for, as they provided critical material for these homeworks. I have tried my best to acknowledge all sources I refer to between the end of this book and NEMV. I end by thanking my collaborators, colleagues, funding agencies and program managers, conference organizers and committee members, journal editorial boards, and finally the book editors and publishers — Sue Fan Law, Joseph Sebastian Ang and Zvi Ruder, for being ultra-patient with delays, understanding of my various other commitments and being great support throughout the exercise.

Avik Ghosh
University of Virginia

Contents

Part 1

Equilibrium Physics and Chemistry

Chapter 1

Fundamentals of Information Processing

1.1 Digital Computing: The Inverter

It is hard to overstate the social and technological impact of digital logic and the Internet of Things — from cell phones to smart cities, laptops to wearables, almost every part of our lives is dominated by tiny objects that switch between two binary states, on and off. While we seem to prefer decimal systems on paper because of our evolutionary inheritance of ten fingers, in solid state electronics where speed and power are dominant issues, it is much more convenient to use binary logic with two states that are well separated and easy to distinguish — for instance the absence or presence of electronic charge, or the up and down orientations (poles) of a magnet.

One of the simplest logic gates is the inverter (NOT) gate, where a large input ('1') gives a small output ('0') or vice versa. Figure 1.1 shows a way to accomplish this with complementary valves controlling water flow. In digital logic, we build an inverter instead by placing two complementary gates between a power supply and a ground (Fig. 1.2). The 'n-switch' or negative-Metal Oxide Semiconductor (nMOS) turns on if we apply a positive gate voltage to it and attracts negatively charged electrons, while the positive-Metal Oxide Semiconductor (pMOS) turns on by applying negative gate voltage to repel electrons. We see later why each element works the way it does. But if we place such an nMOS–pMOS pair to create a complementary MOS (CMOS) element, split an input gate voltage across both and draw an output signal from between them, what do we expect? When the input voltage is positive, it will turn the nMOS on (low resistance, say $k\Omega$) and the pMOS off (high resistance, say $M\Omega$), shorting out the lower branch whereupon the output will register the ground voltage. When the input is negative, we get the reverse. The upper branch shorts out while the lower

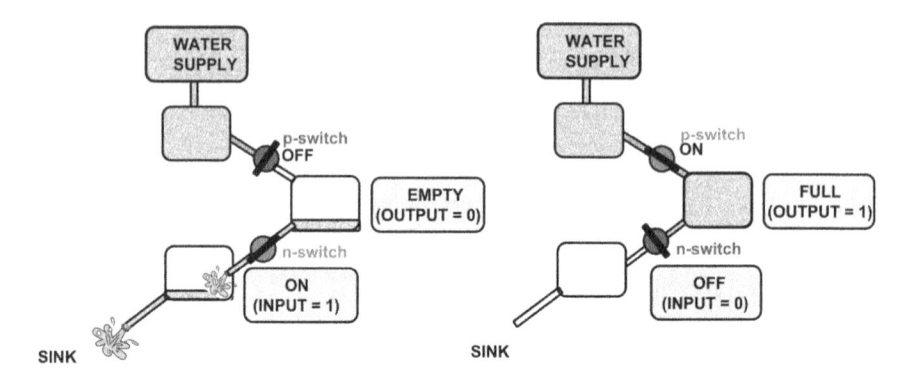

Fig. 1.1 An inverter made of two complementary valves connected to a water supply and a sink. When the lower 'n-switch' valve is on (left), the upper 'p-switch' is off and the output bucket is empty. Turning off the 'n-switch' valve (right) turns on the 'p-switch' and fills the output bucket. The output bucket in turn can activate the lower valve of a downstream inverter by engaging an intermediate water wheel, not shown.

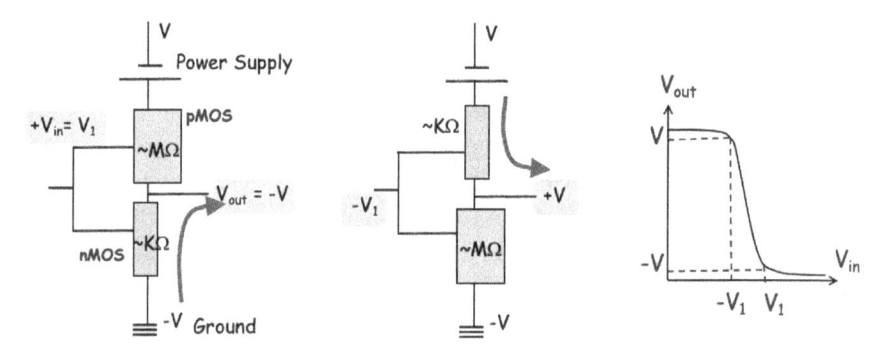

Fig. 1.2 Resistances for CMOS with opposite inputs, and corresponding input–output transfer curve. The slope $\partial V_{\text{out}}/\partial V_{\text{in}}$ is the gain.

branch becomes an open circuit, and the output registers the high power supply voltage. In other words, we have an inverter. By stringing together a set of inverters we can then build higher logic gates such as a NAND or NOR (Fig. 1.3), and thereon to universal Boolean logic.

What has made the CMOS inverter particularly attractive is that at steady-state one of the two CMOS elements is always off, so there is no current between the power supply and the ground. The only current we have is either leakage current (i.e., the transistors are not fully off), or transient current when we are in the process of shorting out one branch or the other. We can estimate the power spent in running such a CMOS inverter. The static leakage power is simply given by Joule heating as $I_{\text{off}}V_P$, the

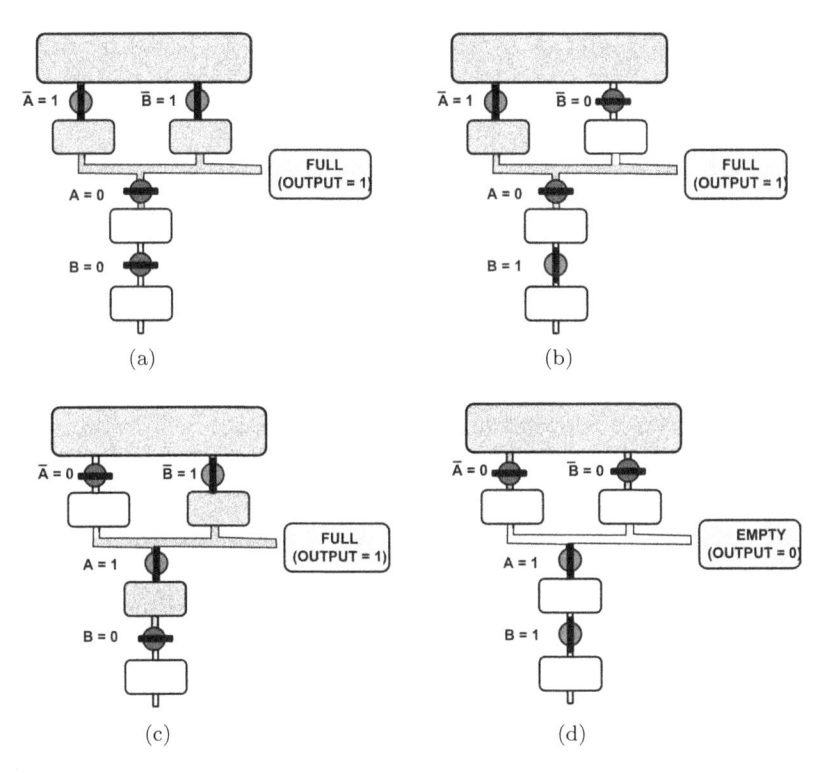

Fig. 1.3 A water-based NOT-AND (or NAND) gate is realized by using two valves A and B, and their complements $\bar{A}$ and $\bar{B}$, so that we get a zero output (NOT) if and only if both valves A *AND* B are on.

off current times the power supply voltage. Ideally, this should be zero, but present day transistors being ultrasmall, they tend to leak. The transient or dynamic power is obtained by estimating the energy wasted while charging up one of the transistor capacitors to completion. The battery delivers charge $Q = CV_P$ at constant voltage and thus delivers an energy $E_B = QV_P = CV_P^2$, but the capacitor voltage ramps up with charge from zero to max so it sees on average only half the voltage V_P, thus storing an energy $E_C = QV_P/2 = CV_P^2/2$ when fully charged $\boxed{\text{P1.1}}$. The rest, $E_B - E_C = CV_P^2/2$ is lost irreversibly as heat in the connecting wires during the charging cycle. Considering both charging and discharging in a cycle, and a clock frequency f for the number of cycles per second, we get

$$P_{\text{CMOS}} = \underbrace{\alpha CV_P^2 f}_{\text{Dynamic}} + \underbrace{I_{\text{off}}V_P}_{\text{Static}} \tag{1.1}$$

where α is the fraction of cycles where the gates are active on average (typically 10%). Plugging in typical numbers from recent semiconductor roadmaps, we find that an ultrasmall planar transistor of dimensions 10 nm by 1 μm (areal density $10^{10}/cm^2$), having a capacitance of 0.35 fF (with a 1 nm SiO_2), switching at 1 V and 1 GHz frequency, would dissipate a whopping 3.5 kW/cm^2 of power. (The same as a rocket nozzle!). Ultimately, this dissipation arises because the minimum voltage for switching is constrained by a target error rate, and we need to charge sizeable capacitors to this minimum voltage $\boxed{P1.2}$. It is this unmanageable generation of heat that led to the ultimate end of the conventional Moore law and Dennard scaling, and the abandonment of the semiconductor roadmap.

In the accompanying reference book NEMV (Sections 2.1–2.5), I explain in detail how the energy loss occurs. In short, the charges have a natural tardiness given by the resistance–capacitance (RC) product, the time for a capacitor to charge and discharge. If we drive a transistor at an angular frequency $\omega = 2\pi f$ much slower than this RC, the charges can easily follow along and the dissipation drops. This is why we drive slower on a bumpy road $\boxed{P1.3}$. However when we turn on a switch abruptly like the step function above, that turn-on profile brings in a lot of Fourier components, many much faster than the RC, and the lagging electrons see a 'friction' leading to the dissipated dynamic power peaking at the above equation.

How can we reduce dissipation? The parameters α and C are set by a particular scaling node and the associated circuit layout. The solution boils down to either deviating from this equation (e.g., by running the circuit slowly or reversibly, as we will see shortly), or else sticking to this equation and reducing the voltage V_P. As we will see, V_P is set by the desired error rate and the thermal energy in a Boltzmann ratio. For certain applications (e.g., neuromorphic pattern recognition), we can live with that error. For high-performance digital applications, we will need to use new physics that relies on non-Boltzmann and/or correlated switching — such as tunnel transistors, nanomechanical relays, or magnetic switching. In all these devices we rely on either the opening of new channels or internal voltage amplification through feedback mechanisms, so that a voltage V_P accomplishes a bigger conductance gain than usual.

1.2 Reducing Heat: Adiabatic Logic

Thus, one way to cut down the heat generated is to drive the circuits *adiabatically*, slowly in stages. $\boxed{P1.3}$ shows the average power dissipated per cycle during the charging–discharging process at a single AC frequency

$$P_\omega = \frac{V_P^2}{2R}\left[\frac{(\omega RC)^2}{1+(\omega RC)^2}\right].$$ (1.2)

This is the dissipation of each single Fourier component of a square wave pulse of frequency f, which all add up to the original expression, $P_{\text{square-wave}} = CV_P^2 f$ $\boxed{\text{P1.4}}$. But going back to a single AC mode above, we see clearly that running a signal slowly compared to the RC time constant of the electron, $\omega RC \ll 1$, we get very low power dissipation, albeit at the expense of delay. This is clearly unsuitable for applications such as data processors that need much faster turnaround time.

1.3 Reducing Heat: Reversible Logic and Information Entropy

Another way to cut down the energy cost is to run circuits *reversibly*. This requires some information gathering. If we knew for instance which well an electron sits in, we can do a 'moonwalk' transformation that raises the other well with an applied voltage, slides the existing well to the site of the other well (e.g., using a superconducting Josephson junction), and finally raises the wall between the wells (Fig. 1.4). This process dissipates zero energy as the electron never leaves the ground state energy. But if we did not know the initial state, then there is a 50–50 chance we raise the electron

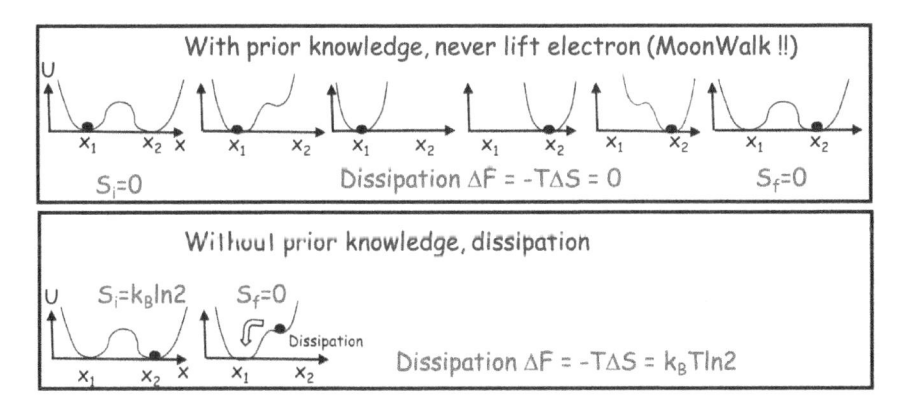

Fig. 1.4 The energy–information trade-off. (Top) Knowing the electron location in one of two wells separated by an energy barrier, we can slowly raise the second well and slide the first laterally before restoring the barrier, thus moving the electron without dissipating energy. (Bottom) Not knowing the electron location *a priori* risks raising the wrong well before a slide, dissipating at least $k_B T \ln 2$ of energy in the process.

when we raise the second well, and this will end up dissipating energy by pushing the electron downhill. To quantify the energy–information relation, we note that the *Free energy* of the electron is given by $F = U - TS$, U the potential energy, S the entropy and T the temperature. The entropy quantifies the disorder, the number of choices available to *a single* electron in its phase space

$$S = k_B \ln \Omega \quad (k_B T \approx 25 \text{ meV at } T = 300 \text{ K}), \tag{1.3}$$

where Ω is the number of microstates available for a given constrained macrostate (e.g., total energy or total number of particles). If we have two non-interacting systems with microstates Ω_1 and Ω_2, then the total number of distinct microstates is their combinatorial product while their entropies must add up $S_1 + S_2$, explaining thus the logarithm. Thus, the change in free energy at the end of computation, i.e., the *minimum* dissipated heat

$$\Delta Q \geq \Delta Q_{\min} = \Delta F = -T\Delta S = k_B T \ln (\Omega_i / \Omega_f) \tag{1.4}$$

(the equality happens if we make the transformation gradually). If the states are equally distributed with probably $p = 1/\Omega$, then we can write $S = k_B \ln (1/p) = -k_B \ln p$. If however we have a distribution of states with a probability distribution as well, then we calculate the average entropy over the various possibilities $\boxed{\text{P1.5–P1.6}}$

$$S = \langle -k_B \ln p \rangle = \sum_i p_i S_i = -k_B \sum_i p_i \ln p_i. \tag{1.5}$$

For instance, for a two-input AND gate $C = A$ AND B, with $C = 1$ if and only if both A and B are 1, we see the output cannot be reversed to generate the inputs uniquely. We have 75% probability of seeing a 0 and 25% of a 1, so the final entropy is $S_f = -k_B \left(\frac{3}{4}\ln\frac{3}{4} + \frac{1}{4}\ln\frac{1}{4}\right)$. The four possible binary inputs (A, B) are all equally likely, so the initial entropy $S_i = -k_B \left(4 \times \frac{1}{4}\ln\frac{1}{4}\right)$, so that the dissipation for each electron in the AND gate is $\Delta F = 3k_B T \ln 3/4 \approx 0.82 k_B T \approx 3.2 zJ$ (for reference, the hydrolysis of an ATP molecule during phosphorylation costs about 20 zJ).

Knowing both the initial and final states corresponds to $\Omega_i = \Omega_f = 1$, or alternately $p_{i,f} = 0, 1$, so that $\Delta S = S_i = S_f = 0$, meaning we can have zero dissipation ($\Delta F = 0$) if we do it cleverly by maintaining $\Delta U = 0$ (e.g., the moonwalk above). An example is a two-input two-output CNOT gate, where we flip the second input if and only if the first is unity. Now each input pair (A, B) and output pair (C, D) are equal probabilities, so that

$S_i = S_f$, and the dissipation is zero. We can see easily that this CNOT gate is reversible. In fact, concatenating two CNOTs one after another will generate the original inputs at the output terminals.

If we erase the final state ($\Omega_f = 1$) while keeping the initial state unknown and multivalued, then $\Delta F = k_B T \ln \Omega_i$, which ultimately limits the voltage $V_P = \Delta F/q$. The overall answer is then

$$\boxed{V_P = (k_B T/q) \ln \left(1/p_{\text{err}}\right)} \tag{1.6}$$

assuming the final erased state has zero error (else we get a ratio of error probabilities). For CMOS, we get $\Delta F = k_B T \ln 2$. The equation is for a single charge, and needs to be supplemented with an extra factor $N = CV_P/q$ to describe the number of independent electrons when evaluating the total dissipation, using $\Omega = 2^N$ and a corresponding error rate $p_{\text{err}} = 2^{-N}$.

If we are constrained to Boolean logic where we need an erase operation ($\Omega_f = 1$) and also yearn for speed, then we're limited by Eq. (1.1). C is usually set by the interconnect capacitance for a given layout, while p_{err} is set by reliability demands for a given technology. However, there can be *non-thermal* ways to excite an electron across a barrier that may bypass the Boltzmann constraints set by Eq. (1.6). This is an emerging area of research, and I devote several sections to some possibilities in NEMV (Ch 26). To understand this however, we need to have at least a preliminary understanding of how an electronic switch works.

1.4 What Makes an Electronic Switch Work?

What is special about electrons? What limits technology based on phonons or photons? An electronic switch relies on a few key attributes. We will develop these concepts fully in the coming chapters.

(i) **Electrons form waves:** As we see in the next chapter, each electron has associated with it a quantum probability wave, one that reveals itself if we choose to repeat an experiment many times under similar conditions and take a histogram of its outcomes. Placing such a wave in a solid, i.e., a grating of atoms arranged periodically, creates standing waves that through destructive interference eliminate certain wavelengths related to the lattice constant. As a result, the available electronic energy levels in solids form bands separated by forbidden energy gaps. These bands can be quite complex, as the electron waves carry additional symmetries associated with their spatial angular momenta, their quantum mechanical spins, interactions with the atomic lattice and with each other. A lot of these symmetries reveal themselves when we plot the energies vs wavevectors $(E - k)$ in 3D and also plot their wavefunctions.

(ii) **Electrons are light:** A 1000-fold compared to the atoms in the lattice supporting them. Ultralow electron masses mean that the bandwidths and bandgaps of semiconducting solids vary widely, from 100 meV to 10s of eV, spanning a large frequency spectrum. As a result, bandgaps can be tuned widely by alloying and band-engineering. In contrast, thermal energies, set by the motion of the heavier atoms constituting a thermal bath, are only around $k_B T \sim$ 25 meV at room temperature (k_B: Boltzmann constant, T: temperature in kelvins). Thus, small vibrations of the lattice do not overturn the destructive interference creating the bandgaps.

(iii) **Electrons are fermions:** Pauli exclusion implies that we can put no more than one electron per spin state. We thus fill up the bands like water in a bathtub, bottom to top until we run out of electrons. The highest energy where we run out of electrons is the Fermi energy E_F. If this happens to lie inside the bandgap, then all valence states are filled and inactive (electrons cannot jump around without double occupancy forbidden by Pauli exclusion). However, if the Fermi energy runs right through a band, as in a metal, we have filled states within a few $k_B T$ of empty states and charges can jump between them to drive a current.

In effect digital electronics is very much like playing with monochromatic electron waves. Altering the electronic or optical properties of a solid amounts to playing with that single Fermi energy. For instance, a gate bias can pull the Fermi energy in and out of a band, changing the current by several orders of magnitude. In practice, we often give the gate an additional nudge by initiating the Fermi energy near one of the bandedges. For n-type semiconductors, we do so with a donor (e.g., trace amounts of pentavalent phosphorus in silicon) that tries to blend in with the tetravalent silicon matrix by releasing its excess electron, liberated easily by thermal fluctuations, aided by dielectric screening. For p-type semiconductors, trace amounts of triavelent boron do the same job, i.e., place the Fermi energy near the valence bandedge and reduce the threshold to turn it on.

It is useful to hold this mental picture of a switch created by the alignment of a few relevant Fermi surface electrons with a band or a gap

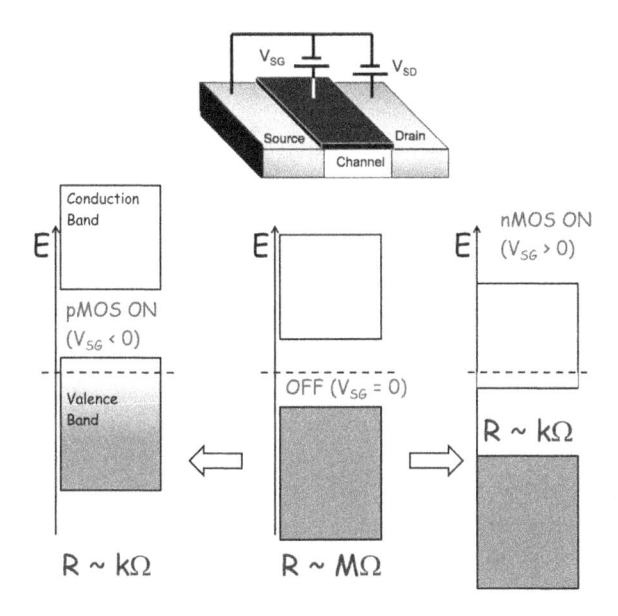

Fig. 1.5 (Center) Transistor band-diagram in the OFF state. By applying a gate voltage V_{SG}, we can move the levels and place the Fermi energy (dashed line) into the empty conduction band (right) or the filled valence band (left), reducing the resistance and turning the transistor on. In practice, we initiate the turn-on by doping the channel, moving the initial placement of the Fermi energy closer to a bandedge.

(Fig. 1.5). This simple picture works provided the electrostatics is in place to prevent the gap itself from moving — requiring a strong gate control that gets hard to maintain for short channels (and is one of the motivations behind growing them in the vertical direction — tall instead of small, to provide adequate room to wrap the channel with gates from all its sides). Similarly, the transistor stays off provided the source-drain separation is long enough that the electrons can't tunnel through the gap in the off state. Finally, the gate oxide must be thick enough and insulating enough so that the electrons don't leak out through the third terminal. All these are precisely the challenges transistor technologies run into upon aggressive downscaling.

1.5 The Importance of 'Gain'

We explained why an nMOS or a pMOS turns on or off the way they do, to create an inverter. What a good inverter needs in addition is a very steep transfer curve or high gain $g = |\Delta V_{\text{out}}/\Delta V_{\text{in}}| \gg 1$ in Fig. 1.2. This way a small change $\Delta V_{\text{in}} = 2V_1$ can translate to a large change $\Delta V_{\text{out}} = 2V$. Why is this gain important and how can we accomplish this?

For analog applications like an amplifier, this is a no brainer. A small signal must translate to a large output. What is often unappreciated is we also need gain for digital logic. Imagine a chain of an odd number of inverters that feeds back at the end to the first inverter. Such a chain acts as a *ring oscillator* — an input 1 ripples down through a sequence of alternating 0s and 1s, ending with a 0 that overrides the input of 1, and then the process continues till we override the 0 and so on, generating an oscillator that is useful for timing circuits. Let us recalibrate so the input voltage swings between V_1 and 0 and the output swings between V and 0, with a maximum gain $g = V/V_1$ along the linear part of the transfer curve. Suppose the first input drifts down from the high value to $V_1 - \Delta$, corrupted by external noise. We can work out the output after N stages $\boxed{\text{P1.7}}$. We find that if $g < 1$, then for large N the outputs change by $(-g)^N \to 0$ and become independent of N, converging to the same value, so we lose all binary information. On the contrary, if $g > 1$, then $(-g)^N \Delta$ blows up around the linear segment of the transfer curve and V_{out} oscillates between a high and low value depending on whether N is odd or even, as we expect the ring oscillator to do. In other words, a high gain g restores a signal corrupted by noise.

But how do we ensure enough gain? As we will see later $\boxed{\text{P18.6}}$, the gain is determined by the degree of *saturation* of the nMOS and pMOS currents as we ramp up the power supply voltage V.

1.6 Le roi est Moore, vive le roi!

The focus of the semiconductor industry over several decades was on keeping Moore's Law, the ability to pack more devices onto a chip, alive by making the electrostatics optimal, keeping the gate capacitance much larger than the source or drain capacitance, in order to force transistor currents to saturate, thereby giving us a high CMOS gain. Numerous structural and material designs such as high-k dielectrics and FinFETs (Section 18.6) were designed to make the gate dominate the electrostatics. With the enormous thermal budget of present day transistors however, the emphasis has moved from one $\mathcal{E}$ to another — from controlling $\mathcal{E}$lectric fields to managing $\mathcal{E}$nergy dissipation. As we saw above, this is limited by the Boltzmann constraint on switching voltage, $p_{\text{err}} = \exp\left[-qV_P/k_B T\right]$ (Eq. (1.6)). Accordingly, one of the most fertile areas of research in device physics is to identify ways to bypass the Boltzmann limit on minimum gate voltage $V_{\min}$ $\boxed{\text{P18.9}}$. I present a few ideas in what follows, with the caveat that they are highly exploratory, and a robust technology based on them may never see the light of day!

We will see later that the current is set by the number of conducting modes within an energy window set by the applied drain bias. As we vary the gate voltage, the Fermi energy moves toward a bandedge, filling that band with charges at a rate set by the thermal voltage $\sim k_B T/q$. We can reduce the voltage by attracting excess charges into the band faster than this rate, by opening new conducting modes with the application of gate voltage. As an example, a *tunnel transistor* $\boxed{\text{P22.4}}$ starts with the conduction band of the channel sitting above the valence band of a source junction. By applying a positive gate voltage to the channel, we push its conduction bandedge down past the valence bandedge of the source region (Fig. 22.4), getting an abrupt onset of band-to-band tunneling between the two bandedges. Ideally, this new conduction mode should create more charges than the Boltzmann limit, so that the voltage V_P needed for a given current target is smaller than usual. In practice, this comes at a price, as the ON current is also low. Attempts to raise it through heterojunction engineering increases the OFF current by invoking trap-assisted band-to-band and Auger tunneling. Similar ideas for creating new conducting modes can lead to relay-based switches and Mott transition switches (NEMV, Ch 26).

Throughout this book, we will discuss several approaches to designing electronic switches — from material to conceptual. But first we need a thorough grasp of quantum physics.

Homework

P1.1 **Energetics.** The work done to bring an incremental charge dQ to a capacitor is given by $V_C dQ$, where $V_C = Q/C$ is the voltage on the capacitor. Find the total energy to charge a capacitor to the battery voltage V_D, and the energy delivered by the battery operating at constant voltage V_D. Find the balance of energy that is dissipated in the wires.

P1.2 **Energy-delay.** Show that the energy-delay product of a switch can be written as $Q^2 R$, where R is the resistance. Use Joule's law for power dissipation in terms of current $I = Q/T$, T being the RC delay.

P1.3 **Energy-delay-error.** Industry often evaluates technologies using the energy-delay product. However, this evaluation must be anchored to a particular error target, since we can always reduce the energy-delay if we are sloppy about error rate. The write error in a two-well one-barrier system is given by the Boltzmann probability of jumping over a barrier,

$$p_{\text{err}} = e^{-qV/k_B T} = e^{-Q/Q_C} \tag{1.7}$$

with $Q = CV$. Work out Q_C. For a write error rate of $p_{\text{err}} = 10^{-12}$ and a circuit with an interconnect capacitance of 1 fF, find V, Q_C, Q and the energy dissipated QV at room temperature.

In nanomagnetic computing, we inject spin polarized current into a magnet to apply a spin torque that flips its magnetization, thereby writing an up or down (1 or 0). The write error rate, obtained by solving a stochastic 1D Fokker*–Planck equation for a perpendicular magnet for low errors, is

$$p_{\text{err}} = 4\Delta e^{-2(Q/Q_c - 1)}, \tag{1.8}$$

where $\Delta = M_S H_K \Omega / 2k_B T$ is the barrier, M_S being the saturation magnetization, H_K is the uniaxial anisotropy and Ω is the volume. Instead of capacitive considerations on Q_C for electronic switching, the requirement here is angular momentum consideration. The number of electrons is given by Q_C/q, and each electron carries a spin angular momentum $P\mu$ (we will see this later) where P is the magnetic polarization and μ the electron magnetic moment. This angular momentum must be adequate to flip a magnetization M_S times its volume Ω, in presence of damping α

$$(Q_C/q)(P\mu) = M_s \Omega (1 + \alpha^2). \tag{1.9}$$

For an ellipsoidal 100 nm $\times$ 20 nm $\times$ 1 nm Fe magnet with about 10^4 spins, $\alpha = 0.1$, $P = 0.7$, show that the number of electrons Q_C/q needed for destabilization is a bit above 10^5. Show that to get an error of 10^{-9} and a switching time of $\tau = 1ns$, we need a current ~ 60 MA/cm^2.

P1.4 Deconstructing dissipation. Let us work out the dissipation for an AC input. For an RC circuit connected to a battery $V_B = V_0 \cos \omega t$, we can write down Kirchhoff's voltage law in phasor notation as

$$RdQ/dt + Q/C = V_0 e^{i\omega t}. \tag{1.10}$$

In phasor notation, we write $Q = Q_0 e^{i\omega t}$ so that derivatives become simple and the differential equation becomes algebraic. Solve for Q, V_C, V_R, V_B in terms of t and show that the energies integrated over one AC period (prove and use the simplification $\langle A(t)B(t)\rangle = \langle \text{Re}[Ae^{i\omega t}] \times \text{Re}[Be^{i(\omega t + \phi)}]\rangle = \frac{1}{2}\text{Re}\langle AB^*\rangle$, where * represents complex conjugation

$$E_C = \int_0^T V_C(t)dQ(t) = \frac{1}{2}\text{Re}\int_0^T V_C dQ^* = 0, \quad T = 2\pi/\omega,$$

$$E_R = \int_0^T V_R(t)dQ(t) = \frac{CV_0^2}{2}\left[\frac{2\pi\omega RC}{1 + (\omega RC)^2}\right] = \int_0^T V_B dQ = E_B, \tag{1.11}$$

showing that after each cycle there is some dissipation in the wire during charging/discharging. Also work out the power dissipated per cycle $P_\omega(T) = E_R(T)/T$ and show that it vanishes if we run the circuit much slower than the RC time constant, while for a fast circuit including a step function turn-on, the dissipation is $V_0^2/2R$ (Eq. (1.2)).

If so, how can the Gigahertz clock act on electrons with picosecond transit time and get the full dissipation we see today? This is because the clock is less like a sinusoid at GHz frequency, and more like a square wave pulse with GHz duty cycle. The abrupt turn ons and offs of the square pulse still bring in fast Fourier components that dissipate energy. To see this, let us go to a square wave pulse of duty cycle $T = 1/f$. Show that the Fourier coefficient $V_\omega = \int_0^T V(t)e^{i\omega t}dt/T$ has an absolute value $\sin(\omega/2f)/(\omega/2f)$. Multiplying the power P_ω (Eq. (1.2)) with the coefficient $|V_\omega^2|$ and integrating over ω normalized by f, we can do a few simplifications. The biggest contributions to P_ω are for $\omega RC \gg 1$, so the integrals are from $1/RC$ to ∞. Assuming $a = 2fRC \ll 1$, show that we recover the square wave result $P_{\text{square-wave}} = CV_0^2 f$. The reason we need small a is because we want to give enough time for the capacitor to fully charge or discharge before flipping the switch, meaning that $T = 1/f \gg RC$. In other words, abrupt turn-on or off with adequate duty cycle. (Check out NEMV Section 2.5 for how a slow LCR turn-on $\sim (1 - e^{-\lambda t})$ can reduce this dissipation by an activation factor $\sim \lambda RC$.)

P1.5 **Combinatorics.** Consider N balls divided into M boxes.

(a) What is the number of ways Ω you can choose N_1 out of N balls to put into the first box, N_2 out of the remaining $N - N_1$ for the second box, all the way to N_M for the Mth box? What is the entropy S?
(b) Using Stirling's formula for large N, $\ln N! \approx N \ln N - N$, simplify S and show that we can get the expression mentioned in the chapter in terms of probabilities p_1, p_2, etc. What is the relation between $p_{1,2,3,...}$ and $N_{1,2,3,...}$?

P1.6 **Entropy and reversibility.** Write down the logic tables and solve for the initial entropy S_i, final entropy S_f and energy dissipation ΔF of NAND gate, NOR gate, XOR gate and XNOR gates.

Let us look at a new two input/two output gate: if the first bit A is 1, erase the second bit B; if the first bit A is 0, do nothing. Write down the logic table and solve for S_i, S_f and ΔF. Is this reversible?

Finally, define a 2-input 2-output gate as shown. Compare its dissipation with the previous gates. Which is more reversible? What if one output (say for input 0,1) is probabilistic with probability p?

A	B	C	D
0	0	0	0
0	1	1	0
1	0	1	0
1	1	1	1

P1.7 **Inverter gain and signal restoration.** Consider an inverter whose input voltage swings between V_1 and 0, and output between V and 0, with gain $g = V/V_1$. Write down the linear equation connecting V_{in} and V_{out}. Assume the input is $V_{\text{in}} = V_1 - \Delta$. Let this output feed into another inverter, and so on, in a cascade of N inverters, and then back in a circular loop. Show that the output voltage after N steps

$$V_{\text{out}} = \left[g/(1+g) \right] V_1 + (-g)^N \Delta. \tag{1.12}$$

For $g < 1$, show that V_{out} converges between successive stages and any binary information is lost. For $g > 1$, show that V_{out} swings between high and low values for odd and even N, as we expect an oscillator to do.

Chapter 2

Quantum Physics: A Whole New World

2.1 The Two-Slit Experiment

At the end of the 19th century, many outstanding mysteries in physics were considered solved. Classical mechanics explained complex spinning tops and simple machines like pulleys and levers, strain and elasticity, motion of planets as well as swirling sea waves, while thermodynamics accounted for collective behavior of large numbers of molecules. In parallel, Maxwell's laws of electromagnetism synthesized our understanding of electric and magnetic fields, with optics emerging as a consequence. However, there was no simple way to explain the stability of an atom, specifically the ability for negatively charged electrons to resist falling into the positively charged nucleus drawn by its attractive forces. Even orbital motion could not account for such resilience (as it does for planets escaping the Sun's pull), because an accelerating charge is expected to radiate energy continuously and aggressively plummet spiraling into the nucleus.

Larmor's equation for radiation yields a differential equation for the shrinking orbit with time, which predicts a lifetime of a few tens of picoseconds for a hydrogen atom! $\boxed{\text{P2.1}}$ If this were indeed true, we would all be cosmic mayflies, instead of being here with the collective history of the billions of years that the universe has existed to date. The discrepancy is several orders of magnitude and quite impossible to reconcile within classical physics. It is noteworthy that electrons in atoms do radiate, but only at discrete spectroscopic frequencies. Looking at these data, Niels Bohr posited that the atom must have some special allowed orbits where the electron must somehow be stationary. Radiation occurs when the electron skips between orbits. In other words, the energy levels are quantized $\boxed{\text{P2.2}}$. Atoms are governed by quantum physics.

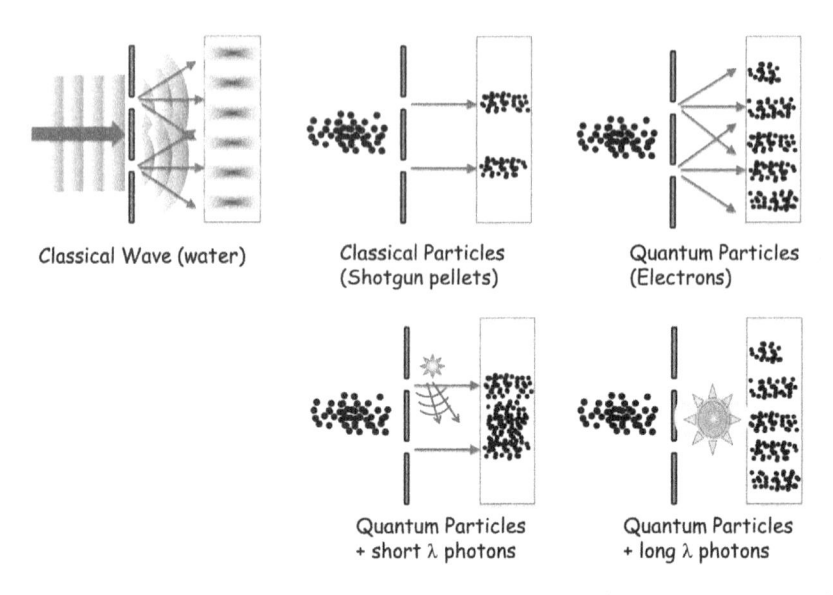

Classical Wave (water)   Classical Particles   Quantum Particles
                            (Shotgun pellets)   (Electrons)

                    Quantum Particles   Quantum Particles
                    + short λ photons   + long λ photons

Fig. 2.1 Wave-particle duality in the two slit experiment. Quantum particles self-interfere and show fringes that build up over time, while registering discrete individual pings. Using light to track them either washes out the fringes or else lacks resolution to locate them accurately.

The counter-intuitiveness of quantum physics can perhaps best be understood by the two-slit experiment (Fig. 2.1). When we generate a wave in a water tank and let it pass through two closely spaced slits cut into a screen, the wave generates simultaneous circular wavefronts at each slit that interfere on the other side. Where the crests of the two wavefronts overlap, a maximum intensity gets recorded on a detector positioned on the far side, while we get a minimum where the crest of one wave sits on the trough of the other upon arrival at the detector, generating an interference pattern. In contrast, if we spray shot gun pellets at the same two slits, they pass through one slit or the other and impinge on the detector right along their lines of sight, creating two clear beams with no trace of interference.

If we now repeat this experiment a third time, using an electron gun that shoots electrons one at a time at the two slits, then the electrons will register on the detector as individual 'pings' much like the pellets did. However, over time, an interference signal builds up, even though the electrons came in one at a time! For instance, if the detector is coated with phosphor, we will see a pattern of fringes, each composed of individual flashy dots.

The quantum viewpoint is that there is a wave associated with each electron. This cloud determines where the electron will end up (i.e., a 'probability wave'). Much like water waves, it can break into wavelets at the two slits and self-interfere. However, at the instant of detection, this wave 'collapses' into a particle-like dot, at a different location each time dictated by the probability distribution embedded in the wave, generating those periodic intensity variations. The wavefunction is a simultaneous superposition of all possible particle-like results — we call them 'eigenstates' — states with definite measurable signatures ('eigenvalues'). Upon interaction with a classical device — in this case the detector, a measurement happens that collapses the wavefunction into one specific eigenstate. Which one it collapses to is determined by the probability distribution — the dots collect densely where the probability amplitude is the largest.

2.2 Particles or Waves?

If we try to trick the system by measuring which slit the electron goes through, say by using a detector that flashes at one slit or the other, then we find the interference fringe disappears. Each photon imparts a momentum p to the electron that kicks it away from its original trajectory, scrambling up the wavelets and washing away the interference pattern. To minimize the impact of the photons, we need to dial down their momentum, by increasing their wavelength (redshifting). Here, we invoke the empirical de Broglie and Planck relations, which relate kinematic, i.e., particulate properties like momentum p and energy E with wave-like properties like wavelength λ and frequency f. These equations were obtained empirically from experiments such as Compton scattering and the photoelectric effect — i.e., longer wavelength (redder) particles have lower energy and momentum.

$$p = \frac{h}{\lambda} = \hbar k, \quad E = hf = \hbar \omega,$$

$h = 6.603 \times 10^{-34} Js$ is Planck's constant, $\hbar = h/2\pi$ is reduced Planck's constant, $k = 2\pi/\lambda$ is the wavevector and $\omega = 2\pi f$ is the angular frequency.

Redshifting photons to a longer wavelength λ reduces their impact p, restoring the electron interference on the detector. By the time this happens however, the wavelength is large enough that we can no longer resolve between the two slits, and see instead a gigantic flash across the whole screen. The electron reverts to its enigmatic wave-like behavior when we don't look, and collapses to a particle like behavior when we look. In other words, we can either measure the position of the electron at the slits, or its momentum away from them — but never both with a high precision.

The above imprecision is baked into the very foundations of quantum physics. Consider an electron wavefunction with a precise momentum $p = h/\lambda$. Having a corresponding precise wavelength, it is in effect a monochromatic probability wave that occupies all space with a constant amplitude, so its position x is completely uncertain. If we now want to confine the electron by sculpting a narrow Gaussian wave packet, we gain a lot more precision on its x value, but in the process of pulse shaping we mix in many Fourier components, each with a different k, so that its momentum is now all over the place. *A signal can be either tightly space limited or tightly band limited but never both.* Since we associate the wavelength with momentum, this means we cannot specify a simultaneous set of precise position and momentum pairs to pinpoint an electron $\boxed{\text{P2.3}}$. The electron cloud does not own a precise trajectory even in principle!

2.3 How Do We Measure?

To quote a song about Sister Maria, *how do we catch a cloud and pin it down?* The probabilistic state of the electron cloud is encoded in its wavefunction $\Psi(x,t)$ as a superposition of eigenstates (of any operator — they all form a complete set to break into, much like a wave can be re-expressed as a sum of Fourier components, but also a sum of polynomials), $\Psi = \sum_n c_n \psi_n(x)$. Each eigenstate ψ_n registers a precise value O_n corresponding to a given measurement operator $\hat{O}$. Because an eigenstate has an unambiguous measurement, the operator gives back ψ_n intact. This is like applying the 'height' operator to a student — who reports the height but stays unaffected by the measurement. If, however, we try to measure the exact height of a class of students, we get a scatter plot, much like the dots on phosphor. The class as an entity is not a suitable eigenstate of the height operator, but rather a superposition of heights.

$$\Psi(x) = \sum_{n=1}^{\infty} c_n \psi_n(x) \quad \text{(superposition of eigenstates } \psi_n\text{)},$$

where

$$\hat{O} \underbrace{\psi_n(x)}_{\text{Eigenstate}} = \underbrace{O_n}_{\text{Eigenvalue}} \psi_n(x)$$

$$\text{(state scaled and restored after operation } \hat{O}\text{).} \qquad (2.1)$$

A unique set of coefficients $\{c_n\}$ designates the shape of Ψ. Upon measuring with $\hat{O}$, it will scale the coefficients by different constants O_n and as a result the shape of Ψ would change. We no longer recover $\Psi(x)$. The superposition is a sum of eigenstates, but not itself an eigenstate.

2.4 Some Prominent Measurables

One needs to appreciate how special an eigenstate is — we operate on the function $\psi_n(x)$ with $\hat{O}$ and typically distort it, but for an eigenstate we recover a scaled version of the same shape, the scale factor being the eigenvalue. For instance, a function that retains shape under a Fourier transform and is thus its eigenstate is a Gaussian (as are all eigenstates of the oscillator potential). For the position operator $\hat{x}$, the eigenstate with a fixed position x_0 is clearly a delta function at a single position x_0, i.e., a blip

$$\hat{x}\delta(x - x_0) = x_0\delta(x - x_0). \tag{2.2}$$

The only operator that does so is $\hat{x} = x$. In fact, we can verify by noting that any function can be constructed out of such spiky delta functions. Recall a delta function is the limiting case of a function that is infinitesimally thin and infinitely tall so that the area under it is unity. Also, multiplying any function with a delta function reproduces that function but replaces the argument by that single point where the delta spikes. Thus,

$$\hat{x}f(x) = \hat{x}\int \delta(y - x)f(y)dy = \underbrace{x\int \delta(y - x)}_{\text{using Eq. (2.2)}} f(y)dy = xf(x) \quad \text{for any } f.$$

$$\tag{2.3}$$

What about momentum? A fixed momentum read implies a fixed wavelength, i.e., the eigenstate is a monochromatic plane wave. It's eigenvalue must be $\hbar k$ as per de Broglie equation (Eq. (2.2)). We must thus have

$$\hat{p}e^{ikx} = \hbar k e^{ikx}, \tag{2.4}$$

which indicates that the momentum operator $\hat{p}$ is ultimately the slope of the wavefunction (plane waves and more, generally exponentials, preserve shape under slopes or derivatives). So to summarize in 1D,

$$\boxed{\hat{x} = x, \quad \hat{p} = -i\hbar\frac{\partial}{\partial x}}. \tag{2.5}$$

Note that the momentum eigenstate is also a blip in k space, since the Fourier transform of a plane wave is a delta function. We can see that it's impossible to measure both position and momentum at the same time. A clear position recording requires a sharply localized function whose slope (momentum) varies widely, while a clear momentum recording requires a function with a fixed amplitude whose position is all over the place.

By extension, we can write down the kinetic and potential energy

$$\hat{T}(\hat{p}) = \frac{\hat{p}^2}{2m} = -\frac{\hbar^2}{2m}\frac{\partial^2}{\partial x^2}, \quad \hat{U}(\hat{x}) = U(x) \tag{2.6}$$

The eigenstate for total energy E would look a lot like that for momentum (constant energy is constant frequency, monochromatic), except in time domain. A positive moving plane wave is $e^{i(kx-\omega t)}$ with $E = \hbar\omega$ and the operator $\hat{\mathcal{H}}$ is called the Hamiltonian

$$\hat{\mathcal{H}}e^{-iEt/\hbar} = Ee^{-iEt/\hbar}, \tag{2.7}$$

so that the Hamiltonian operator $\boxed{\text{P2.4}}$

$$\boxed{\hat{\mathcal{H}} = i\hbar\frac{\partial}{\partial t}}. \tag{2.8}$$

2.5 How We Measure Matters!

If we take two measurements and expect a clear result (say height and weight of a student), this can only happen if the target was a *simultaneous eigenstate* of both operators. In that case, we can pull the eigenstate through each operator, yielding the same unambiguous answer since products of eigenvalues (numbers) are interchangeable — they 'commute'. For position and momentum, such a simultaneous eigenstate cannot exist, since

$$\underbrace{\left[\hat{x}, \hat{p}\right]}_{\text{Commutator}} F(x) = \left[\hat{x}\hat{p} - \hat{p}\hat{x}\right] F(x)$$

$$= -i\hbar x\frac{\partial F}{\partial x} + i\hbar\frac{\partial(xF)}{\partial x} = i\hbar F \quad \text{for all } F \implies \boxed{[\hat{x}, \hat{p}] = i\hbar} \tag{2.9}$$

meaning the commutator $[\hat{x}, \hat{p}] = i\hbar$ is non-zero, and the order of the measurements matters! Measuring position first and momentum next ends up with a plane wave eigenstate, while doing it the other way around ends with a delta function — a different answer. Each measurement in a sequence of non-commuting operators wipes out memory of the previous one. Depending on which we measure last — position or momentum, we leave behind a wavefunction collapsed to a different looking eigenstate — a blip in position vs a plane wave. In fact, we will see shortly $\boxed{\text{P3.6}}$ that the uncertainty principle arises directly as a consequence of this ambiguity of measurement.

Homework

P2.1 Lifetime of a classical atom. Here's a fun problem! How long would an atom last were it not for quantum mechanics? Larmor's theorem gives us the power radiated by an electron — the rate of decrease of its electronic energy, proportional to its acceleration squared

$$\frac{dE}{dt} = -\frac{\mu_0 q^2 a^2}{6\pi c}, \tag{2.10}$$

where $a = v^2/r$ is the centrifugal acceleration for a rotating electron, $\mu_0 = 4\pi \times 10^{-7}\, H$ is the permeability of free space, $q = 1.6 \times 10^{-19}\, C$ is the electronic charge and $c = 3 \times 10^8\, \mathrm{m/s}$ is the free space speed of light. To get the energy as a function of radius, we use energy conservation and force balance between Coulomb and centrifugal forces to eliminate v

$$E = \underbrace{\frac{mv^2}{2}}_{\text{Kinetic Energy}} - \underbrace{\frac{q^2}{4\pi\epsilon_0 r}}_{\text{Coulomb Energy}}$$

$$\underbrace{\frac{mv^2}{r}}_{\text{Centrifugal Force}} = \underbrace{\frac{q^2}{4\pi\epsilon_0 r^2}}_{\text{Coulomb Force}}, \tag{2.11}$$

where $\epsilon_0 = 8.854 \times 10^{-12}$ F/m is the permittivity of free space and $m = 9.1 \times 10^{-31}$ kg is the mass of an electron. Use the above two equations to replace E with r, and show that we end up with a differential equation that shows a constant rate of shrinkage of a sphere in terms of the size of an electron where the Coulomb attraction into the nucleus equals its relativistic rest energy

$$\frac{d}{dt}\left(\frac{4\pi r^3}{3}\right) = -4\pi r_0^2 \times \frac{4c}{3}, \quad \text{where} \quad \frac{q^2}{4\pi\epsilon_0 r_0} = mc^2. \tag{2.12}$$

Solve the differential equation to find the trajectory of the electron and show that a hydrogen atom of radius 0.529 Å $\boxed{\text{P2.2}}$ would only last $\sim$15 picoseconds before it radiates all its energy and falls into the nucleus!!

P2.2 Bohr theory of the atom. While we are at it, let us estimate the energy and momentum of a hydrogen atom by imposing an added quantization rule on the angular momentum

$$mvr = n\hbar, \quad n = 1, 2, 3, \ldots, \tag{2.13}$$

where $\hbar = h/2\pi \approx 1.1606 \times 10^{-34}$ Js. Combined with Eq. (2.11), show that the radius of an electron in the hydrogen atom and energy are given by

$$r_n = n^2 a_0, \quad a_0 = \frac{h^2 \epsilon_0}{\pi m q^2} \approx 0.529\,\text{Å} \quad \text{(Bohr radius)},$$

$$E_n = \left(\frac{1}{n^2}\right) E_0, \quad E_0 = -\frac{m q^4}{8 h^2 \epsilon_0^2} \approx -13.6\,\text{eV} \quad \text{(1 Rydberg)}. \tag{2.14}$$

These are actually accurate!

P2.3 **A feel for QM.** Let us get a feel for quantum mechanical numbers. The de Broglie wavelength describing the spread of an electron wave (its spatial uncertainty) is given by Eq. (2.2), where the momentum $p = \sqrt{2mE}$ with $E \approx k_B T = 25\,\text{meV}$ is a rough estimate for the kinetic energy of a vibrating atom of mass m at room temperature T (300 K). Estimate the de Broglie wavelengths for the following systems:

(a) An electron with mass 9.1×10^{-31} kg at room temperature.
(b) A hydrogen atom with a proton that is about 2000 times heavier, at room temperature.
(c) A helium atom with two protons and two neutrons at 2 K (this should explain why atomic helium can show quantum effects at low T).
(d) A carbon atom with six protons and six neutrons at room temperature.
(e) A small gyroscope of mass 100 g at room temperature.

P2.4 **Operators as generators.** Show that the momentum operator $\hat{p}$ is the *generator* of spatial translation, and the Hamiltonian operator $\hat{\mathcal{H}}$ is the generator of time translation. To do so, Taylor expand the wavefunction at a displaced coordinate and show it can be written as an exponential operator acting on ψ,

$$\psi(x + \Delta x) = \underbrace{e^{i\hat{p}\Delta x/\hbar}}_{\text{Spatial translation}} \psi(x), \quad \text{where } \hat{p} = -i\hbar\partial/\partial x. \tag{2.15}$$

Similarly, argue that a time translation operator

$$\psi(t + \Delta t) = \underbrace{e^{-i\hat{\mathcal{H}}\Delta t/\hbar}}_{\text{Time translation}} \psi(t) \qquad (2.16)$$

Explain why for a free electron (and only a free electron!) ψ can be a simultaneous eigenstate of $\hat{\mathcal{H}}$ and $\hat{p}$. In that case, we get a plane wave $\psi(\Delta x, \Delta t) = e^{i(p\Delta x - E\Delta t)/\hbar}$, E and p being the eigenvalues of $\hat{\mathcal{H}}$ and $\hat{p}$.

Chapter 3

Interpreting Quantum Probabilities

While nature seems to be inherently probabilistic, with all possible outcomes encoded into the wavefunction in the form of a superposition, we can extract useful and relevant information from it. Gamblers and actuaries build their careers on probability distributions. What is the relation between probability and wavefunction? Note that Ψ is in general, complex — given the imaginary i sitting in front of the Schrödinger equation. We just saw plane wave solutions e^{ikx} for free electrons. Probability, however, is a non-negative number, and the relation is obtained by taking the absolute value squared to yield the probability density

$$P = |\Psi|^2, \tag{3.1}$$

with an added normalization rule $\int P(x, t)dx = 1$, since probabilities should add up to unity. The electrons must be somewhere!

3.1 Averages and Standard Deviations

What else can we extract from the probability distribution? Given a histogram of exam scores, for instance, we can estimate the average and standard deviation. The average is obtained by weighting each outcome with its probability. For instance, the average number of dots on a die is 3.5, obtained by summing each possible number of dots on a face, one through six, with its individual probability of 1/6. We can similarly evaluate the average of any operator $\hat{O}$, using the probability $P = \Psi^*\Psi$

$$\langle O \rangle = \sum_i O_i P_i \implies \int dx \Psi^*(x)\hat{O}\Psi(x). \tag{3.2}$$

Note the order! For a scalar $\hat{O}$ like position, the order doesn't matter and we see $P(x)$ straightaway. For more complicated operators such as derivatives (momentum), the sequence of terms in the integrand matters!

Operators corresponding to real measurables turn out to be *Hermitian*, which satisfy a condition under a sequence swap (more accurately complex conjugation followed by transpose). Define the *Adjoint* of an operator $A^\dagger$

$$\int dx \left(\Phi^* \hat{A} \Psi \right)^\dagger = \int dx \Psi^* \hat{A}^\dagger \Phi. \tag{3.3}$$

Quantum operators that correspond to real measurements are Hermitian, meaning they are identical to their adjoint $A = A^\dagger$. Under this condition, it is straightforward to verify that their eigenvalues λ from the eigen equation $A\Phi = \lambda\Phi$ equal their complex conjugate, meaning we get real and thus physically meaningful measurements $\boxed{\text{P3.1, P3.2}}$.

One operator of some significance is the current density operator $\hat{J}$, Loosely speaking, the average electron density times its velocity (thus momentum). But remember that for operators, not just the average itself, but the individual operands and their order matter. We can obtain the operator by using the equation of continuity for charge conservation (increase of charge implies an influx or negative divergence of current density)

$$\frac{\partial \hat{n}}{\partial t} = -\frac{\partial \hat{J}}{\partial x}. \tag{3.4}$$

Using $\hat{n} = \Psi^* \Psi$ and the Schrödinger equation for $\partial\Psi/\partial t$ and $\partial\Psi^*/\partial t$, we can show $\boxed{\text{P3.3}}$ that the current density is given by

$$\hat{J} = \frac{\hbar}{2mi} \left(\Psi^* \frac{\partial\Psi}{\partial x} - \frac{\partial\Psi^*}{\partial x} \Psi \right). \tag{3.5}$$

On a 1D grid of points separated by a lattice constant a, we can use the finite difference approximation for a derivative, $d\Psi/dx|_{x=na} = (\Psi_{n+1} - \Psi_n)/a$, whereupon $J = (\hbar/2mia)[\Psi_{n+1}^* \Psi_n - \Psi_n^* \Psi_{n+1}]$. For a free electron with $\Psi \sim e^{ikx}$, this amounts to $(\hbar k/m)|\Psi|^2$, i.e., charge density times velocity. For a bound electron forming a standing wave $\Psi \sim \sin kx$, there are two counter-propagating currents so the current density predictably is zero.

3.2 Schrödinger Equation

Combining the operators above, the expression for total energy and its eigenstate $\Psi(x,t)$ can be written as

$$\underbrace{i\hbar\frac{\partial}{\partial t}}_{\hat{\mathcal{H}}}\Psi(x,t) = \left[-\underbrace{\frac{\hbar^2}{2m}\frac{\partial^2}{\partial x^2}}_{\hat{T}} + \underbrace{U(x,t)}_{\hat{U}}\right]\Psi(x,t) \qquad (3.6)$$

which is the celebrated non-relativistic Schrödinger equation in 1D. For a free electron ($U = 0$), the solution, easily verified by substitution, is the aforementioned plane wave, $\exp i(kx - Et/\hbar)$. For a time-static potential, we expect the time-dependent part of the solution to look like a free electron, i.e., $\Psi(x,t) = \psi(x)e^{-iEt/\hbar}$. Substituting, we get the time-independent Schrödinger equation, which essentially shows energy conservation,

$$\hat{\mathcal{H}}\psi(x) = \left[-\frac{\hbar^2}{2m}\frac{\partial^2}{\partial x^2} + U(x)\right]\psi(x) = E\psi(x), \qquad (3.7)$$

and the eigenstates $\psi(x)$ are essentially energy eigenstates.

3.3 Quantum-Classical Relations and Uncertainty

Now that we know how to extract averages, it is straightforward to show, by simply manipulating the Schrödinger equation, the origin of Ehrenfest's equation, that *quantum averages follow classical physics* $\boxed{P3.4, P3.5}$

$$\frac{d}{dt}\langle x\rangle = \frac{\langle p\rangle}{m},$$
$$\frac{d}{dt}\langle p\rangle = -\left\langle\frac{\partial U}{\partial x}\right\rangle \qquad \text{(Newton's law).} \qquad (3.8)$$

As an exercise, let us prove the first equation. From the definition of averages as well as Schrödinger's equation,

$$\frac{d\langle x\rangle}{dt} = \frac{d}{dt}\int \psi^*\hat{x}\psi dx = \int\left[\underbrace{\frac{\partial\psi^*}{\partial t}}_{-\psi^*\hat{\mathcal{H}}/i\hbar}\hat{x}\psi + \psi^*\hat{x}\underbrace{\frac{\partial\psi}{\partial t}}_{\hat{\mathcal{H}}\psi/i\hbar}\right]dx$$
$$= \frac{1}{i\hbar}\int \psi^*[\hat{x},\hat{\mathcal{H}}]\psi dx = \frac{1}{i\hbar}\left\langle\left[\hat{x},\hat{\mathcal{H}}\right]\right\rangle. \qquad (3.9)$$

From $\hat{\mathcal{H}}$ the potential energy operator is also a function of x alone, so it commutes with $\hat{x}$ (they are simultaneously measurable). This leaves the kinetic energy operator $\hat{p}^2/2m$ in the commutator with $\hat{x}$. Using $[\hat{x}, \hat{p}] = i\hbar$, we get $[\hat{x}, \hat{p}^2/2m] = i\hat{p}\hbar/m$, and recover the first Ehrenfest equation.

The Ehrenfest result is fairly profound! Although the averages $\langle x \rangle$ and $\langle p \rangle$ are calculated using completely separate ensembles of incompatible x and p values, they are related the way we expect them to be classically. The first gives us the average velocity, while the second is Newton's law — the rate of change of momentum is the average Force, $F = -\partial U/\partial x$. Where classical and quantum mechanics will deviate is the size of the individual departures from these averages.

This result would certainly explain the intuitive idea behind Bohr's allowed states that don't radiate. Those states are energy eigenstates of the hydrogen atom, and since each eigenstate has a single frequency $\Psi(x, t) = \psi(x)e^{-iEt/\hbar}$, the probability $|\Psi(x, t)|^2 = |\psi(x)|^2$ is static — an electron in this state cannot radiate. Accordingly the average position $\langle x \rangle$ is also static, and thus the average acceleration is zero. Since classical physics deals with the average, Larmor's theorem relating radiation to acceleration $\boxed{\text{P2.1}}$ would also predict zero radiation classically. If we now have a superposition of orbits, $\Psi(x, t) = a_1\psi_1(x)e^{-i\omega_1 t} + a_2\psi_2(x)e^{-i\omega_2 t}$, the probability will show a sloshing between the two states at frequency $|\omega_1 - \omega_2|$ like a shapeshifter, and we will get discrete radiations driven by their energy difference $|E_1 - E_2|/\hbar$, as we see with atomic spectra.

The difference between classical and quantum objects arises then in their deviations from their average. Remember standard deviation is the root mean squared (RMS) of the deviation $D = O - \langle O \rangle$, an average of the absolute values of the deviations of a sample of entries from their mean, squared to prevent accidental cancellations in sign, square rooted to keep the unit same as the mean.

$$\sigma_O = \sqrt{\langle D^2 \rangle} = \sqrt{\langle [O - \langle O \rangle]^2 \rangle} = \sqrt{\langle O^2 \rangle - \langle O \rangle^2} \quad \text{(expand and verify)}, \tag{3.10}$$

i.e., average of squares minus the square of the average. Since we know how to calculate averages of any quantum operator, we can calculate that for position and momentum as well. What we can then show $\boxed{\text{P3.6}}$ is

$$\sigma_x \sigma_p \geq \hbar/2, \tag{3.11}$$

which is the precise definition of the uncertainty principle. The product of the uncertainties of position and momentum has a lower bound, so that precision in one inevitably implies ignorance of the other. In the absence of knowledge of both position and momentum, we can no longer define a trajectory for the electron, even in principle! Furthermore, it suggests that given a quantum well, an electron cannot sit still at its bottom like a classical object would do. Putting an electron at rest at the bottom would mean we know its position and also its momentum (in this case, zero) with absolute certainty. The best an electron can do is hover and shuttle somewhat above the bottom with a non-vanishing *zero point energy*.

We can verify the uncertainty principle with various eigenvalue solutions that we will encounter going forward $\boxed{\text{P3.7}}$. From symmetry, we expect the lowest uncertainty product (equal to $\hbar/2$) to arise for a Hamiltonian that is symmetric in x and p. Since the Hamiltonian is always quadratic in p from the kinetic energy, this means the minimum uncertainty solution must arise for a potential that is quadratic in x, in other words, for a harmonic oscillator potential. The reader can easily verify it by calculating $\langle x \rangle$, $\langle x^2 \rangle$, $\langle p \rangle$ and $\langle p^2 \rangle$ for the ground state of a harmonic oscillator — a function that retains shape under a Fourier transform ($x \leftrightarrow k$), i.e., a Gaussian $\boxed{\text{P3.7}}$.

3.4 Particle in a Box

Let us start by confining an electron into a box with infinitely high walls

$$U(x) = \begin{cases} 0, & \text{for } 0 < x < L, \\ \infty, & \text{for } x < 0 \text{ and } x > L. \end{cases} \tag{3.12}$$

Each eigenstate $\psi_n(x)$ must vanish at the ends, since the probability $|\psi_n(x)|^2$ of the electrons is zero beyond the barriers. The only energy is the kinetic energy inside the box since $U = 0$. Recall that kinetic energy $\hat{T}$ is proportional to the *curvature* of the wavefunction, so the lowest energy solution must vanish at the ends with minimum curvature, while still maintaining a certain area to enforce normalization. The solution then is a simple bump — maximum in the middle. This is markedly different from a free classical particle which would spread uniformly across the box. The next solution must also minimize its curvature, vanish at the ends, but more importantly be *orthogonal* to the first eigenstate, since the various modes must be independent of each other. In other words, the average of their overlaps $\psi_1(x)\psi_2(x)$ must vanish. This means ψ_2 must have one zero in the middle to enforce orthogonality (only one to keep the curvature low). Similarly, ψ_3 must have two zeros, and so on, until the high energy solutions — the ones with lower wavelengths, start resembling the uniform distribution that classical physics predicts. We thus get a set of quantized solutions that look like the modes of a guitar string clamped at its ends, namely $L = n\lambda_n/2$, fitting an integer number of half-wavelengths (Fig. 5.1). This means $k_n = 2\pi/\lambda_n = n\pi/L$ — *the wavevectors are equally spaced.*

We can get these answers mathematically. Inside the wall, we expect counter-propagating free electrons setting up a standing wave, so each eigenstate must be a mix of $e^{\pm ikx}$, i.e., sines and cosines. Since cosines do not vanish at $x = 0$, we retain only the sines. To vanish at $x = L$, we must now have $\sin kL = 0$, which corresponds to $k_n = n\pi/L$ as above. Enforcing normalization $\int_0^L |\psi_n(x)|^2 dx = 1$, and using the quantized k values in the parabolic $E - k$ for a free electron, we now get

$$\Psi_n(x,t) = \psi_n(x)e^{-iE_nt/\hbar}, \quad \psi_n(x) = \sqrt{\frac{2}{L}}\sin k_n x, \quad E_n = \frac{\hbar^2 k_n^2}{2m}, \quad k_n = \frac{n\pi}{L}. \tag{3.13}$$

3.5 Other Kinds of 'Boxes'

Let us play with the box a bit. What if we want to get a precise location of the particle instead of the probabilistic one currently? We need to make the box thinner. Since the area under the probabilities $|\psi_n(x)|^2$ must still add to unity, each solution must now vary faster. The increase in curvature will push up all their kinetic energies, causing them to spread apart. This is how continuum bands in metallic gold for instance start to spread out and discretize as we go to nanostructured gold, changing color in the process as the transition frequencies $|E_m - E_n|/\hbar$ alter as a result. Furthermore, if you look at each eigenstate, the slope is now wildly varying from positive to zero to negative, meaning the consequence of higher certainty in position is to make the momentum a lot more uncertain.

What if we made the wall heights finite at U_0? We can solve this too mathematically, except the solution is not a closed form expression but graphical. However, it's not hard to see what would happen. For infinite walls, we had an infinite potential energy at the barrier, but notice that the wavefunction abruptly vanished to zero, meaning its curvature (slope of slope) was infinity. This means the kinetic energy $\propto -\nabla^2\psi$ was negative infinity there, and the two infinities canceled out to give a finite total energy E. For finite walls, the kinetic energy and thus the curvature must now be finite, meaning the wavefunctions must tend to zero gradually, with 'infinite support' as mathematicians would say. This inevitably means $\psi_n(x)$ must seep out of the box, with a corresponding non-zero probability in the forbidden negative kinetic energy, barrier region. If we made the walls thinner than this tail, the electron would actually emerge at the other side with a reduced but non-zero probability — a process called quantum tunneling.

Viewed as a wave, this is not surprising — we are familiar with skin depths and frustrated transmissions for EM waves evanescent into opaque metallic films, for instance. For indivisible electrons this indeed seems spooky — out of many incident, indivisible and non-interacting electrons, a small fraction would probabilistically tunnel right through a barrier, for instance across vacuum gap at an STM tip.

Finally, what if we modified the shape of the box — say use a harmonic oscillator potential $U(x) = m\omega^2 x^2/2$? We would expect that the wavefunctions modify a bit to accommodate a finite barrier by tailing out of the well, starting with a Gaussian for the lowest energy, and then picking up additional zeros for orthogonality by pre-multiplying with a polymonial (a Hermite function as it stands). Unlike the square well, we are no longer trying to simply minimize the curvature (i.e., kinetic energy) but the total energy. In fact, as we soar in energy and shrink in wavelength, the probability gets larger away from the center. Lower wavelength modes act classically, where the electron has maximum probability at the two ends where the kinetic energy is lowest and the electron slows down, much like we tend to dither near the turning points of a swing. Finally, as we go up in total energy, the width of the box also increases, meaning the levels get pushed down relative to a square well, making them ultimately vary linearly rather than quadratically with index (Fig. 5.1). Spectroscopists identify vibrational modes by looking for such a linearly spaced ladder.

The above intuition can be succinctly summarized with the exact solution for a harmonic oscillator

$$\psi_n(x) \propto \underbrace{H_n(x/a_0)}_{\text{Hermite}} \underbrace{e^{-x^2/2a_0^2}}_{\text{Gaussian}}, \quad E_n = \left(n + \underbrace{1/2}_{\text{Zero Point Energy}}\right)\hbar\omega \quad , \quad n = 0, 1, 2, \ldots,$$

$$(3.14)$$

where the size of the wavefunction $a_0 = \sqrt{\hbar/m\omega}$ is given by the turning points where the initial kinetic energy $\hbar^2/2ma_0^2$ at the bottom of the well has turned entirely into potential energy $m\omega^2 a_0^2/2$. Instead of demoing the algebra, we will later solve this problem numerically (Fig. 5.1).

3.6 Dirac Equation

The equation for electromagnetic fields is given by the wave equation

$$\frac{\partial^2 \Psi}{\partial t^2} = c^2 \frac{\partial^2 \Psi}{\partial x^2} \tag{3.15}$$

resulting in a linear dispersion $E = |p|c$ for a plane wave $\Psi \propto \exp i[(px - Et)/\hbar]$. In contrast, the equation for a massive free non-relativistic electron is given by the Schrödinger equation $i\hbar\partial\Psi/\partial t = -(\hbar^2/2m)\partial^2\Psi/\partial x^2$. This equation gives us a parabolic dispersion $E = p^2/2m$ for a plane wave solution. The Dirac equation gives us a way to combine these two equations so we can describe relativistic particles with linear dispersions as well as slower non-relativistic particles with parabolic dispersions. The trick is to realize that quadratic equations can be written as two coupled linear equations. We'll work with 1 spatial dimension for simplicity. It is easy to check that the Dirac equation

$$E \begin{bmatrix} \Psi^+ \\ \Psi^- \end{bmatrix} = \begin{bmatrix} mc^2 & -i\hbar c\partial/\partial x \\ -i\hbar c\partial/\partial x & -mc^2 \end{bmatrix} \begin{bmatrix} \Psi^+ \\ \Psi^- \end{bmatrix} \tag{3.16}$$

involves quadratic derivatives for each component, but linear for the overall two component vector, leading to eigenvalues that yield a band dispersion that is a hybrid between the photonic and electronic versions

$$E^\pm = \pm\sqrt{\hat{p}^2c^2 + m^2c^4} = \begin{cases} \pm\hat{p}c, & \text{high speed} \\ \pm\left\{ \underbrace{mc^2}_{\text{rest}} + \dfrac{\hat{p}^2}{2m} + O(\hat{p}^4/m^4c^4) \right\}, & \text{low speed} \end{cases} \tag{3.17}$$

This is easy to show since in 1D $\hat{p} = -i\hbar\partial/\partial x$. For small momenta $p \ll mc$ the $E - p$ reduces to that of the kinetic energy of a free particle plus a rest energy term, i.e, the non-relativistic Schrödinger equation we've been discussing so far. This rest energy term is arguably the most famous equation in physics.

The equations change a bit in 3D, with added terms to describe the y and z derivatives. Specifically, the p-dependent part of the matrix becomes $\vec{\sigma} \cdot \vec{p}$ where the Pauli spin matrices, with very intriguing commutation properties P3.8 , are given by

$$\sigma_x = \begin{pmatrix} 0 & 1 \\ 1 & 0 \end{pmatrix}, \quad \sigma_y = \begin{pmatrix} 0 & -i \\ i & 0 \end{pmatrix}, \quad \sigma_z = \begin{pmatrix} 1 & 0 \\ 0 & -1 \end{pmatrix},$$

$$\vec{\sigma} \cdot \vec{p} = \begin{pmatrix} \hat{p}_z & \hat{p}_x - i\hat{p}_y \\ \hat{p}_x + i\hat{p}_y & -\hat{p}_z \end{pmatrix}. \tag{3.18}$$

Since each of these terms multiplies one of the Ψ components, each of those components $\Psi^\pm$ now becomes a 2×1 vector, and the overall wavefunction picks up four components.

Dirac's great contribution was to develop a relativistic quantum equation for electrons, and make several predictions from it. For starters, you will notice he obtained a set of negative energy solutions, which led to the prediction of antiparticles! Furthermore, if you work out the solutions to the above equation in presence of an atomic field with potential $V(r)$ sitting on the diagonals, you will see the emergence of a few terms beyond the regular Schrödinger equation, one of which is the spin–orbit interaction. We refer the reader to the homework problems to work this out $\boxed{\text{P3.9}}$, and NEMV Section 15.3 for consequences such as *Klein Tunneling*.

Homework

P3.1 **Hermiticity and real eigenvalues.** Consider the eigenvalue equation $\hat{O}\psi_n = \lambda_n \psi_n$ for a Hermitian operator, $\hat{O}^\dagger = \hat{O}$. Apply the adjoint operation to the equation, and show that $\lambda_n = \lambda_n^*$, meaning λ_n is a real eigenvalue. Also show that for two distinct eigenvalues λ_a and λ_b, the corresponding eigenstates ψ_a and ψ_b are orthogonal, $\int \psi_a^* \psi_b dx = 0$.

P3.2 **Hermiticity.** Show that the momentum operator $\hat{p}$ is Hermitian.

P3.3 **Charge conservation and the equation of continuity.** Show that the electron density operator and the current density operator discussed in this chapter satisfy the equation of continuity, $\partial \hat{n}/\partial t = -\partial \hat{J}/\partial x$.

P3.4 **Ehrenfest theorem.** Prove the Ehrenfest theorem by working with the definitions of the averages as well as Schrödinger equation.

P3.5 **Evolution.** Similar to the evolution of $\langle \hat{x} \rangle$ in Ehrenfest's theorem, show that the evolution of the expectation of *any* operator $\langle \hat{O} \rangle$ is

$$\frac{d\langle \hat{O} \rangle}{dt} = \left\langle \frac{\partial \hat{O}}{\partial t} \right\rangle + \frac{1}{i\hbar} \left\langle \left[\hat{O}, \hat{\mathcal{H}} \right] \right\rangle. \tag{3.19}$$

Show that if the potential U is independent of time, the expected energy E is conserved. If it is independent of x, the momentum $\langle p \rangle$ is conserved.

P3.6 **Proving uncertainty.** Let us derive the uncertainty relation for non-commuting operators $\hat{x}$ and $\hat{p}$. The RMS is defined by the deviations from the mean, $\hat{D}_x = \hat{x} - \langle x \rangle$ and $\hat{D}_p = \hat{p} - \langle p \rangle$. Now, the function

$$F(\lambda) = \langle |\hat{D}_x + i\lambda \hat{D}_p|^2 \rangle \tag{3.20}$$

must be non-negative for all λs, assumed real and positive. Expand the expression, and re-express the D_x, D_p terms in terms of σ_x, σ_p and $[\hat{x}, \hat{p}]$, the commutator (keep in mind when expanding that the orders matter!!). Also, note that the commutator of an operator like $\hat{x}$ or $\hat{p}$ with a number (e.g. $\langle x \rangle$ or $\langle p \rangle$) must be zero. The only time the commutator is non-zero is when both are operators.

Since $F \geq 0$ for all λ, let's minimize $F(\lambda)$ by setting the derivative with λ to zero, and find the minimum value of F. Now, requiring that minimum F be non-negative will lead to a new inequality. If you do this correctly, this should be the uncertainty principle in its obvious recognizable form.

P3.7 **Verifying uncertainty.** Prove the uncertainty principle by using

(1) The lowest eigenstate of the particle in a box between $x = -L/2$ to $L/2$. For $\psi = \sqrt{2/L} \sin \pi x/L$, it is easy to argue that $\langle x \rangle = \langle p \rangle = 0$, simply from symmetry arguments (think through it!): Find $\langle x^2 \rangle$, $\langle p^2 \rangle$, thence σ_x and σ_p, and show that $\sigma_x \sigma_p > \hbar/2$.
(2) Repeat for the lowest eigenstate for a particle in an oscillator potential, $\psi = e^{-x^2/2a_0^2}/(\pi a_0^2)^{1/4}$.

P3.8 **Pauli matrix maths.** Using the forms for the Pauli matrices (Eq. (3.18)), show that their determinants are -1, traces 0 and eigenvalues are ± 1. Also prove their commutation and anti-commutation rules

$$\{\sigma_x, \sigma_y\} = \sigma_x \sigma_y + \sigma_x \sigma_y = 2i\sigma_z, \quad \text{(and cyclic permutations)},$$
$$[\sigma_a, \sigma_b] = \sigma_a \sigma_b - \sigma_b \sigma_a = 2\delta_{ab}I, \quad (a, b = x, y, z). \tag{3.21}$$

Also show that

$$(\vec{\sigma} \cdot \vec{p})^2 = p^2,$$
$$(\vec{\sigma} \cdot \vec{p})V(\vec{\sigma} \cdot \vec{p}) = V(\vec{\sigma} \cdot \vec{p})^2 - (i\hbar\vec{\nabla}V \cdot \vec{\sigma})(\vec{\sigma} \cdot \vec{p}). \tag{3.22}$$

P3.9 **Dirac maths.** The Dirac equation can be written as

$$E \begin{bmatrix} \Psi^+ \\ \Psi^- \end{bmatrix} = \begin{bmatrix} mc^2 + qV & \vec{\sigma} \cdot \vec{p}c \\ \vec{\sigma} \cdot \vec{p}c & -mc^2 + qV \end{bmatrix} \begin{bmatrix} \Psi^+ \\ \Psi^- \end{bmatrix}. \tag{3.23}$$

3.9.1 Expand the second line, assuming low energy ($E, V \ll mc^2$), show that

$$\Psi^- \approx \frac{1}{2mc^2} \left(1 - \frac{W - qV}{2mc^2} \right) \vec{\sigma}.\vec{p}c\Psi^+, \tag{3.24}$$

where $W = E - mc^2$.

3.9.2 Substitute in the first equation, expand using the identities from part 3.3.1. Show that we get the non-relativistic Hamiltonian

$$\mathcal{H} \approx \underbrace{\frac{p^2}{2m} + V}_{} \underbrace{- \frac{p^4}{8m^3c^2}}_{\text{relativistic mass term}} + \underbrace{\frac{1}{2m^2c^2}\vec{\sigma}\cdot(\vec{\nabla}V)\times\vec{p}}_{\text{spin-orbit}} + \underbrace{\frac{\hbar^2}{8m^2c^2}\nabla^2 V}_{\text{Darwin term}}$$

(3.25)

3.9.3 For an atom whose V depends only on r (central potential), show that the spin–orbit term looks like $H_{\text{so}} = A(dV/dr)\vec{\sigma}\cdot(\vec{r}\times\vec{p})$. Also, using the commutative properties of scalar triple product show that this term can be written as $\xi(r)\vec{L}\cdot\vec{S}$, where $\vec{L} = \hat{r}\times\hat{p}$ is the angular momentum operator and $\vec{S} = \hbar\vec{\sigma}/2$ is the spin operator.

Chapter 4

Bumps Along the Quantum Path: Steps, Wells and Barriers

4.1 Particle on a Step

Continuing with our exploration of wave properties of electrons, let us imagine an electron incident on a step function potential (Fig. 4.1), $U(x) = U_0\Theta(x)$, much like a ball thrown at a brick wall. Classical mechanics would predict that the electron would have zero probability of being transmitted to the other side if its kinetic energy is less than the barrier height U_0 and 100% (unit transmission) otherwise. What does quantum physics predict? If you're not familiar with this, the answer might surprise you. To solve this problem, let us write down the piece-meal solutions in each region. For a free electron, we expect plane waves, so the wavefunction can be written as (note direction of decay so wavefunctions are finite at $x = \pm\infty$)

$$\psi(x) = \begin{cases} e^{ik^*x} + re^{-ikx}, & x \le 0 \quad (\text{*means complex conjugate}) \\ te^{ik'x}, & x > 0, \end{cases} \tag{4.1}$$

where $k = \sqrt{2mE/\hbar^2}$ and $k' = \sqrt{2m(E - U_0)/\hbar^2}$. The barrier keeps the total energy of the electron fixed (there is no mechanism to funnel energy out), but the kinetic energy goes down when the electron goes over the barrier, so its wavelength increases. The change in wavelength creates a partial reflection, much like water waves moving into a shallow region.

We now match boundary conditions at $x = 0$ $\boxed{\text{P4.1, P4.2}}$, $\psi(0^-) = \psi(0^+)$, and $d\psi/dx|_{0^-} = d\psi/dx|_{0^+}$. The first automatically conserves charge, the second conserves current density. Applied to ψ above, allowing k to be complex, we get $1 + r = t$ and $k^* - rk = k't$, whose solutions are

$$r = (k^* - k')/(k + k'), \quad t = (k^* + k)/(k + k') = 2k_R/(k + k') \tag{4.2}$$

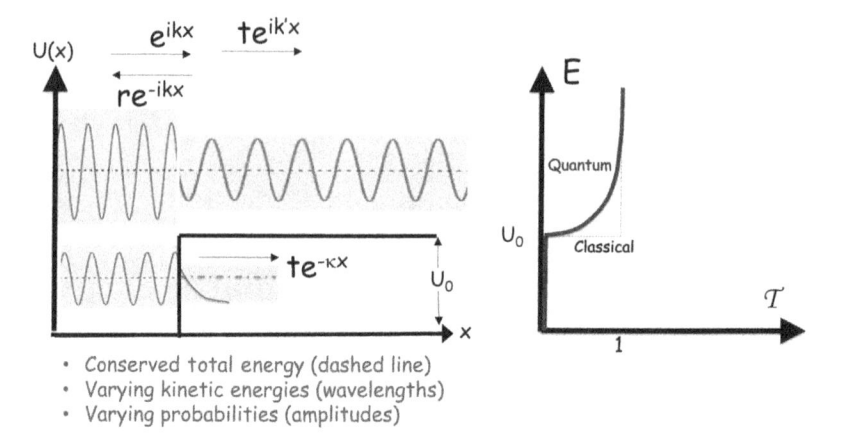

- Conserved total energy (dashed line)
- Varying kinetic energies (wavelengths)
- Varying probabilities (amplitudes)

Fig. 4.1 Waves at a step and their transmission probability vs. energy.

with R denoting real component and I imaginary. We wish to calculate the transmission $\mathcal{T}$, which is the ratio of *current densities* J (particle flow per second per unit area) of the transmitted and incident components. The complement is the reflection $\mathcal{R} = 1 - \mathcal{T}$, given by current density ratio of reflected and incident components. For a free electron using Eq. (3.5), $J_t(x \geq 0) = q(\hbar k'_R/m)|t|^2 e^{-2k'_I x}$, $J_r(x \leq 0) = q(\hbar k_R/m)|r|^2 e^{2k_I x}$ and $J_i(x \leq 0) = q(\hbar k_R/m)|1|^2 e^{2k_I x}$, we get

$$\mathcal{R} = |r|^2 = \frac{|k^* - k'|^2}{|k + k'|^2} \quad \text{(step function potential)},$$

$$\mathcal{T} = |t|^2 \left(\frac{k'_R}{k_R}\right) = \frac{4k_R k'_R}{|k + k'|^2},$$

$$\mathcal{R} + \mathcal{T} = 1 \quad \text{(check it!}\ \boxed{\text{P4.1}}\text{)}. \tag{4.3}$$

For energies $E > U_0$, k and k' are both real, and the reflectivity depends (predictably) on the fractional change in wavelength across the junction. For energies $E < U_0$, k' becomes imaginary, $k' = i\kappa$ with $\kappa = \sqrt{2m(U_0 - E)/\hbar^2}$, whereupon $\mathcal{R} = 1$ and $\mathcal{T} = 0$. Nothing transmits!

We can plot this transmission over the entire energy range by using the energy dependences of k and k'. The counter-intuitive part is that even when the electron clears the barrier, there is a finite probability of reflection. This is actually what we see for water waves that move into a shallow region, but we certainly don't expect it for bullets. We now expect a few electrons with energy above the barrier to reflect back magically from a seemingly invisible barrier because the electron wavelength is large enough to 'sense' it and react.

To recover the classical result of a step-function variation in transmission, we need to make the fractional change in wavelength small (see equation for $\mathcal{R}$ above), meaning we need a slowly varying potential. The phase picked up by an electron over one wavelength must be correspondingly small. We can see that by solving for a slowly varying potential between $x = \pm\infty$, $U(x) = U_0/(1 + e^{-x/x_0})$. For this potential, we can actually solve the Schrödinger equation exactly (see Landau–Lifshitz), although the algebra is daunting and involves in-depth knowledge of hypergeometric functions. The result however is instructive (compare with expression above):

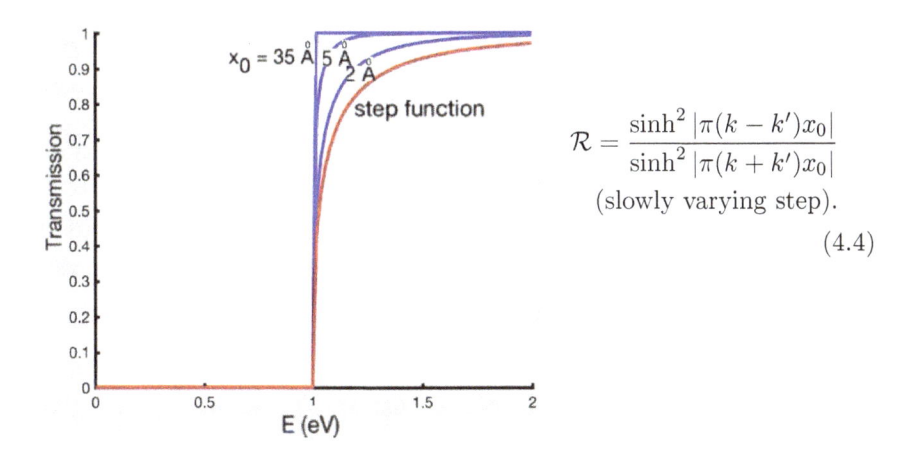

$$\mathcal{R} = \frac{\sinh^2 |\pi(k - k')x_0|}{\sinh^2 |\pi(k + k')x_0|}$$
(slowly varying step).

$$(4.4)$$

It is straightforward to see that if we keep x_0 finite and reduce the de Broglie wavelength so that $k'x_0 \gg 1$, the reflection coefficient decays as $\sim \exp[-2\pi x_0 k'] = \exp[-x_0/\lambda']$, vanishing immediately upon clearing the barrier as a classical particle is expected to do.

The exercise above explains the distinction between the quantum and the classical. Quantum properties are ultimately stored in the phase variations of the various eigenstates under an external potential. When the potential varies slowly enough so that the fractional change in phase over one wavelength is minimal, the electron acts classically, as in the slowly varying step above. This also explains how phase-breaking processes at room temperature cancel any phase accumulated over one wavelength and tend to keep quantum objects classical.

4.2   Particle on a Barrier: Resonance and Tunneling

Things get more complicated (and richer!) P4.3–P4.5 if we now use a finite barrier of width L. Classically, the finite width does not matter — once you've cleared the wall, you've cleared the wall (again don't think of a physical wall and gravity, but an energy barrier). Wave dynamics is more involved! Figure 4.2 shows what happens. Even when we don't clear the barrier, there is an evanescent tail of the wavefunction that penetrates the barrier. If the barrier is thin enough, the tail will make it out to the other side and the particle will emerge with a finite, albeit reduced probability (lower right inset, Fig. 4.2). One out of several incident electrons will actually disappear from one side of the barrier, tunnel right through and emerge at the other.

For electrons above barrier, there is more intrigue! The transmission is nominally less than unity, as there is inevitable back reflection due to a sudden change in potential, same as a step (middle right inset, Fig. 4.2). However, at specific magic energies, the transmission probability shoots up to unity (resonances), and the particle makes it through with absolute certainty. Looking at the wavefunction at that energy (upper right inset in Fig. 4.2), we see what happens. When the width of the barrier exactly matches an integer number of half-wavelengths, the value and slope of the wavefunction are identical in absolute value at the two ends. Since

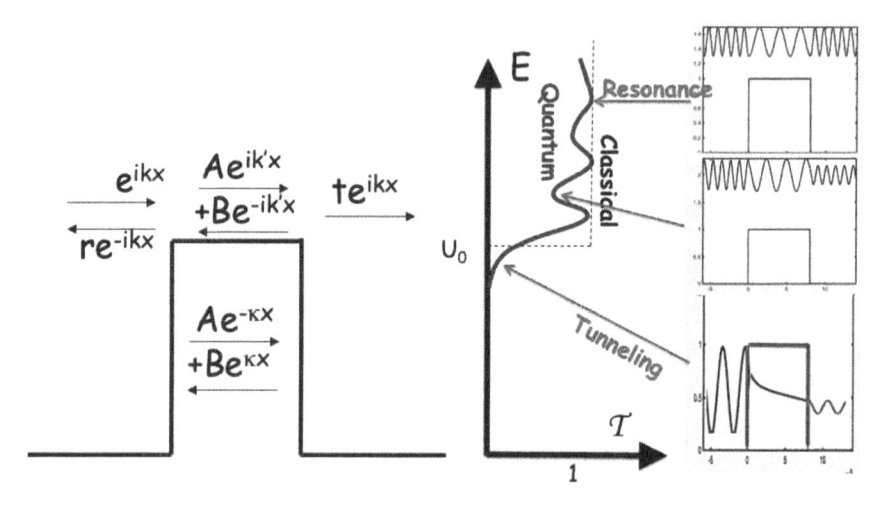

Fig. 4.2 Transmission across a barrier shows quantum tunneling at low energies and resonances (perfect transmission) at select higher energies.

we match value and slope of the plane waves at the two ends, the symmetric structure implies symmetric plane waves going away from the barrier on both sides with identical amplitudes, so that their ratio, i.e., the transmission is unity (strictly speaking, this ratio is $(1 - r)/t$, but since charge is also conserved, i.e., $1 + r = t$, the combination ensures $r = 0$ and $t = 1$).

Once the physics makes sense, it is much easier to get the algebra in place. The wavefunction for $E > U_0$

$$\psi(x) = \begin{cases} e^{ikx} + re^{-ikx}, & x \le 0, \\ Ae^{ik'x} + Be^{-ik'x}, & 0 < x \le L, \\ te^{ikx}, & x > L. \end{cases} \qquad (4.5)$$

Matching the wavefunction and derivatives at the two edges of the barrier $x = 0$ and L, we get

$$\left. \begin{aligned} 1 + r &= A + B \\ k(1 - r) &= k'(A - B) \end{aligned} \right\} \Rightarrow 2 = \left(1 + \frac{k'}{k}\right)A + \left(1 - \frac{k'}{k}\right)B,$$

$$\left. \begin{aligned} Ae^{ik'L} + Be^{-ik'L} &= te^{ikL} \\ k'(Ae^{ik'L} - Be^{-ik'L}) &= kte^{ikL} \end{aligned} \right\} \Rightarrow 0 = \left(1 - \frac{k'}{k}\right)Ae^{ik'L} + \left(1 + \frac{k'}{k}\right) \times Be^{-ik'L}.$$

$$(4.6)$$

Solving for A and B from the two right equations, and then plugging back into the last equation to get t, we finally get the transmission $\mathcal{T} = |t|^2$:

$$\mathcal{T} = \frac{4k^2(k')^2}{\left[4k^2(k')^2 \cos^2 k'L + \left(k^2 + (k')^2\right)^2 \sin^2 k'L\right]}, \quad k' = \sqrt{\frac{2m(E - U_0)}{\hbar^2}}.$$

$$(4.7)$$

We can readily see that at resonances when $k'L = n\pi$ (i.e., $L = n\lambda'/2$, in other words, fitting an integer number of half-wavelengths), the sine term vanishes and the cosine becomes unity, so that $\mathcal{T} = 1$. Once again for a slow, rounded potential, we eventually recover the clasical result $\boxed{\text{P22.2}}$.

For electron energies lower than the barrier height, $U_0 > E$, $k' = i\kappa$ becomes imaginary, and

$$\mathcal{T} = \frac{4k^2\kappa^2}{\left[4k^2\kappa^2 \cosh^2 \kappa L + \left(k^2 - \kappa^2\right)^2 \sinh^2 \kappa L\right]}, \quad \kappa = \sqrt{\frac{2m(U_0 - E)}{\hbar^2}}$$

$$\approx \frac{16E(U_0 - E)}{U_0^2}e^{-2\kappa L} \quad \text{(when } \kappa L \gg 1\text{)}. \qquad (4.8)$$

At this time, we see an electron tunneling through, since its probability is non-zero below the barrier height. The exponential sensitivity of tunneling current on barrier length is used in scanning tunneling microscopy (STM) to image the atomic structure of surfaces — as the STM tip travels over the atomically corrugated surface at a constant height, the tunnel barrier changes width L and so does the current, exponentially, so that the current map reveals the topography of the surface (on occasion, we switch to constant current mode with a feedback, whereupon the tip dances up and down to trace out the atomic topography).

The above equation works for a constant potential barrier U_0. For slowly varying potentials $U(x)$, we can generalize the exponential $\sim e^{-2\kappa L}$ to get the Wentzel–Kramers–Brillouin (WKB) tunneling formula

$$\mathcal{T}_{\text{WKB}} \approx e^{\displaystyle -2 \int_{x_1}^{x_2} \kappa(x) dx} \quad , \quad \kappa(x) = \sqrt{\frac{2m[U(x) - E]}{\hbar^2}}, \tag{4.9}$$

where $x_{1,2}$ are the turning points between which the electronic energy E is below the barrier height, in other words, the end points are set by $E = U(x_{1,2})$. Note that we dropped a prefactor term $\sim 16E[U_0 - E]/U_0^2$. That term comes from the mismatch in kinetic energy at the barrier. It is a small term compared to the exponential variation, but can well end up being important. For instance, the difference in current between the parallel and antiparallel states of a magnetic tunnel junction (the so-called tunnel magnetoresistance, TMR) is mainly set by the ratio of this pre-factor for the parallel and antiparallel configurations.

4.3 Point Scattering

We just worked out the transmission through a barrier. If we make the barrier thin but increase its height so its impact is still finite, we will reach the scattering from a point particle — a delta function, with $U(x) = U_0 \delta(x)$. Since a delta function integrates to unity and U has units of volts, U_0 has units of volts-meter. Like before, we wish to calculate the transmission of an electron incident as a plane wave e^{ikx}, plotted as a function of energy. By matching the wavefunction values ψ at $x = 0$, we get

$$\psi(0) = 1 + r = t. \tag{4.10}$$

This part is the same as before. Normally, we match derivatives as well at this point, but since the potential is infinite at the delta function, a finite total energy E will require an infinite negative kinetic energy (curvature), meaning there must be a discontinuity in the derivative (recall a similar discontinuity at the infinitely tall walls of a particle in a box). We can extract this discontinuity by integrating the Schrödinger equation over a small distance around the spike between $x = \pm\epsilon$ with $\epsilon \to 0$:

$$\frac{-\hbar^2}{2m} \underbrace{\int_{-\epsilon}^{\epsilon} dx \frac{\partial^2 \psi}{\partial x^2}}_{[\partial\psi/\partial x|_\epsilon - \partial\psi/\partial x|_{-\epsilon}]} + \underbrace{\int_{-\epsilon}^{\epsilon} dx U_0 \delta(x)\psi(x)}_{U_0\psi(0)} = E \underbrace{\int_{-\epsilon}^{\epsilon} dx \psi(x)}_{2\epsilon\psi(0) \to 0}.$$

Substituting $\psi(0)$ from Eq. (4.10)

$$\frac{-\hbar^2}{2m}\left[ik(t - 1 + r)\right] + U_0 t = 0. \tag{4.11}$$

From these two equations, we can solve for t and get the transmission

$$\mathcal{T} = |t|^2 = \frac{1}{1 + \dfrac{m^2 U_0^2}{\hbar^4 k^2}} = \frac{1}{1 + \dfrac{m U_0^2}{2\hbar^2 E}}. \tag{4.12}$$

Note that the sign of U_0 does not matter, meaning the scattering happens whether we have a barrier ($U_0 > 0$) or a well ($U_0 < 0$). The electron perceives any difference in potential that varies abruptly over a distance shorter than its de Broglie wavelength and reacts by reflecting back. In the case of the delta function, even though it is infinitely high, it is also infinitely thin, and what matters ultimately is the integrated potential (Eq. (4.11)).

To convert this to a finite energy, we introduce a thickness to the delta function, $U_1 = U_0 a$, converting U_1 to units of volts. We then get

$$\mathcal{T}_{\text{defect}} = \frac{1}{1 + \dfrac{U_1^2}{\Gamma^2}}, \tag{4.13}$$

where

$$\Gamma = \frac{\hbar^2 k}{ma} = \frac{\hbar v}{a}, \quad v = \frac{\hbar k}{m} \tag{4.14}$$

is the level broadening, inversely proportional to the transit time, $\Gamma = \hbar/\tau$, $\tau = a/v$. We can also get this equation from the limiting case of a rectangular barrier $\boxed{\text{P4.6}}$.

Homework

P4.1 **Conservation:** For plane waves at a step with complex ks, r and t were obtained in Eq. (4.2) and $\mathcal{R}$ and $\mathcal{T}$ in Eq. (4.3). Show that $\mathcal{R}+\mathcal{T} = 1$.

P4.2 **Transmission with mass variation:** The second boundary condition for a particle at a step was current conservation — continuity of $d(\psi/m)/dx$. Repeat for a step where the masses $m_{1,2}$ of the electron are different on the two sides, and show that the reflection coefficient is given by mismatch of *velocities k/m*. This would explain why for materials with constant band velocity (e.g., graphene), there is no reflection in 1D, a piece of physics called *Klein tunneling* (see NEMV, Section 15.3).

P4.3 **Transfer matrix method:**

(a) Consider an electron going through a sequence of wells and barriers. We can decompose the propagation and transfer of electron waves into four pieces: (a) well to barrier transmission, (b) barrier to well transmission, (c) propagation in barrier and (d) propagation in well. For each process, we can connect the coefficients of forward and backward waves to the right and left ends using a 2×2 matrix. The four processes can be described by two matrices: one describing transmission across an

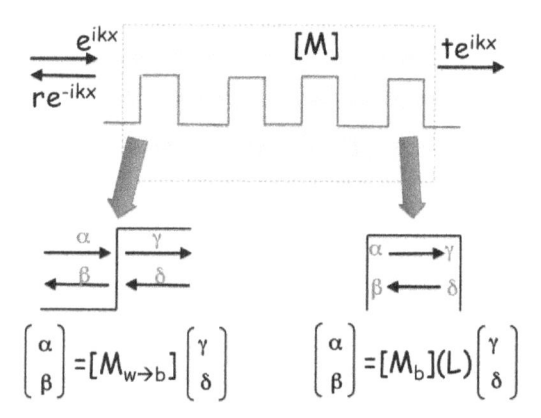

abrupt interface and the other propagation through a medium. Show that these are

$$[M]_{A \to B} = \begin{pmatrix} \dfrac{1 + \eta_{AB}}{2} & \dfrac{1 - \eta_{AB}}{2} \\[2mm] \dfrac{1 - \eta_{AB}}{2} & \dfrac{1 + \eta_{AB}}{2} \end{pmatrix}, \quad \eta_{AB} = k_B/k_A \text{ (material A to B)},$$

$$[M]_A(L) = \begin{pmatrix} e^{-ik_A L} & 0 \\ 0 & e^{ik_A L} \end{pmatrix} \quad \text{(material A width L).}$$

$$(4.15)$$

(b) Starting from the right end outputs, we can now move to the left end inputs to sequence the matrices. For instance, transfer through a barrier can be written as

$$[M_B](L) = [M_{w \to b}][M_b(L)][M_{b \to w}].$$ $$(4.16)$$

(c) Show that $M_{b \to w} = M_{w \to b}^{-1}$, and using that show that when $k_b L = n\pi$, i.e., resonance, the transfer matrix becomes the identity matrix and transmission is thus unity. Also, show that

$$[M_B] = \begin{bmatrix} \cos\theta_1 - i\alpha \sin\theta_1 & -i\beta \sin\theta_1 \\ i\beta \sin\theta_1 & \cos\theta_1 + i\alpha \sin\theta_1 \end{bmatrix},$$

$$\theta_1 = k_b L, \quad \alpha = \frac{\eta_{wb} + \eta_{wb}^{-1}}{2}, \quad \beta = \frac{\eta_{wb} - \eta_{wb}^{-1}}{2}. \quad (4.17)$$

Also, show that $det(M_B) = 1$, and thus, $|[M_B]_{11}| = \sqrt{1 - \beta^2 \sin^2\theta_1}$.

(d) From $[M_B]$, we can calculate transfer through two barriers one well:

$$[M] = [M_B(L_1)][M_w(L_2)][M_B(L_3)].$$ $$(4.18)$$

At the ends, we can put the input coefficients as 1 and r, and output coefficients as t and 0 for the forward and reverse waves

$$\begin{pmatrix} 1 \\ r \end{pmatrix} = [M] \begin{pmatrix} t \\ 0 \end{pmatrix}$$ $$(4.19)$$

so that

$$\mathcal{T} = |t|^2 = \frac{1}{|[M]_{11}|^2}.$$ $$(4.20)$$

Define $[M_B]_{11} = |[M_B]_{11}| e^{-i\phi}$, in other words, $\phi = \tan^{-1}(\alpha \tan \theta_1)$. Using this expression, show that

$$\mathcal{T} = |t|^2 = \frac{1}{1 + 4\beta^4 \sin^4 \theta_1 \sin^2 (\theta_2 + \phi)}, \tag{4.21}$$

where $\theta_2 = k_w W$ is the phase across the well. We thus see resonances when $\sin(\theta_2 + \phi) = 0$, and when $\sin \theta_1 = 0$. Interpret these two results.

P4.4

Breit–Wigner resonance: Going back to the transmission expression $\mathcal{T}$ for the barrier, show that near the nth resonance $E \approx E_n + \delta E$, we can Taylor expand and get the single impurity result, with the broadening Γ_n given by the rate of change of energy with phase ϕ.

$$\mathcal{T}_{\text{Barrier}}(E) \approx \frac{1}{1 + \delta E_n^2/\Gamma_n^2}, \quad \text{where } \Gamma_n = dE/d\phi = dE/d(kL). \tag{4.22}$$

Since $\Gamma_n = \hbar v/L$ (Eq. (4.14)), we show that the velocity of the electron cans be written as $v = dE/\hbar dk$. We will revisit this relation in Section 8.2.

P4.5

Quantum-classical correspondence: Input a trial wavefunction $\psi(x) = A(x)e^{i\phi(x)}$, with real A, ϕ in Schrödinger equation, and match real and imaginary parts to get two coupled differential equations. Show that we recover the WKB expression assuming $|d^2U(x)/dx^2|/|U(x)| \ll 1/\lambda^2$, i.e., a potential $U(x)$ with slow variation compared to the electron wavelength over which the fractional change of phase is negligible. Prove that the slow varying expression for a smooth step connects the classical and quantum limits (Eq. (4.4)). For a smooth barrier, we will need numerical tools $\boxed{\text{P22.2}}$.

P4.6

Single particle scattering: Simplify the transmission $\mathcal{T}(E)$ for a barrier of height U_0 and width a in terms of E and U_0, assuming $E < U_0$ (tunneling regime). Keep the arguments of the sinh and cosh functions in terms of κa. The m and $\hbar$ terms should cancel out. Now, work out the results for a 'delta' function (a spike). Following the property of a delta function, we make its height $U_0 \to \infty$ and width $a \to 0$ such that the area, i.e., the product $U_0 a = U_1$, and the energy E stay constant. Using Taylor expansion and simplification (keep in mind as $U_0 \gg E$, $ka, \kappa a \to 0$) show that you get the expression above for a delta function.

The transit time τ for a free electron of mass m to cross a barrier of width a is given by a/v. Write v in terms of the kinetic energy E. The corresponding broadening Γ is obtained from the energy–time uncertainty principle, $\Delta E \Delta t = \Gamma \tau = \hbar$. Show that the transmission above has the shape of a Lorenzian with broadening Γ, i.e.,

$$\mathcal{T} = \frac{1}{1 + \dfrac{U_0^2}{\Gamma^2}}.$$

Chapter 5

Numerical Methods and Approximations

Few eigenvalue problems can be solved exactly (particle in an infinite box, oscillator, hydrogen atom). For most potentials, we develop approximation techniques like variational principle $\boxed{\text{P5.1}}$ and perturbation theory $\boxed{5.2}$, or numerical methods $\boxed{\text{P5.3 and P5.4}}$, discussed in what follows.

5.1 Time-Independent Problems: Finite Difference in 1D and 2D

Let us first set up a way to calculate the eigenspectrum of a Hamiltonian in 1D. We resolve the desired eigenfunction onto a uniform grid of points $\{x\} = (x_1, x_2, \ldots, x_N)$ separated by a distance a_0, and write down the discrete representation of $\psi(x)$ as an $N \times 1$ vector

$$\psi(x) \Longrightarrow \underbrace{\begin{pmatrix} \psi_1 \\ \psi_2 \\ \vdots \\ \psi_N \end{pmatrix}}_{[\psi]}. \tag{5.1}$$

The potential term is local, $U(x)\psi(x)$, and can thus be written as a matrix multiplication with a diagonal $[U]$ matrix

$$U(x)\psi(x) \Longrightarrow \begin{pmatrix} U_1\psi_1 \\ U_2\psi_2 \\ \vdots \\ U_N\psi_N \end{pmatrix} = \underbrace{\begin{pmatrix} U_1 & 0 & \ldots & \ldots & 0 \\ 0 & U_2 & 0 & \ldots & 0 \\ \vdots & \vdots & \vdots & \vdots & \vdots \\ \ldots & \ldots & \ldots & 0 & U_N \end{pmatrix}}_{[U]} \begin{pmatrix} \psi_1 \\ \psi_2 \\ \vdots \\ \psi_N \end{pmatrix}. \tag{5.2}$$

Finally, we reach the kinetic energy term, which involves second derivatives and is thus nonlocal in x. A first derivative on the discrete grid can be written variously as (prime meaning derivative)

$$\frac{d\psi}{dx}\bigg|_{x_n} = \psi'_{x_n} \approx \underbrace{\frac{\psi_{n+1} - \psi_n}{a_0}}_{\text{forward difference}} = \underbrace{\frac{\psi_n - \psi_{n-1}}{a_0}}_{\text{backward difference}} = \underbrace{\frac{\psi_{n+1/2} - \psi_{n-1/2}}{a_0}}_{\text{symmetric difference}}, \tag{5.3}$$

all of which should converge for $a_0 \to 0$. Using symmetric difference of a symmetric difference, we can then get the kinetic energy

$$\frac{-\hbar^2}{2m}\frac{d^2\psi}{dx^2}\bigg|_{x_n} = \frac{-\hbar^2}{2m}\left(\frac{\psi'_{n+1/2} - \psi'_{n-1/2}}{a_0}\right) = \frac{-\hbar^2}{2m}\left(\frac{\psi_{n+1} + \psi_{n-1} - 2\psi_n}{a_0^2}\right)$$

$$= 2t_0\psi_n - t_0\psi_{n+1} - t_0\psi_{n-1}, \quad \text{where } t_0 = \hbar^2/2ma_0^2. \tag{5.4}$$

This means the kinetic energy term can be written as

$$\hat{T}\psi(x) \Longrightarrow \underbrace{\begin{pmatrix} 2t_0 & -t_0 & \ldots & \ldots & 0 \\ -t_0 & 2t_0 & -t_0 & \ldots & 0 \\ \vdots & \vdots & \vdots & \vdots & \vdots \\ \ldots & \ldots & \ldots & -t_0 & 2t_0 \end{pmatrix}}_{[T]} \begin{pmatrix} \psi_1 \\ \psi_2 \\ \vdots \\ \psi_N \end{pmatrix}. \tag{5.5}$$

In MATLAB, we can then set up the Hamiltonian matrix $[T+U]$ using a single line command over an N-atom grid, and then find the eigenvalues. This is obtained with one line, $[\,V, D\,]=\mathrm{eig}(H)$. The result gives us an $N \times N$ diagonal matrix $[D]$ whose diagonal entries are the separate eigenvalues (in fact, the process of finding eigenvalues is called *diagonalization* — the corresponding eigenvectors no longer couple through the Hamiltonian and are independent modes — much like the normal modes of a vibrating solid). The $N \times N$ eigenvector matrix $[V]$ has each eigenvector ψ_p, $p = 1, 2, \ldots, N$ displayed vertically along the pth column, the qth row entries showing the value of ψ_p at the qth spatial point x_q.

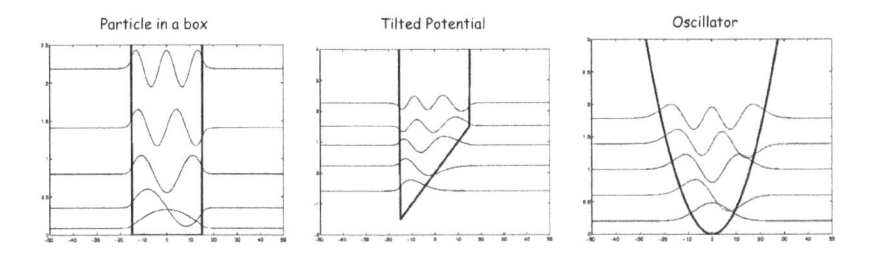

Fig. 5.1 Potential wells and their numerically extracted eigenspectra.

In Fig. 5.1, we demonstrate it for a rectangular box, a tilted box and an oscillator potential (the code line in use is uncommented).

```
%% Set parameters and grids in mks units and eV
t=1; Nx=101; x=linspace(-5,15,Nx);

%% Write down Hamiltonian
% U = [ 100*ones(1,11) zeros(1,79) 100*ones(1,11) ]; % Particle in a box
% U=[ 100*ones(1,11) linspace(0,5,79) 100*ones(1,11) ]; % Tilted box
U=x.^2; U=U/max(U); U = diag(U); % Oscillator
T=2*t*eye(Nx)-t*diag(ones(1,Nx-1),1)-t*diag(ones(1,Nx-1),-1);
H = T+U;

%% Find eigenspectrum and plot vertically
[ V, D ]=eig(H);

%% Plot
for k=1:5
plot(x,V(:,k)+10*D(k),'r','linewidth',3) % Plot at height propto energy
hold on
grid on
end
plot(x,U,'k','linewidth',3); % Zoom if needed
axis([-5 5 -2 10 ])
```

Let us now consider a 2D square lattice, with couplings $-t_0$ from every site to its left and right $(x, y \pm 1)$, top and bottom $(x \pm 1, y)$. The 2D Hamiltonian can be reduced to a pentadiagonal problem, by lumping the two indices into a single running index (one that rasters line by line). For an $N \times N$ array of rectangular grid points, we can entangle and disentangle the coordinates as (Matlab commands)

$$(x, y) \to z = (x - 1) * N + y$$
$$z \to y = \mathrm{mod}(z, N); x = \mathrm{floor}(z/N) + 1 = (z - y)/N + 1. \quad (5.6)$$

The nearest neighbor couplings $(x, y \pm 1)$, $(x \pm 1, y)$ then boil down to $z \pm 1$, $z \pm N$. On a $N^2 \times N^2$ matrix that lists z entries along the rows and columns, this then boils down to entries at the first principle off diagonals and another set at Nth principle off diagonals. At the end we can disentangle the representation back onto the (x, y) grid. The 2D kinetic energy term now simplifies to a pentadiagonal $N^2 \times N^2$ matrix

$$[T] = \begin{pmatrix} 4t_0 & -t_0 & 0 & \dots & -t_0 & \dots & 0 \\ -t_0 & 4t_0 & -t_0 & \dots & \dots & -t_0 & 0 \\ 0 & \dots & 4t_0 & \dots & \dots & \dots & -t_0 \\ \dots & \dots & \dots & \dots & \dots & \dots & \dots \\ -t_0 & \dots & \dots & \dots & \dots & \dots & \dots \\ \dots & -t_0 & \dots & \dots & \dots & 4t_0 & -t_0 \\ \dots & \dots & -t_0 & \dots & \dots & -t_0 & 4t_0 \end{pmatrix}. \quad (5.7)$$

Furthermore, *in a magnetic field* the second set of off diagonal elements pick up a complex phase arising from the electromagnetic momentum $\boxed{\text{P11.8}}$ (see NEMV Section 21.3 for the physics)

$$T_{i,i\pm N} = -t_0 e^{\mp 2\pi i m \Phi/\Phi_0}, \quad m = 1, 2, \dots, N^2 - N, \quad (5.8)$$

where Φ is the magnetic flux (field times area) across a unit cell, and $\Phi_0 = h/2q$ is the flux quantum. In presence of a magnetic field, the eigenvalues of this 2D structure show discrete Landau levels at low fields and a more intricate Hofstadter butterfly at larger fields when the cyclotron orbits are tight enough to sample the underlying discrete lattice potential $\boxed{\text{P11.8}}$.

5.2 Time-Dependent Problems: Crank–Nicolson

Let us try to solve a 1D problem with a rectangular barrier, and track the waves as a function of time. We use the Crank–Nicolson method to solve the Schrödinger equation on a time grid of points

$$i\hbar\frac{\partial\psi}{\partial t} = \hat{\mathcal{H}}\psi \quad\Longrightarrow\quad \psi(t) = \underbrace{e^{-i\int_0^t dt'\,\hat{\mathcal{H}}(t')/\hbar}}_{U(t)}\psi(0),$$

$$\psi(t+\Delta t) = \left(1 - \frac{i\hat{\mathcal{H}}(t)\Delta t}{2\hbar}\right)\left(1 + \frac{i\hat{\mathcal{H}}(t)\Delta t}{2\hbar}\right)^{-1}\psi(t). \tag{5.9}$$

The second equation follows by approximating the unitary operator $U = e^{-i\hat{\mathcal{H}}\Delta t/\hbar}$ in a way that preserves the probability $|\psi|^2$ (a set of finite difference forward Euler approximations), as you can readily check. We can now step through a time grid to calculate the wavefunction at every time based on its previous result. We choose an initial condition given by a plane wave, multiplied by a Gaussian for easy visualization.

```
%% Time-dependent SE showing scattering of particle from potential
Nx=101; x=linspace(0,100,Nx); dx=x(2)-x(1); % real space grid
sig=8; x0=20; k=1; % parameters for wavepacket distribution
Nb=20; k=2*pi/(Nb+1); % for resonance in RTD
psi=exp(-i*k*x).*1/(sig*sqrt(2*pi)).*exp(-(x-x0).^2)/(2*sig^2));
% Initial condition — plane wave times Gaussian for visualization
H=2*eye(Nx)-diag(ones(1,Nx-1),1)-diag(ones(1,Nx-1),-1);

%% Create potential — comment/uncomment relevant line
% Nd=3;V=blkdiag(zeros(40),eye(Nx-40)); % Step
% Nd=30;V=blkdiag(zeros(40),-eye(Nd),zeros(Nx-40-Nd)); % Well
Nd=30;V=blkdiag(zeros(40),0.5*eye(Nd),zeros(Nx-40-Nd)); % Barrier
% Nb=20;Nw=20;
%V=blkdiag(zeros(20),eye(Nb),zeros(Nw),eye(Nb),zeros(Nx-20-Nw-
2*Nb)); %RTD
% V=V+diag(linspace(0,1,Nx)); %Field
H=H+V;

%% Boundary condition
sig=zeros(Nx); % Hard wall
%s1=-exp(i*k); s2=s1; sig1=zeros(Nx);
%sig(1,1)=s1; sig(Nx,Nx)=s1; % Open boundary conditions
```

```
%sig=zeros(Nx); sig(1,Nx)=-1; sig(Nx,1)=-1; % Periodic BCs
H=H+sig;
Nt=301;t=linspace(1,300,Nt);dt=t(2)-t(1);
p=zeros(Nx,Nt);p(:,1)=psi';
for kt=1:Nt-1
M=eye(Nx)-i*H*dt/2;N=eye(Nx)+i*H*dt/2;
p(:,kt+1)=M*inv(N)*p(:,kt);
end
surf(t,x,p.*conj(p));
view(2)
shading interp
```

5.3 Boundary Conditions: Hard Wall, Periodic and Open

When we solved the particle on a barrier (Section 4.2), we already specified the asymptotic solutions for $x \to \pm\infty$, namely, plane waves $e^{\pm ikx}$. However, if we try to solve this numerically, we need to build these asymptotic structures into the numerics, so an electron can escape. To solve the $N \times N$ matrix Hamiltonian in real-space, how should we describe the two ends? Remember that the tridiagonal $[H]$ matrix has each diagonal entry $2t_0 + U_n$ straddled by identical off-diagonal elements $-t_0$, except the first and last sites — elements $(1,1)$ and (N,N) which are missing their left and right neighbors, respectively. If we keep those partners unfilled, then we get a hard wall problem where the electron reflects at $x = 0$ and L. Alternately, we can impose periodic boundary conditions where site N couples back onto site 1, as in atoms on a ring. This means the missing Hamiltonian element $H(1,0)$ to the left of site 1 now gets folded onto element $(1,N)$ as $-t_0$, while the Hamiltonian $H(N,N+1)$ is similarly folded onto element $(N,1)$. This gives us an electron escaping (Fig. 5.2) but reinjected back on the other side.

To get the open boundary conditions supporting plane waves at the two ends, we need to add a *complex potential* at the end atoms, whose imaginary parts will endow a plane wave with a complex wavevector $k + i\kappa$, so that $e^{i(k+i\kappa)x} = e^{ikx}e^{-\kappa x}$ will have a decaying term describing the electron's escape into the contacts. The complex Hamiltonians are called self-energy $\Sigma_{1,2}$, for the two ends, and describe Schrödinger equation with open

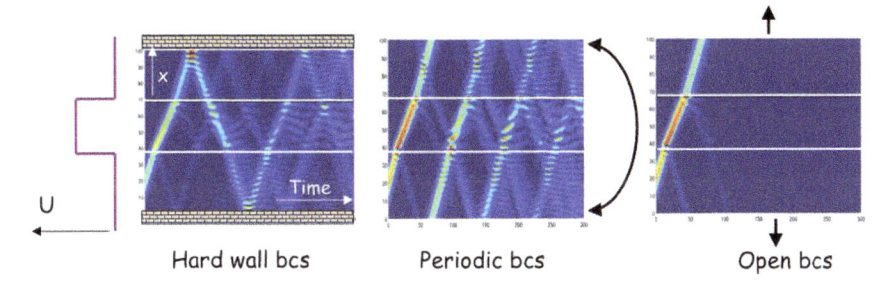

Fig. 5.2 Barrier U, and colorplots showing electron trajectories (x vertical axis, time horizontal). For hard wall boundary conditions, electrons bounce at edges, for periodic, they reemerge at the bottom, and for open boundary conditions, they arrive from $x = -\infty$ and disappear into $x = +\infty$.

boundary conditions for the outgoing electrons escaping into the contacts

$$\Big(EI - [H] - \Sigma_1(E) - \Sigma_2(E)\Big)\psi = S_1 + S_2, \qquad (5.10)$$

where $S_{1,2}$ are the incoming plane waves. In our Crank–Nicolson code, we built S_1 through a plane wave initial condition times a Gaussian (for visibility), while $S_2 = 0$. We will work out the self-energies later, but let us reveal their forms at the moment

$$\big[\Sigma_1\big] = \begin{bmatrix} -te^{ika} & 0 & \dots & \dots & 0 \\ \vdots & \vdots & \vdots & \vdots & \vdots \\ \dots & & \dots & 0 & 0 \end{bmatrix}, \quad \big[\Sigma_2\big] = \begin{bmatrix} 0 & 0 & \dots & \dots & 0 \\ \vdots & \vdots & \vdots & \vdots & \vdots \\ \dots & \dots & \dots & 0 & -te^{ika} \end{bmatrix}, \quad (5.11)$$

where E and k are related by the 1D eigenvalue equation $E - 2t(1 - \cos ka)$. For 3D with multiple orbitals, we will need to solve a matrix recursion equation to determine the self energies.

5.4 Basis Sets, Basis Transformations, Vectors and Matrices

We can decompose a wavefunction $\Psi(x)$ into any superposition of shape functions or *basis sets* $\phi_n(x)$, $\Psi(x) \approx \sum_n a_n \phi_n(x)$. The basis sets must be *complete* so we can express any function in terms of the set $\{\phi_n(x)\}$. We also prefer them to be orthogonal, and normalized, which helps make the coefficients unique. In other words

$$\int dx\phi_m^*(x)\phi_n(x) = \delta_{mn}, \quad \sum_n \int dx\phi_n(x)\phi_n^*(x) = 1. \tag{5.12}$$

$\underbrace{\qquad}_{\text{Orthonormality}} \qquad \underbrace{\qquad}_{\text{Completeness}}$

Completeness helps us extract the expansion coefficients easily. Given a basis set $\{\phi_n(x)\}$ and a target function $\Psi(x) = \sum_n a_n\phi_n(x)$,

$$\int dx\phi_m^*(x)\Psi(x) = \sum_n a_n \underbrace{\int dx\phi_m^*(x)\phi_n(x)}_{\delta_{mn}} = a_m. \tag{5.13}$$

The expansion coefficients $\{a_n\}$ can then be collected as an $N \times 1$ vector

$$\{a_n\} \Longrightarrow \begin{pmatrix} a_1 \\ a_2 \\ \vdots \\ a_N \end{pmatrix} = \begin{bmatrix} V \end{bmatrix}. \tag{5.14}$$

For operator averages that involve bilinear products of Ψ, we get corresponding bilinear combinations of expansion coefficients that can be written as a matrix. For instance, the expectation value of the energy

$$\langle E \rangle = \int dx\Psi^*(x)\hat{\mathcal{H}}\Psi(x) = \sum_{m,n} a_m^* a_n \underbrace{\int dx\phi_m^*(x)\hat{\mathcal{H}}\phi_n(x)}_{H_{mn}}. \tag{5.15}$$

We can collect the matrix elements as an $N \times N$ matrix

$$\begin{bmatrix} H \end{bmatrix} = \begin{pmatrix} H_{11} & H_{12} & \cdots & \cdots & \cdots \\ H_{21} & H_{22} & H_{23} & \cdots & \cdots \\ \vdots & \vdots & \vdots & \vdots & \vdots \\ \cdots & \cdots & \cdots & H_{N,N-1} & H_{NN} \end{pmatrix}. \tag{5.16}$$

The average can now be written as

$$\langle E \rangle = V^\dagger H V. \tag{5.17}$$

Suppose we now want to change to a different basis set $\{\chi_\alpha(x)\}$, $\Psi(x) = \sum_\alpha b_\alpha \chi_\alpha(x)$, changing say from a set of polynomials $\{x^n\}$ to a set of sinusoids $\{\sin n\pi x/L\}$. We first define the expansion coefficients

$$\{b_\alpha\} \Longrightarrow \begin{pmatrix} b_1 \\ b_2 \\ \vdots \\ b_N \end{pmatrix} = \begin{bmatrix} W \end{bmatrix}. \tag{5.18}$$

Expanding the new basis set in terms of the old, using Eq. (5.13), we get

$$\chi_\alpha(x) = \sum_m U_{\alpha m}\phi_m(x) \Longrightarrow V = UW. \tag{5.19}$$

In our above example, U comes from the Taylor expansion of sinusoids, $\sin kx \approx kx - (kx)^3/3! + \cdots$. U is a *unitary* matrix that satisfies $UU^\dagger = U^\dagger U = I$, in order to preserve the orthonormality of the two vector sets. Under this condition, it is straightforward to work out the transformation rules between the two representations for the vectors $V \longleftrightarrow W$ and matrix elements $H' \longleftrightarrow H$ in these two separate basis sets.

$$V = UW \Longleftrightarrow W = U^\dagger V,$$
$$H' = UHU^\dagger \Longleftrightarrow H = U^\dagger H'U, \tag{5.20}$$

giving us a recipe to jump between basis sets and representations, for instance Cartesian to spherical, Bloch to Wannier representations, plane wave to Gaussian, even incomplete non-orthogonal sets $\boxed{\text{P5.5 and P5.6}}$.

Homework

P5.1 **Variational principle.** We can often guess the shape of the lowest eigenstate and tweak the parameters to get the lowest energy solution. That lowest energy solution should be a good approximation to the exact solution, depending on how smart our guess is. The exact solution, of course, will be the lowest energy solution.

We start with a parabolic potential $U(x) = m\omega^2 x^2/2$. Let us guess the solution to be a normalized Gaussian, $\psi(x) = Ae^{-x^2/4\sigma^2}$. Find A. σ is our variational parameter that we adjust for lowest energy.

A couple of algebraic identities also help — that the average of x should be zero (the Gaussian is symmetrically disposed around zero), and the variance is σ^2, so $\langle x \rangle = 0$, $\langle x^2 \rangle = \sigma^2$. Make sure you verify them!!

(a) What is the average kinetic energy $\langle T \rangle = -\dfrac{\hbar^2}{2m} \displaystyle\int_{-\infty}^{\infty} \psi^*(x) \dfrac{\partial^2 \psi(x)}{\partial x^2} dx$? Use the identities above so you don't need to do any integrals!!

(b) What is the average potential energy $\langle U \rangle = \displaystyle\int_{-\infty}^{\infty} \psi^*(x) \dfrac{m\omega^2 x^2}{2} \psi(x) dx$? Again, use the above identities.

(c) Find the lowest energy solution $E = \langle T \rangle + \langle U \rangle$ with respect to the variational parameter σ, i.e., $\partial E(\sigma)/\partial\sigma = 0$. Find both σ and E.

(d) Repeat with a wrong but approximate solution, a Lorentzian function $\psi(x) = A/(x^2 + a^2)$.

(e) For fun, try a truncated sinusoid, i.e., $\psi = \sqrt{\dfrac{2}{L}}\sin\left(\dfrac{\pi[x - L/2]}{L}\right)$, $-L/2 < x < L/2$, zero otherwise, with L being the variational parameter. I'm plotting the answers — you still have to get to them!

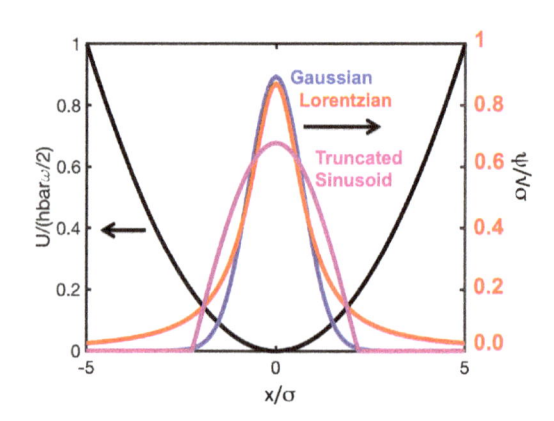

(f) Finally, do a variational solution for a particle in a box of width L with variational solution $Ax^\alpha(L-x)^\alpha$, α being the variational parameter.

P5.2 **Perturbation theory.** Consider a particle in a box with a small burr in it. Common sense dictates that if the burr is small, the eigenfunctions without the burr will now see a dip where the burr sits, pushing the electrons away from the middle. If the burr grows, at one point it may sequester the box into two wells, with the electron sitting largely on one side or the other, like the *sombrero* potential that follows P5.4. Exact eigenstates will show this for sure, but we can do a fair bit of approximation.

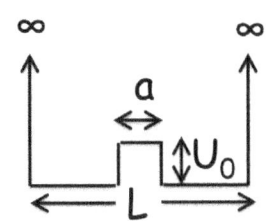

We know the solutions to the burr-free box, $\hat{\mathcal{H}}_0\psi_n^{(0)} = E_n^{(0)}\psi_n^{(0)}$. We want to solve $(\hat{\mathcal{H}}_0 + V)\psi_n = E_n\psi_n$, with a small V from the burr above. We expect the results to evolve continuously from the unperturbed solutions as we turn on the perturbation V with a pre-factor λ that goes from zero to one. Let us use the Hamiltonian $\hat{\mathcal{H}}_0 + \lambda V(x)$. The solution can also be written as an expansion

$$\psi_n = \psi_n^{(0)} + \lambda\psi_n^{(1)} + \lambda^2\psi^{(2)} + \cdots ,$$
$$E_n = E_n^{(0)} + \lambda E_n^{(1)} + \lambda^2 E^{(2)} + \cdots . \tag{5.21}$$

We know the zeroth-order terms $\{\psi_n^{(0)}, E_n^{(0)}\}$ but not the higher-order shifts. (a) Using $(\hat{\mathcal{H}}_0 + V)\psi_n - E_n\psi_n$, equate equal powers of λ. From the first order in λ equation, left multiplying and integrating $\int dx\psi_n^{(0)*}$, using $\hat{\mathcal{H}}_0\psi_n^{(0)} = E_n^{(0)}\psi_n^{(0)}$ and assuming orthonormality $\int dx\psi_p^{(0)*}(x)\psi_m^{(0)}(x) = \delta_{mn}$, show

$$\boxed{E_n^{(1)} = \int dx\psi_n^{(0)*}(x)V(x)\psi_n^{(0)}(x) = V_{nn}} . \tag{5.22}$$

In other words, the lowest order shift in eigenvalue is the expectation value of the perturbation potential in the original state.

(b) Write the dominant corrective term $\boxed{\psi_n^{(1)} = \sum_{m \neq n} c_{mn} \psi_m^{(0)}}$. Substitute

in Schrödinger equation, left multiply by $\int dx \psi_p^{(0)*}(x)$, use orthonormality, to show

$$\boxed{c_{pn} = \int \psi_p^{(0)}(x) V(x) \psi_n^{(0)}(x)/(E_n^{(0)} - E_p^{(0)}) = V_{pn}/\Delta E_{np}^{(0)}}. \qquad (5.23)$$

(c) Use sinusoids for the particle in a box, $\psi_n^{(0)}(x) \propto \sin(n\pi x/L)$ to find the perturbative eigenfunctions and eigenvalues in presence of a small burr, and compare with numerical solutions.

| P5.3 | **Half oscillator.**

Let us solve an oscillator numerically. We'll also do a half-oscillator, and see if we can build some intuition. The energy levels and wavefunctions of the regular parabolic oscilla-

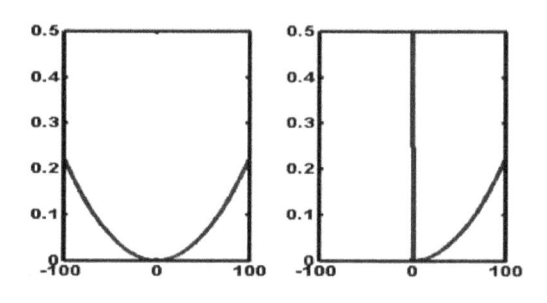

tor potential on the left can in fact be analytically calculated. The energy levels are given by $E_n = (n+1/2)\hbar\omega$, where $\omega = \sqrt{k/m}$ and $n = 0, 1, 2, \ldots$. The corresponding wavefunctions are given by (to within an overall sign)

$$\psi_n(x) = (1/2^n n!)^{1/2} (1/\pi a_0^2)^{1/4} H_n(x/a_0) exp[-x^2/2a_0^2],$$

where $a_0 = \sqrt{\hbar/m\omega}$ is a measure of the oscillator amplitude (obtained by equating kinetic energy $\sim \hbar^2/2ma_0^2$ and potential energy $m\omega^2 a_0^2/2$), while H_n is the nth Hermite polynomial.

Compare your numerical results with the analytical ones and see if they agree (if the agreement isn't exact, explain a possible origin of this discrepancy). You can find a generalized definition online on Hermite polynomials, but for the purpose of this HW, just look at the first six Hermites: $H_0(x) = 1, H_1(x) = 2x, H_2(x) = 4x^2 - 2, H_3(x) = 8x^3 - 12x, H_4(x) = 16x^4 - 48x^2 + 12, H_5(x) = 32x^5 - 160x^3 + 120x, H_6(x) = 64x^6 - 480x^4 + 720x^2 - 120.$

Guided by your numerical results for the semi-parabola, guess the analytical solutions of the half-parabola on the right, and explain the logic behind your guess.

P5.4 **Double-well potential.** We now solve for the *sombrero* potential $U(x) = -x^2/2 + x^4/4$, where $-1.5 < x < 1.5$ in some length units. Find the sorted eigenvalues [D] as a vector. Plot the eigenvalues vertically (One way to do this is to create a vector $X = [-1.5;0;1.5]$ which creates the horizontal lines between -1.5 and 1.5; then you repeat the D vector as 3 columns (you do that by typing 'DD = repmat(D,1,3)', which tiles the matrix D once along the rows and thrice along the columns. Then you plot DD vs. X)

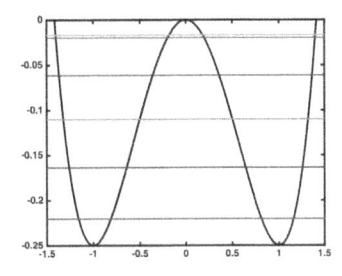

Note that the eigenvalues near the bottom of the potential are 'doubly degenerate' (i.e., come in identical pairs — you will see this better if you just print out the D values instead of eyeballing them on a plot). But the pairs near the top of the central barrier start separating. Explain why.

This is a pre-cursor to what happens for a solid (there we have many wells and barriers, and we see correspondingly many levels, not just pairs, separate to form closely spaced bands separated by gaps).

P5.5 **Basis transformation** Write down the transformation matrix U between sinusoids and polynomials, and use that to transform $[H]$ for a particle in a box on a polynomial basis.

P5.6 **Incomplete, non-orthogonal basis sets.** Consider two *non orthogonal, incomplete* basis sets, a $N_d \times M$ set $\{\phi_m(x)\} = [V]$ (V denotes the matrix representation of $\phi(\vec{r})$ with rows representing N_d real space points and columns representing M orbitals) and an $N_d \times N$ set $\{\chi_\alpha(x)\} = [W]$. We wish to reconstruct a matrix $[G]$ derived in basis $[V]$ in the new basis-set $[W]$ such that their real space representations, $G_{RV} \approx V G_V V^T$ and $G_{RW} \approx W G_W W^T$ match as closely as possible. By forcing a real-space equality $G_{RV} = G_{RW}$, show that $G_W = X^T G_V X$, where the $M \times N$ matrix $X = S^{-1}W^{T}V$ is the normalized overlap between bases, and $S = W^{T}W$.

The expansion coefficients $a_{m\alpha}$ of X correspond to the best fit between the basis sets. Show that minimizing the error

$$Err_m(\{a\}) = \left[\int d^3\vec{r}\left|\phi_m(\vec{r}) - \sum_{\alpha} a_{m\alpha}\chi_{\alpha}(\vec{r})\right|^2\right] / \int d^3\vec{r}\left|\phi_m(\vec{r})\right|^2, \quad (5.24)$$

with respect to the coefficients $\{a\}$ gives us a solution

$$\sum_{\alpha} a_{m\alpha}S_{\alpha\beta} = \int d^3\vec{r}\phi_m(\vec{r})\psi_{\beta}(\vec{r}), \quad (5.25)$$

which in matrix form gives $A = X$.

Chapter 6

Angular Momentum and the Hydrogen Atom

6.1 From 1D to 2D

We have seen the solutions to the particle in a box in 1D, which are sinusoids. For a grid of 100 points, we have 100 eigenstates, each decomposed onto a 100 point spatial grid. If we now extend this to a 2D box on a 100×100 grid with infinite walls, we expect 10,000 eigenstates. There is a convenient way to get at these solutions, by combining 1D eigenstates. This works if the Hamiltonian is decomposable into separable terms in x and y. In Cartesian coordinates, a separable Hamiltonian looks like

$$\left[\underbrace{\left(-\frac{\hbar^2 \partial^2}{2m \partial x^2} + U_x(x) \right)}_{H_x(x)} + \underbrace{\left(-\frac{\hbar^2 \partial^2}{2m \partial y^2} + U_y(y) \right)}_{H_y(y)} \right] \Psi(x,y) = E\Psi(x,y). \quad (6.1)$$

For a square lattice, the potential is zero inside and infinite outside, and is thus trivially separable as above. It is easy to verify that each eigenfunction for a separable potential is simply the product of the individual 1D eigenfunctions, while each eigenvalue is the sum of the 1D eigenvalues

$$\Psi_{m,n}(x,y) = \phi_m(x)\chi_n(y), \qquad E_{m,n} = v_m + \epsilon_n$$

where

$$H_x(x)\phi_m(x) = v_m \phi_m(x) \qquad (m = 1, \ldots, 100),$$
$$H_y(y)\chi_n(y) = \epsilon_n \chi_n(y) \qquad (n = 1, \ldots, 100). \qquad (6.2)$$

We can then construct the 10,000 eigensolutions by combining each of the 100 $\phi_m(x)$s with each of the 100 $\chi_n(y)$s. For a square box of sides $L \times L$,

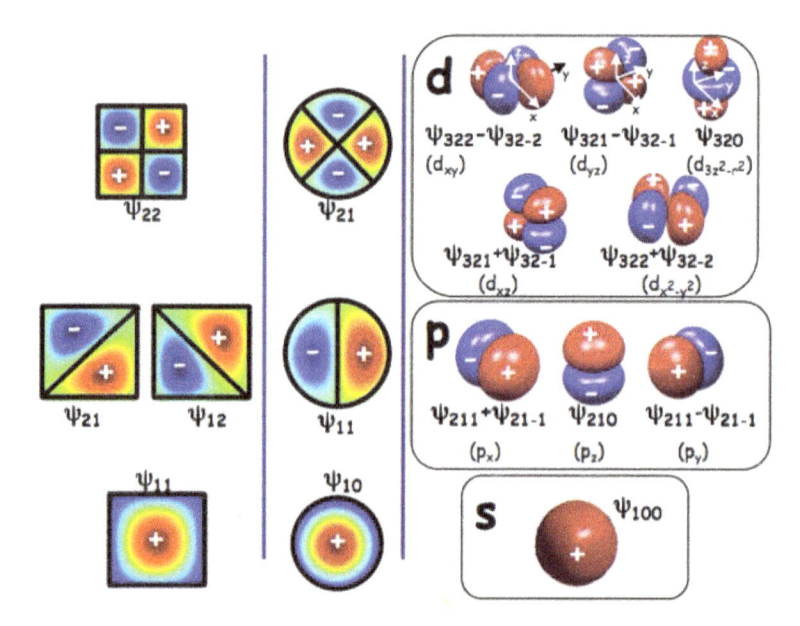

Fig. 6.1 (Left) Eigenstates (top view) for a particle in a 2D square box, showing nodal lines. (Center) 2D circle. (Right) 3D sphere sees nodal planes, leading to orbital eigenstates (see text).

the solutions combine 1D results (Eq. (3.13)) as

$$\Psi_{m,n}(x,y) = \frac{2}{L} \sin\left(\frac{m\pi x}{L}\right) \sin\left(\frac{n\pi y}{L}\right), \quad E_{m,n} = \frac{\hbar^2 \pi^2 \left(m^2 + n^2\right)}{2mL^2}. \quad (6.3)$$

We plot them in Fig. 6.1. The point to notice is that instead of nodes (zeros) in 1D that need to appear in the solutions to enforce orthogonality, we now have nodal lines — the lowest solution with $m = n = 1$ has zero nodal lines (lowest curvature gives lowest kinetic energy), the next has 1 nodal line but with two possibilities ($m = 1, n = 2$ and $m = 2, n = 1$), the third has two nodal lines with $m = n = 2$ and so on.

Let us now deform the square well into a circular well. We expect the solutions to look similar as far as nodal lines are concerned, although we are better off solving this in 2D radial coordinates. The lowest solution again has zero nodal lines and will eventually be designated an s orbital. The next has one nodal line like a dumbbell (a p orbital), while the third has two nodal lines like a four-leaved clover (a d orbital). Essentially, what we just got are the excitation modes of a drum.

When we now extend this to 3D, we expect to get very similar results, albeit stretched out over a sphere. The detailed maths is left as a homework.

But because we have an extra dimension, we expect to see more features — for instance, we see three flavors of p orbitals that we call p_x, p_y and p_z. There are 5d orbitals. 2D nodal lines turn into 3D nodal planes. In general, when we have l nodal planes ($l = 0,1,2,\ldots$), we will have $2l + 1$ orbitals $\boxed{\text{P6.1}}$. The orbitals are labeled by the eigenvalues of their angular momentum. Let us work this out now.

6.2 Angular Momentum Operator

Since position and momentum are incompatible measurements, their cross-product, i.e., the angular momentum, cannot be measured with absolute precision. However, certain components can, as $\vec{r}$ and $\vec{p}$ commute along orthogonal directions, giving us partial information of $\vec{L}$. Using $\vec{L} = \hat{r} \times \hat{p} = -i\hbar\vec{r} \times \vec{\nabla}$, it is straightforward to work out the commutators

$$\left[\hat{L}_x, \hat{L}_y\right] = i\hbar\hat{L}_z, \qquad \left[\hat{L}_y, \hat{L}_z\right] = i\hbar\hat{L}_x, \qquad \left[\hat{L}_z, \hat{L}_x\right] = i\hbar\hat{L}_y,$$

$$\left[\hat{L}^2, \hat{L}_x\right] = \left[\hat{L}^2, \hat{L}_y\right] = \left[\hat{L}^2, \hat{L}_z\right] = 0. \tag{6.4}$$

So the various components of $\vec{L}$ do not play nice with each other, but the overall magnitude L^2 does. In other words, the most complete angular information about the electron that we can garner at a time is the magnitude of the angular momentum and one of its components. The simultaneous eigenstates are called *orbitals*, denoted as $Y_{lm}(\theta, \phi)$. L^2 and L_z are conserved quantities in the absence of a torque that would arise for a directional, angle-dependent potential.

For electrons in a *central* (i.e., directionless) potential such as Coulomb, the Hamiltonian is separable into r, θ, ϕ, which means the eigenstates are also separable into products of lower-dimensional eigenstates

$$\mathcal{H}\Psi = \underbrace{\frac{-\hbar^2}{2mr}\frac{\partial^2}{\partial r^2}(r\Psi) + \frac{\hat{L}^2(\theta, \phi)}{2mr^2}\Psi}_{\hat{T}\Psi = -\hbar^2\nabla^2\Psi/2m} + U(r)\Psi,$$

$$\Psi_{nlm} = \frac{R_{nl}(r)}{r}Y_{lm}(\theta, \phi) \tag{6.5}$$

$\hat{L}^2/2mr^2$ is the rotational (centrifugal) potential — more on that later. Y is the solution to the Schrödinger equation in the θ, ϕ plane, when the potential U does not vary and there is thus no force. In other words, Y *is the solution to a free electron in 2D.* It would look like a plane wave (and indeed it does along the ϕ direction), but for the fact that the axis $\hat{\theta}$ varies from point to point and there is thus a penalty in terms of complexity that arises as a result. $R(r)$ alone takes care of $U(r)$.

Since the ϕ motion of the electron is along a line of latitude whose normal $\hat{z}$ is fixed in space, it looks Cartesian (think of a racetrack). From our experience with Cartesian plane waves where $p_x = -i\hbar\partial/\partial x$ and $\psi \propto \exp[ikx]$, we can readily infer that

$$\hat{L}_z = -i\hbar\partial/\partial\phi, \quad Y_{lm}(\theta,\phi) \propto e^{im_l\phi}, \quad \text{say } Y_{lm}(\theta,\phi) = P_{lm}(\cos\theta)e^{im_l\phi},$$
(6.6)

where P_{lm} is some function of θ, independent of ϕ. Since the wavefunction must be single-valued, this means upon varying ϕ from 0 back to 2π, we should get the same answer, which means $\exp[2\pi im_l] = 1$ and m_l must be an integer. Applying the ϕ derivative on the plane wave, we then get

$$\hat{L}_z Y_{lm}(\theta,\phi) = m_l\hbar Y_{lm}(\theta,\phi), \quad m_l = -l, -l+1, \ldots, l-1, l \qquad (6.7)$$

with $\pm|l|$ being the maximum values that m_l can reach on both sides. There are thus $2l+1$ possible values for m_l.

We can picture the angular momentum vector $\vec{L}$ tracing out the surface of a sphere, since the radius that sets L^2 is fixed. It may be tempting to think this radius equals $l\hbar$, the maximum eigenvalue of $\hat{L}_z$ corresponding to $\vec{L}$ pointing along the $\pm z$ direction, but that would give us $L_x = L_y = 0$ and $L_z = l\hbar$, in other words, a complete information of $\vec{L}$ that is explicitly forbidden by the uncertainty principle and the commutation rules. The maximum L_z can only hover below the north pole much like the minimum ground state wavefunction in a 1D box hovers above its bottom. The maximum eigenvalue of L^2 ends up a bit larger, at $l(l+1)\hbar^2$ P6.2

$$\hat{L}^2 Y_{lm}(\theta,\phi) = l(l+1)\hbar^2 Y_{lm}(\theta,\phi), \quad l \geq 0. \qquad (6.8)$$

The expression for $\hat{L}^2$ is a bit more complicated, and follows from Eq. (6.5) and the definition of ∇^2 in spherical coordinates P6.3

$$\hat{L}^2 = -\hbar^2\left[\frac{1}{\sin\theta}\frac{\partial}{\partial\theta}\left(\sin\theta\frac{\partial}{\partial\theta}\right) + \frac{1}{\sin^2\theta}\frac{\partial^2}{\partial\phi^2}\right]. \qquad (6.9)$$

It has the requisite $\partial^2/\partial\theta^2, \partial^2/\partial\phi^2$ terms, but for the added $\sin\theta$ factors that arise because $\hat{\theta}$ and $\hat{\phi}$ change orientation all over the sphere.

6.3 Orbital Eigenstates for Angular Momentum

Let us start with the smallest value of angular momentum, $l = 0$. Since m_l varies over integers from $-l$ to l, this means $m_l = 0$. The eigenvalues are all zero, thus so are all the angular derivatives. The solution is therefore an angular constant

$$Y_{00}(\theta, \phi) = \text{constant} \quad \text{(s orbital).} \tag{6.10}$$

As angular momentum increases, we get more oscillations and the results look more 'lobey', in order to maintain angular orthogonality to the lower modes. For $l = 1$, we get $2l + 1$, i.e., three orbitals, which can be verified by direct substitution in the equations for $\hat{L}^2$ and $\hat{L}_z$ $\boxed{\text{P6.4}}$

$$\left. \begin{array}{c} Y_{1,1} \propto \sin\theta e^{i\phi} \\ Y_{1,0} \propto \cos\theta \\ Y_{1,-1} \propto \sin\theta e^{-i\phi} \end{array} \right\} \quad \text{(p orbitals).} \tag{6.11}$$

We can then take real and imaginary parts of the complex functions to yield the real p orbitals $Y_{1,1} + Y_{1,-1} \propto \sin\theta\cos\phi = x/r$, $Y_{1,1} - Y_{1,-1} \propto \sin\theta\sin\phi = y/r$, and $Y_{10} \propto \cos\theta = z/r$, leading to the popular names of $p_{x,y,z}$. More generally, we can invoke the expression above (Eq. (6.6)), and work out the differential equation involving $\hat{L}^2 P_{lm}$, leading to an *associated Legendre polynomial* $\boxed{\text{P6.5}}$.

6.4 Radial Eigenstates in a Coulomb Potential

Substituting the equation for Ψ and the eigenstates of $\hat{L}^2$ in the Hamiltonian for a Coulomb potential, we get the effective 1D Schrödinger equation for R $\boxed{\text{P6.6 and P6.8}}$ with a 1D normalization (the 1D normalization over dr instead over the 3D radial 'volume' $r^2 dr$ is why we pulled the $1/r$ factor out in Eq. (6.5))

$$\left[-\frac{\hbar^2}{2m}\frac{\partial^2}{\partial r^2} + \underbrace{\frac{-Zq^2}{4\pi\epsilon_0 r} + \frac{l(l+1)\hbar^2}{2mr^2}}_{U_{\text{eff}}(r)} \right] R = ER, \quad \int_0^\infty R^2 dr = 1, \tag{6.12}$$

where U_{eff} suggests that there is an attractive Coulomb potential that dominates for larger r and a repulsive centrifugal potential $L^2/2I$, with $I = mr^2$ being the moment of inertia, which dominates at small r (differentiating the potential with respect to r, we get the well-known centrifugal force $F_c = -\partial U/\partial r = L^2/mr^3 = mv^2/r$ where $L = mvr$).

The easiest way to solve this is to use our numerical tools for the 1D potential U_{eff}. We can also solve it piece by piece — first, we can solve it for $r \to \infty$ whereupon the entire U_{eff} drops out and we get

$$-(\hbar^2/2m)\partial^2 R/\partial r^2 \approx ER \quad \Longrightarrow \quad R(r \to \infty) \propto \exp\left[-\kappa r\right], \qquad (6.13)$$

with $\kappa = \sqrt{2m|E|/\hbar^2}$. This term shows the overall exponential tail of the electrons tunneling out of the Coulomb barrier. Next, we solve it for small r, where only the repulsive centrifugal potential survives and both the Coulomb and the total energy terms become miniscule, whereupon

$$-(\hbar^2/2m)\partial^2 R/\partial r^2 + l(l+1)\hbar^2 R/2mr^2 \approx 0 \quad \Longrightarrow \quad R(r \to 0) \propto r^{l+1}, \tag{6.14}$$

as easily checked by substitution. This term shows that the electrons stay away from the nucleus except the s orbitals ($l = 0$) for which R/r does not vanish because they do not have any angular momentum and thus no centrifugal barrier. The radial wavefunctions for s orbitals peak at the nucleus and show a prominent cusp (Fig. 6.2).

Finally, at intermediate radii R gathers a number of oscillations to make the radial functions, for instance 1s, 2s, 3s, orthogonal to each other (having enforced orthogonality between s and p or d by the $2l+1$ lobes for each Y_{lm} orbital). For the s series, the lowest R_{10} will not show any zeros and maintain the lowest curvature, starting with a cusp at $r = 0$ and then dropping exponentially. R_{20} will need 1 zero, R_{n0} will need $n-1$ zeros, and in general R_{nl} will need $n-l-1$ zeros. In other words, the set l must go from $0, 1, \ldots, n-1$, with $n \geq 1$, while the R becomes a polynomial with $n-l-1$ zeros connecting the power low and exponential limits (Eqs. (6.13) and (6.14))

$$\Psi_{nlm} = \frac{R_{nl}}{r} Y_{lm}(\theta, \phi), \quad R_{nl} \propto r^{l+1} L_{n-l-1}^{2l+1}(r)e^{-\kappa r}, \quad \kappa = \sqrt{2m|E|/\hbar^2}. \tag{6.15}$$

The polynomial L_a^b can be calculated by substituting the above form in the Schrödinger equation and then crosslisting the resulting second-order differential equation against well-known solutions. The defining feature of the polynomial ends up being the ratio between terms with successive indices.

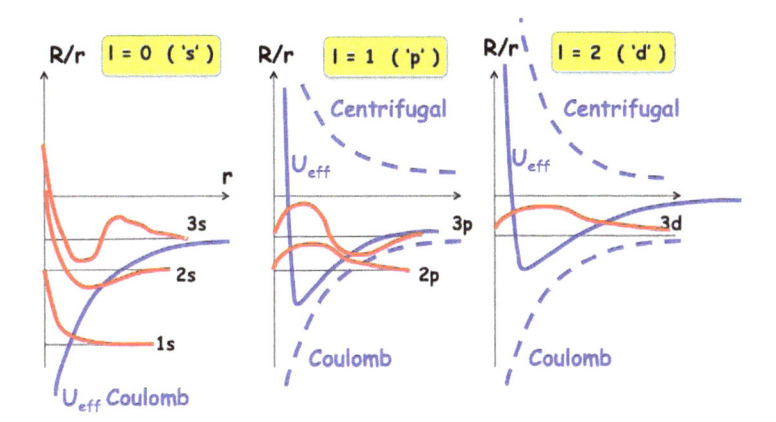

Fig. 6.2 Radial eigenstates in hydrogen orthogonalize due to radial nodes. s orbitals show a cusp at the origin due to a missing centrifugal barrier.

In this case, it turns out to be an *associated Laguerre polynomial* of the form L_a^b which has a zeros.

6.5 Total Energy and Accidental Degeneracy

We can find the energy by using our numerical tools to find the eigenstates in the effective potential U_{eff}. Much like the 1D quantum oscillator, the lengthscale a_0 is given by balancing kinetic and potential energies. To get the exact ratio, we invoke the *virial theorem* $\boxed{P6.9}$, which states that when $U(r) \propto r^n$, the expectation values of kinetic and potential energies satisfy $\langle T \rangle = (n/2)\langle U \rangle$. For an oscillator with $n = 2$, these two are equal (again, the $x-k$ symmetry!). For a hydrogen atom with $n = -1$, the kinetic energy is negative half of the potential energy, giving us the Bohr radius $\boxed{P2.2}$

$$\underbrace{\frac{\hbar^2}{2mu_0^2}}_{T} = \underbrace{\frac{q^2}{8\pi\epsilon_0 a_0}}_{-U/2} \implies a_0 = \frac{\hbar^2(4\pi\epsilon_0)}{mq^2}. \tag{6.16}$$

Unlike an oscillator where the tunnel barrier is energy-independent, for the Coulomb potential the tunnel barrier decreases with energy, meaning the size of the tunneling tail increases with energy, $1/\kappa = na_0$. Substituting the expression for a_0, we get the total energy $\boxed{P2.2}$

$$E_n = -\frac{1}{n^2}\left(\frac{\hbar^2}{2ma_0^2}\right), \quad n = 1, 2, 3, \ldots. \tag{6.17}$$

Note that the energy is independent of angular momentum l. This is an *accidental* degeneracy (degeneracy means equal energy). Angular momentum plays a role when electron–electron interactions start to dominate (multi-electron atoms) and then the configuration of the other electrons matters. For a single electron, there is perfect spherical symmetry $\boxed{\text{P6.10}}$ and E depends not on l but on n alone.

Homework

P6.1

Orbital degeneracy. Why do we have $2l+1$ orbitals for a given l? It turns out that any lth-order polynomial can be described as a superposition of just these $2l+1$ orbitals and r^2 (which is conserved). This is because $(x+y+z)^l$, when expanded, generates $2l+1$ orthogonal terms (not counting powers of r^2 that can be factored out at any point). Show that the second-order polynomial $x^2 - y^2 + xz$ can be written as a superposition of Y_{2m}s. Also show this for the polynomial $z^2 - xy$.

P6.2

Ladder operators. We never really proved that $\hat{L}^2$ has eigenvalues of $l(l+1)\hbar^2$. We motivated it by saying it must be greater than $l^2\hbar^2$ to keep uncertainty alive. We can prove it purely by manipulating so-called *ladder operators*, $\hat{L}_\pm = \hat{L}_x \pm i\hat{L}_y$. Since these terms don't commute with $\hat{L}_z$, they will act on $Y_{lm}(\theta,\phi)$ and alter their m. As it turns out, they act like a ladder in that $\hat{L}_+$ increases m by 1 and $\hat{L}_-$ decreases it by 1. Use the angular momentum commutation relations to prove that

$$\left[\hat{L}_z, \hat{L}_\pm\right] = \pm\hbar\hat{L}_\pm, \quad \left[\hat{L}_+, \hat{L}_-\right] = 2\hbar\hat{L}_z \tag{6.18}$$

Now, let us interrogate the angular momentum states of $\hat{L}_\pm$ acting on $Y_{lm}(\theta,\phi)$. By operating on it by $\hat{L}_z$ and using the first ladder commutation rule in Eq. (6.18), show that

$$\hat{L}_z\left[\hat{L}_\pm Y_{lm}(\theta,\phi)\right] = (m \pm 1)\hbar\left[\hat{L}_\pm Y_{lm}(\theta,\phi)\right] \tag{6.19}$$

meaning $\hat{L}_\pm Y_{lm}(\theta,\phi)$ is an eigenstate of $\hat{L}_z$ with an eigenvalue of $(m \pm 1)\hbar$ (and not a mixture of m states), in other words, proportional to $Y_{l,m\pm1}$.

$$\hat{L}_\pm Y_{lm}(\theta,\phi) \propto Y_{l,m\pm1}(\theta,\phi). \tag{6.20}$$

This shows that $\hat{L}_\pm$ act like *ladder operators*.

Now, we know that the maximum m equals l, meaning *the ladder must terminate there*! In other words, $\hat{L}^+ Y_{ll} = 0$. We can now find the eigenvalue of $\hat{L}^2$ by applying it to this highest state. First, show that

$$\hat{L}_x^2 + \hat{L}_y^2 = \frac{1}{2}\left(\hat{L}_+\hat{L}_- + \hat{L}_-\hat{L}_+\right) = \hat{L}_-\hat{L}_+ + \frac{1}{2}\left[\hat{L}_+, \hat{L}_-\right]. \tag{6.21}$$

From these results and the second rule in Eq. (6.18), show

$$\hat{L}^2 Y_{ll}(\theta, \phi) = \left(\hat{L}_x^2 + \hat{L}_y^2 + \hat{L}_z^2\right) Y_{ll}(\theta, \phi) = \underbrace{l(l+1)\hbar^2}_{\text{eigenvalue of } Y} Y_{ll}(\theta, \phi). \tag{6.22}$$

Finally, using Eq. (6.21) for $\hat{L}_-\hat{L}_+$ and the result you just got, find the proportionality constant of Eq. (6.20) and show it is $\sqrt{(l \mp m)(l \pm m + 1)}\hbar$.

P6.3 **Operators.** From Cartesian to radial coordinate transformation, show

$$\hat{L}^2 = -\hbar^2\left[\frac{1}{\sin\theta}\frac{\partial}{\partial\theta}\left(\sin\theta\frac{\partial}{\partial\theta}\right) + \frac{1}{\sin^2\theta}\frac{\partial^2}{\partial\phi^2}\right] \text{ and } \hat{L}_z = -i\hbar\frac{\partial}{\partial\phi}.$$

P6.4 **p Orbitals.**

(a) The p orbitals correspond to $l = 1$. From the above list of allowed m values, we get three kinds of p orbitals corresponding to $m = -1, 0, 1$, giving us Y_{11}, Y_{10} and $Y_{1,-1}$ listed in Eq. (6.11). Show that they satisfy the eigenvalue equations (6.7), (6.9) with the correct eigenvalues.

(b) Plot the real and imaginary parts of the functions in polar coordinates (look under the plot option 'polar' in Matlab), then show that the orbitals point along the cubic axes.

(c) Repeat for the d_{xy} orbital which looks like $\sin^2\theta\sin\phi\cos\phi$ for $l = 2$. Work it out for the two eigenvalue equations and plot it.

P6.5 **f Orbitals.** Just for fun, let's work out the f-orbital names. The wavefunctions

$$Y_{3,\pm3} = e^{\pm3i\phi}\sin^3\theta, \quad Y_{3,\pm2} = e^{\pm2i\phi}\sin^2\theta\cos\theta,$$
$$Y_{3,\pm1} = e^{\pm i\phi}\sin\theta(5\cos^2\theta - 1), \quad Y_{3,0} = (5\cos^3\theta - 3\cos\theta). \tag{6.23}$$

Take real and imaginary parts $Y_{3,m} \pm Y_{3,-m}$ and re-express in terms of $x = \sin\theta\cos\phi$, $y = \sin\theta\sin\phi$ and $z = \cos\theta$ to label the seven f orbitals.

P6.6 Guessing orbitals. The plot shows the radial and angular wavefunctions for two specific orbitals that an electron in carbon occupies. By looking at their radial and angular dependences, identify the (n, l, m) for both orbitals. Explain the logic behind your choices.

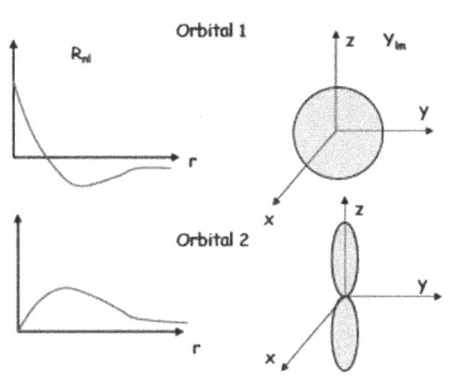

P6.7 Modes of a drum. Let us work out the modes of a 2D particle in a box of radius a. The 2D Schrödinger equation is $-\hbar^2 \nabla^2 \Psi/2m = E\Psi$, with $\nabla^2 = (1/r)\partial/\partial r(r\partial/\partial r) + (1/r^2)\partial^2/\partial\theta^2$. Assuming a solution $\Psi_{np}(r, \theta) = R_{np}(r)e^{ip\theta}$, show that the radial Schrödinger equation is

$$r^2\partial^2 R/\partial r^2 + r\partial R/\partial r + (r^2/\lambda^2 - p^2)R = 0, \qquad (6.24)$$

where $\lambda = \hbar/\sqrt{2mE}$ is the de Broglie wavelength. The solutions to this (look it up!) are Bessel functions, $R_{np}(r) = J_p(r/\lambda_n)$, with the quantization n determined by the boundary conditions. For instance, if the wavefunction vanishes at $r = a$ (as in a drum), λ_n is given by the nth root of the pth Bessel function, $J_p(a/\lambda_n) = 0, n = 1, 2, 3, \ldots$.

Repeat in 3D with $\psi_{nlm}(\vec{r}) = R_{nl}(r)Y_{lm}(\theta, \phi)$, and show that the radial term is given by $R_{nl}(r) = j_l(r/\lambda_n)$, the spherical Bessel function.

P6.8 Hydrogen atom. Repeat for the 2D and 3D Hydrogen atoms. We now have a potential! For the radial solution, assume a polynomial function to enforce orthogonality, times the short range $(r = 0)$ power law and long range $(r = \infty)$ exponential solutions, $R(r) = r^p e^{-r/|\lambda|}f(r)$, $f(r) = \sum_n c_n r^n$, and extract the ratio of successive terms c_n/c_{n-1}. For the series to be finite, you need one of the c_n terms to vanish, which should give you the quantized energy. Show that for the 2D Hydrogen atom, $E_n = E_1/(n - 1/2)^2$, where E_1 is the result of the 3D hydrogen ground state (Eq. (6.17)). Thus, the lowest energy for 2D is four times the 3D result (the energy rise is due to quantum confinement in the third, out of plane dimension). This explains why in 2D excitons bind so well!

P6.9 **Virial theorem.** Let us prove the virial theorem in 1D. The *virial* is the quantity

$$\hat{Q} = \sum_i \hat{r}_i \cdot \hat{p}_i. \tag{6.25}$$

The Hamiltonian $\hat{\mathcal{H}} = \sum_i \dfrac{\hat{p}_i^2}{2m_i} + \underbrace{U(\vec{r})}_{Ar^n}$. Use Eq. (3.19) to evaluate $d\hat{Q}/dt$ in terms of commutator of $\hat{\mathcal{H}}$ with the virial, and show that $d\hat{Q}/dt = \langle 2T \rangle - n\langle U \rangle$. For steady-state, $d\hat{Q}/dt$ vanishes, yielding the virial theorem.

P6.10 **Angular coordinates.** Starting with Cartesian to spherical polar transformation geometry, show that a particle acceleration $d^2\vec{r}/dt^2$ naturally breaks into radial, tangential, centrifugal and Coriolis acceleration terms simply because the polar unit vectors change orientation from point to point. In other words, their origin is purely geometrical.

Chapter 7

Atomic and Molecular Structure

7.1 Aufbau Principle and Hund's Rules

The final wavefunction shows $2l + 1$ zeros from the orbitals and $n - l - 1$ zeros from the radial functions, so that the total number of nodes equals their sum, $n + l$. Energetically, we must therefore start with the lowest n and l (curvature) to minimize the kinetic energy, and then fill along each line of constant $n + l$, so that the sequence of filling orbitals is $1s, 2s, 2p, 3s, 3p, 4s, 3d$ and so on. This is called the *Aufbau* (German, meaning 'build-up') principle.

Within each orbital set, how should we choose preferentially — say between $2p_{x,y,z}$? At this stage, since we are jamming multiple electrons into the same orbit, minimizing Coulomb repulsion becomes a priority, which brings us to Hund's rules for the ground states of lighter atoms.

Hund's Rule 1: Maximize m_s — We will shortly see that electron spins act like angular momentum, with

$$S^2 \chi_{s,m_s} = s(s+1)\hbar^2 \chi_{s,m_s},$$

$$S_z \chi_{s,m_s} = m_s \hbar \chi_{s,m_s}, \quad m_s = -s, \ldots, s \qquad (7.1)$$

with $s = 1/2$, so that we get only $2s + 1 = 2$ values for m_s, namely $\pm 1/2$ (up and down spins). Hund's first rule tells us to maximize m_s when we build up the electronic configuration of an atom. This rule is driven by the need to minimize Coulomb repulsion. In order to reduce electron overlap, we need to spread them far apart and this is easier for parallel spins that stay away because of Pauli exclusion (we will describe the corresponding 'exchange' force shortly). Furthermore, singly occupied states are screened poorly and thus contract more under the pull of the nucleus, increasing the attractive energy. Thus, if we have four electrons that need to go into three p orbitals $p_{1,2,3}$, we *maximize m_s*, the spin angular momentum, by going $p_{1\uparrow}$, $p_{2\uparrow}$, $p_{3\uparrow}$ until $m_s = 3/2$ is at its maximum, and only then bringing the 4th in to pair up as $p_{1\uparrow\downarrow}$ for a net $m_s = 1$.

Hund's Rule 2: Maximize m_l — The next priority is to maximize m_l, the magnetic quantum number. It is easy to show again that the electron overlap is minimum if m_l is maximum, because this represents two electrons rotating in the same direction. Oppositely rotating electrons are guaranteed to cross each other over each period, while co-rotating electrons will only do so once in a while. If the angular frequencies of the two electrons are ω_1 and ω_2, then their relative speed is $\omega_1 - \omega_2$ for co-rotating electrons (slower) and $\omega_1 + \omega_2$ for oppositely rotating electrons (faster). The speed determines how quickly two electrons will come into overlap with each other.

Going back to the p orbitals, Hund's second rule implies that we must first put the first up electron into state $p_1 = Y_{11}$, next into $p_2 = Y_{10}$ and finally into $p_3 = Y_{1,-1}$.

Hund's Rule 3: Minimize m_J for less than half-filled, maximize otherwise — The last rule of Hund has to do with how you combine m_l and m_s to create eigenstates of the vector $\hat{J} = \hat{L} + \hat{S}$. This term becomes important since the spin and orbital angular momenta tend to couple in materials like InAs with spin–orbit interaction (recall second term in Dirac equation in $\boxed{\text{P3.9.3}}$). In presence of such interactions $\propto \vec{L} \cdot \vec{S}$, neither L_z nor S_z are preserved and one needs to go to $\vec{J}$. It is easy to show that $|L - S| \leq J \leq L + S$. For less than half filling, $J = |L - S|$, while for greater than half-filled, $J = L + S$. This is because the quantity $\vec{L} \cdot \vec{S}$ is negative when the angular momentum and spin are opposite to each other, which saves energy. Beyond half-full, we need to double up spins opposite to those already in place, and the rule reverses.

7.2 Exchange, Anisotropy and Dzyaloshinskii–Moriya

Note that Hund's rules are ultimately approximate and work for non-interacting atoms. In presence of crystal environments, interfaces and boundaries, interatomic forces start to pick up that can dominate over these forces. The general Hamiltonian for a set of interacting spins in a magnetic structure can be written using symmetry considerations. For a set of two spins, a typical Hamiltonian is

$$\hat{\mathcal{H}} = -\underbrace{J\vec{S}_1 \cdot \vec{S}_2}_{\text{Exchange}} + \underbrace{K\left(S_{1z}^2 + S_{2z}^2\right)}_{\text{Anisotropy}} + \underbrace{D\hat{z} \cdot \left(\vec{S}_1 \times \vec{S}_2\right)}_{\text{Dzyaloshinskii–Moriya (DMI)}} . \quad (7.2)$$

7.3 Heavier Atoms: Coulomb Repulsion

When we go to heavier atoms, the main challenge is the presence of multiple electrons. To get the Coulomb repulsion on an electron in an N-electron atom, we need to mentally gouge out that electron from the charge distribution and then calculate the effect of the remaining non-spherical charge distribution on the missing electron. For lighter atoms, we can ignore this 'correlation' effect, and smear the remaining $N - 1$ electrons back into a sphere, leading to the *self-consistent, mean field* or *Hartree* theory. Accordingly, we can assume a hydrogen-like decomposition of the overall wavefunction, with a radial equation that includes an added Coulomb repulsion.

$$\Psi_{nlm} = \frac{R_{nlm}(r)}{r} Y_{lm}(\theta, \phi)$$

$$\left[-\frac{\hbar^2}{2m}\frac{\partial^2}{\partial r^2} + \underbrace{\frac{-Zq^2}{4\pi\epsilon_0 r} + \frac{l(l+1)\hbar^2}{2mr^2} + U_{ee}^H(r)}_{U_{\text{eff}}(r)}\right] R_{nlm}(r) = E R_{nlm}(r). \quad (7.3)$$

The Coulomb repulsion (Hartree term) U_{ee}^H can be obtained from classical electrodynamics, i.e., Gauss' law. First we work out the charge distribution $n(r) = q \sum_{nlm} |R_{nlm}(r)|^2 \times (N - 1)/N$ over the filled states, i.e., the ground state orbital configuration for the atom. The factor $(N - 1)/N$ accounts for the fact that each electron should only feel the potential due to the other $(N - 1)$ electrons. Gauss's Law tells us that the flux of the electric field is proportional to the charge enclosed (for the charge, remember that the angular integral over $d(\cos\theta)d\phi$ is already accounted for in the $Y_{lm}(\theta, \phi)$ normalization and the r^2 volume term through the $R_{nlm}(r)/r$ term)

$$\underbrace{\mathcal{E} \times 4\pi r^2}_{\text{Flux}} = \int_0^r dr' n(r')/\epsilon_0 \implies \mathcal{E}(r) = \frac{1}{4\pi\epsilon_0 r^2} \int_0^r n(r')dr'. \quad (7.4)$$

The potential $U_{ee}^H(r) = -qV(r)$, where by definition $-\vec{\nabla}V = \mathcal{E}$. Assuming the potential is zero at $r = \infty$, we get

$$U_{ee}^H(r) = q \int_\infty^r \mathcal{E}(r')dr' = q \int_\infty^r dr' \frac{1}{4\pi\epsilon_0(r')^2} \int_0^{r'} n(r'')dr''$$

$$= -\frac{q}{4\pi\epsilon_0 r} \int_0^r n(r')dr' - \int_\infty^r \frac{qn(r')dr'}{4\pi\epsilon_0 r'} \quad \text{(integration by parts)},$$

$$n(r) = q \sum_{nlm} |R_{nlm}(r)|^2 \left(\frac{N-1}{N}\right). \tag{7.5}$$

The two equations need to be solved self-consistently $\boxed{\text{P7.1}}$. We start with a guess for U_{ee}^H and solve for R_{nlm} using Eq. (7.3), then plug that back into Eq. (7.5) to recompute U_{ee}^H and cycle back and forth till two subsequent outputs of n or U_{ee}^H converge to within a small preset tolerance limit.

<u>**Matlab code:**</u>
```
%% Code to calculate and plot Radial functions and energies of silicon
clear all
%% Constants (all MKS, except energy which is in eV)
hbar=1.055e-34;m=9.110e-31;epsil=8.854e-12;q=1.602e-19;

%% Lattice
Np=200;a=(5e-10Np);R=a*[1:1:Np];
t0=(hbar^2)/(2*m*(a^2))/q;

%% Hamiltonian,H = T + UN + Ul + UH, Eq. 7.3

T=(2*t0*diag(ones(1,Np)))-(t0*diag(ones(1,Np-1),1))-
(t0*diag(ones(1,Np-1),- 1));
%% Use Eq. ?? for UN and Ul
UN=(-q*14/(4*pi*epsil))./R; % Nuclear, Z=14 for silicon
l=1;Ul=(l*(l+1)*hbar*hbar/(2*m*q))./(R.*R); % El-El for p orbitals
UH=zeros(1,Np);change=1;
while change>0.1

%% SCF Hartree potential for s orbitals, no Ul
[V,D]=eig(T+diag(UN+UH));
```

```
D=diag(D);[DD,ind]=sort(D);
E1s=D(ind(1));psi=V(:,ind(1));P1s=psi.*conj(psi);P1s=P1s';
E2s=D(ind(2));psi=V(:,ind(2));P2s=psi.*conj(psi);P2s=P2s';
E3s=D(ind(3));psi=V(:,ind(3));P3s=psi.*conj(psi);P3s=P3s';

%% SCF Hartree potential for p orbitals, Ul with l=1
[V,D]=eig(T+diag(UN+Ul+UH));
D=diag(D);[DD,ind]=sort(D);
E2p=D(ind(1));psi=V(:,ind(1));P2p=psi.*conj(psi);P2p=P2p';
E3p=D(ind(2));psi=V(:,ind(2));P3p=psi.*conj(psi);P3p=P3p';

%% Use Eq. 7.5 for n and Unew replacing UH
n0=2*(P1s+P2s+P3s)+(6*P2p+2*P3p); % Only count upto 3p2
n=n0*(13/14); % Count 13 out of 14 electrons
Unew=(q/(4*pi*epsil))*((sum(n./R)-cumsum(n./R))+(cumsum(n)./R));
change=sum(abs(Unew-UH))/Np,UH=Unew;
end


[E1s E2s E3s E2p E3p]

%% Plot Orbitals
[V1,D1]=eig(T+diag(UN+Uscf));D1=diag(D1);[DD1,ind1]=sort(D1);
psi2s=V1(:,ind1(2)); % This gives f(r) whose square was the probability
R2s=psi2s./R'; % True radial function
[V2,D2]=eig(T+diag(UN+Ul+Uscf));D2=diag(D2);[DD2,ind2]=sort(D2);
psi2p=V2(:,ind2(1));
R2p=psi2p./R';
subplot(1,2,1)
h=plot(R,R2s,'linewidth',3); % Plotting R2s
set(gca,'linewidth',3,'fontsize',15);
xlabel('r(m)','fontsize',15);
ylabel('R_2s','fontsize',15);
grid on
axis([0 5e-10 -5e9 5e9])
subplot(1,2,2)
h=plot(R,R2p,'linewidth',3); % Plotting R2p
set(gca,'linewidth',3,'fontsize',15);
xlabel('r(m)','fontsize',15);
```

```
ylabel('R_2p','fontsize',15);
grid on
axis([0 5e-10 -5e9 5e9])
```

7.4 Exchange and Correlation

In calculating the charges, we have only partly accounted for Pauli exclusion forces that keep two electrons of identical configuration apart from double occupancy. We took care of that during the build-up of electronic configuration in an atom, but its contribution to the energy cost (as part of Coulomb) was never explicit. The electron density $n \propto \sum |R|^2$ simply adds single electron probabilities, but does not account for exchange, the fact that any two electrons under exchange will yield a net negative sign for the overall many-body wavefunction. That exchange is ultimately responsible for Pauli exclusion (identical electrons under exchange give a negative of the joint wavefunction which is only possible if their wavefunction product was zero to begin with). The way to enforce the exchange is by dictating that the many-body wavefunction is no longer the simple product of the individual orbital terms, but Slater determinants, so that any exchange amounts to a swap of row or column that creates a minus sign.

$$\Psi(\vec{r_1}, \vec{r_2}, \ldots, \vec{r_N}) \neq \psi_1(\vec{r_1})\psi_2(\vec{r_2})\ldots\psi_N(\vec{r_N})$$

$$\rightarrow \begin{vmatrix} \psi_1(\vec{r_1}) & \psi_2(\vec{r_1}) & \cdot & \cdot & \psi_N(\vec{r_1}) \\ \psi_1(\vec{r_2}) & \psi_2(\vec{r_2}) & \cdot & \cdot & \psi_N(\vec{r_2}) \\ \cdot & & \cdot & \cdot & \cdot \\ \cdot & & & \cdot & \cdot \\ \psi_1(\vec{r_N}) & \psi_2(\vec{r_N}) & \cdot & \cdot & \psi_N(\vec{r_N}) \end{vmatrix} \tag{7.6}$$

Using this many-body operator in Schrödinger equation to simplify the Coulomb, we will then encounter the normal (Hartree) terms described above, but added non-local terms describing exchange (Fock) between two electrons, giving a modified equation (refer NEMV Section 6.2.1)

$$\left[-\frac{\hbar^2}{2m}\frac{\partial^2}{\partial r^2} + \frac{-Zq^2}{4\pi\epsilon_0 r} + \frac{l(l+1)\hbar^2}{2mr^2} + U_{ee}^H(r) \right] R_{nlm}(r)$$

$$\underbrace{- \int U_{ee}^F(r,r')R_{nlm}(r')dr'}_{\text{Exchange}} = ER_{nlm}(r)$$

$$U_{ee}^F(r, r') = -\frac{q}{4\pi\epsilon_0 r}\int_0^r n_{ex}(r, r')dr' - \int_\infty^r \frac{qn_{ex}(r, r')dr'}{4\pi\epsilon_0 r'},$$

$$n_{ex}(r, r') = q\sum_{nlm} R_{nlm}^*(r)R_{nlm}(r')\left(\frac{N-1}{N}\right). \tag{7.7}$$

The non-local exchange term described above creates a reduction in Coulomb energy for parallel spin electrons and is responsible ultimately for Hund's rule and ferromagnetism. This is easier to see for highly localized orbitals, where only $r = r'$ terms contribute. It is straightforward to show that if all the indices (including spin, not shown here) match, then the Hartree and Fock terms precisely cancel, so that parallel spins reduce the Coulomb repulsion by staying away from each other. A little bit of algebra shows that the exchange term exists around each electron over a 'sphere' of influence equal to the electron size. For an electron in a metal for instance, this size is related to the electron density, the radius r_0 satisfying $(4\pi r_0^3/3) \times n = 1$. In metals with a high electron density, the electron size is small and Pauli exclusion doesn't extend very far. For insulators, this term persists and can have a significant impact on the orbital energies.

While the non-locality makes chemistry complex and accounts for Hund's rule, it is not yet the full picture! What Hartree–Fock misses is electron *correlation*. The ultimate aim of the electron is to reduce its overall *free energy* as much as possible, by any means possible, subject to symmetry constraints. The Hatree–Fock equation would accomplish this, if the electrons were restricted to their individual ground state configurations (e.g., the terms over which the summation runs in $\sum_{nlm} R_{nlm}^* R_{nlm}$). But it is entirely possible for the electron to mix higher energy unoccupied (virtual) states if that leads to an overall energy reduction, and the mixture of these higher states would also influence the ground state orbitals. For instance, antiferromagnetic exchange or hopping arises due to electrons spreading their wavefunctions to minimize their curvature and thus the kinetic energy. For a half-filled system (one electron per site), the hopping will momentarily pack two opposite spins into a single orbital — allowed by Pauli exclusion, but this is energetically quite costly. However, as long as this energy spike happens over a small time (set by the uncertainty principle, $\Delta E \Delta t \sim \hbar$), the eventual reduction in kinetic energy ends up with a net gain, giving rise to antiferromagnetism.

A large number of complicated structures arise from the ability for electrons to correlate themselves, to do a complex tango, in order to cut down the overall energy. Superconductivity, Luttinger liquids, Coulomb Blockade, Kondo effect and Mott transitions are just a small sample of such complexity. We discuss details of such processes in NEMV. For the purpose of this lecture, it is sufficient to say that we have only scratched the tip (albeit a fairly complex tip) of a giant electronic iceberg.

7.5   From Orbitals to Bonds: The Chemistry of Molecules

We found that electrons in atoms distribute themselves into orbitals to enforce angular orthogonality, and radial nodes to enforce radial orthogonality. The angular part acts like a free electron, while the radial part settles into a combined landscape set by nuclear Coulomb attraction, centrifugal repulsion and electronic repulsion. What happens if we bring two of these atoms together until their electron clouds overlap? What we will find is that Coulomb potentials will start to dictate their behavior for ionic bonds, and kinetic energy considerations set the stage for covalent bonds.

7.6   Ionic Bonds

Ionic bonds arise due to the partnering of an electropositive element like Na and an electronegative element like Cl, from opposite columns of the periodic table. The conventional wisdom is that Na with 11 electrons wants to lose an electron and reach an inert gas (octet) configuration like neon, while Cl with 17 electrons wants to gain an electron and reach argon. Thus, an electron transitions from Na to Cl, leaving behind a Na^+Cl^- ionic pair. But this ignores the fact that Na still has 11 protons while Cl has 17 protons, and the lack of neutrality comes at a large energy penalty. As we will see, NaCl relies on other partner molecules to stabilize its configuration.

   We can modify our atomic code in the grey box (Section 7.3), written there for Si, to calculate the energy levels for Na and Cl, shown in Fig. 7.1 in the left and middle subplots $\boxed{\text{P7.2}}$. They naively suggest that the electron from Na 3s at -5.72 eV can drop to Cl 3p at -12.8 eV. However, note that these are the *ionization potentials* for the atoms, i.e., the energy to remove an electron $(Na \rightarrow Na^+,\ Cl \rightarrow Cl^+)$. Since we need to add an electron to Cl, we need to calculate the *electron affinity* $(Cl \rightarrow Cl^-)$, which involves the energy cost of an extra electron (going between 17 and 18, instead of 17 and 16). We can calculate that by noting the electron affinity for Cl is the same as the ionization potential of Cl^-, shown on the right subplot. We then see the EA of Cl is -2.9 eV, higher than the IP of Na by $5.72 - 2.9 = 2.32$ eV, substantially larger than the IP of neutral Cl by the charging cost of $-2.9 + 12.8 = 9.9$ eV. Na^+Cl^- bonds are not energetically favorable!

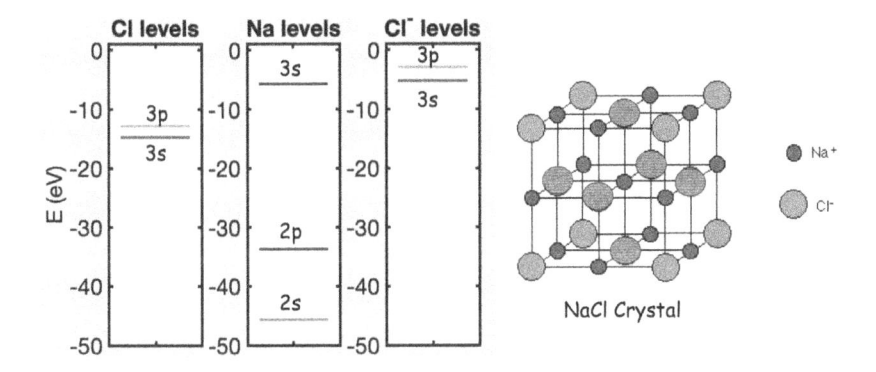

Fig. 7.1 See text. Energetics of individual atoms and NaCl ionic structure.

What stabilizes NaCl is its Coulomb interactions with other pairs. NaCl forms a 3D chessboard structure, shown earlier. Each central Na atom (red) has six nearest neighbor oppositely charged Cl atoms (green shaded), 12 next nearest neighbor similarly charged Na partners in red, and 8 next to next nearest neighbor oppsittely charged Cl atoms (green unshaded). The lattice constant is $a = 0.24$ nm, with a net gain in Coulomb energy by $\mathcal{M}q^2/4\pi\epsilon_0 a \sim -7.32$ eV ($\mathcal{M} \approx -1.74756$ is called the Madelung constant P7.3), which offsets the loss of 2.32 eV and stabilizes the entire structure.

This analysis also explains why crystals like salt (NaCl) or sugar ($C_6H_{12}O_6$) dissolve in water. Water is a polar molecule, as we will see shortly, mimicking the sp^3 tetrahedral structure of methane CH_4, but with a slightly suppressed angle of $\sim 104^0$. The electropositive H atom and electronegative O atoms create a spatial separation of charges, creating dipole moments along the $H - O$ bonds (~ 1.8546 Debye ~ 0.386 e-Å) that do not cancel vectorially because of the angle. This polarity tears away the attracting NaCl atoms, dissolving the salt (or sugar) crystals. We can estimate this by realizing that the dielectric constant of salt water is ~ 80, and this means the Coulomb crystal gain is now ~ 7.32 eV/80 ≈ 0.0925 eV.

7.7 Covalent Bonds

For two elements of comparable valency, a covalent bond provides a way for each to reach inert gas configurations without jettisoning its charge completely and having then to find ways to offset the Coulomb penalty. These elements reach their goal by *sharing* electrons, leading to directional covalent bonds P7.4 . The main mechanism to do so is by spreading the electron cloud, the kinetic energy (proportional to the curvature of the wavefunction) is reduced, as long as there is a sizeable overlap between the atoms. The overlap splits the original atomic energy levels into a bonding-antibonding pair, much like the sum and difference of two mixed frequencies in a nonlinear network, $\sim \cos \omega_1 t \times \cos \omega_2 t \propto \cos(\omega_1 + \omega_2)t + \cos(\omega_1 - \omega_2)t$.

We know how to write down the Hamiltonian for a system, such as a H_2^+ molecular ion with one electron and two protons

$$
\hat{\mathcal{H}} = \underbrace{-\sum_{R_{A,B}} \frac{\hbar^2 \nabla^2_{R_{A,B}}}{2M_{A,B}}}_{\text{Nuclear kinetic energy}} + \underbrace{\frac{q^2}{4\pi\epsilon_0 |\vec{R}_A - \vec{R}_B|}}_{\text{Nuclear–nuclear repulsion } (\mathrm{U_R})} \quad \text{(nuclear)}
$$

$$
\underbrace{-\frac{\hbar^2 \nabla^2}{2m}}_{\text{Electron kinetic energy } (\mathrm{T})} \underbrace{-\sum_{R_{A,B}} \frac{q^2}{4\pi\epsilon_0 |\vec{R}_{A,B} - \vec{r}|}}_{\text{Electron–nuclear atttraction } (\mathrm{U}_{A,B})} \quad \text{(electronic).}
$$

$$\tag{7.8}$$

We can simplify this a bit, since the nuclei are thousands of times heavier than the electron, so we freeze their coordinates $\vec{R}_{A,B}$ and ignore their dynamics, focusing thus on only the last two terms constructing the electronic Hamiltonian $\hat{\mathcal{H}}_{el}$. Even so, the problem is tricky, as we can no longer use spherical mean-field approximations, given that we have two nuclei and thus two potential centers of any sphere. However, we can use the concept of basis sets, by postulating that the eigenfunction of the molecular orbital (MO) is a linear combination of atomic orbitals (LCAOs), i.e., the eigenfunctions of the individual atoms (the LCAO-MO approximation). For this single electron multisystem, we can postulate that

$$
\Psi_{H_2^+}(\vec{r}) \approx \psi_A u_{1s,A}(\vec{r}) + \psi_B u_{1s,B}(\vec{r}), \tag{7.9}
$$

in other words, a linear combination of 1s hydrogenic orbitals sitting at atomic sites A and B, with $u_{1s,A,B}(\vec{r})$ having the form $Ne^{-|\vec{r}-\vec{R}_{A,B}|/a_0}/\sqrt{\pi}$. What we want to determine are just the two mixing coefficients $\psi_{A,B}$.

We can write down Schrödinger's equation for the MO wavefunction

$$
\hat{\mathcal{H}}_{el} \Psi_{H_2^+}(\vec{r}) = E \Psi_{H_2^+}(\vec{r}). \tag{7.10}
$$

We can left multiply in turn by $u_{1s,A,B}$ and integrate over all space, to get a matrix equation for the coeffiicients

$$\begin{pmatrix} H_{AA} & H_{AB} \\ H_{BA} & H_{BB} \end{pmatrix} \begin{pmatrix} \psi_A \\ \psi_B \end{pmatrix} = E \begin{pmatrix} S_{AA} & S_{AB} \\ S_{BA} & S_{BB} \end{pmatrix} \begin{pmatrix} \psi_A \\ \psi_B \end{pmatrix},$$

$$H_{mn} = \int d^3\vec{r}\, u_{1s,m}(\vec{r})\hat{\mathcal{H}}_{el} u_{1s,n}(\vec{r}), \quad S_{mn} = \int d^3\vec{r}\, u_{1s,m}(\vec{r}) u_{1s,n}(\vec{r}).$$

$$(7.11)$$

The matrix elements can actually be calculated analytically using elliptical coordinates $\boxed{\text{P7.5}}$. But for now, we can use our intuition. The diagonal entries of H involve both basis sets on one atom, so the Hamiltonian matrix should be close to the energy of a single atom of hydrogen, E_{1s} with some corrections. The off-diagonal entries are the bonding energy between two neighboring atoms, which we can write as $-t_0$. Normalization implies $S_{AA} = S_{BB} = 1$, while the off-diagonal overlap terms are s. In other words,

$$\begin{pmatrix} E_{1s} & -t_0 \\ -t_0 & E_{1s} \end{pmatrix} \begin{pmatrix} \psi_A \\ \psi_B \end{pmatrix} = E \begin{pmatrix} 1 & s \\ s & 1 \end{pmatrix} \begin{pmatrix} \psi_A \\ \psi_B \end{pmatrix}. \qquad (7.12)$$

Ignoring the overlap term for now, we can get the eigenvalues and normalized eigenvector coefficients (with corrections in $\boxed{\text{P7.5}}$)

$$E_{\pm} = E_{1s} \pm t_0, \quad \psi_A = \mp\psi_B = 1/\sqrt{2}, \quad \Psi^{\pm}_{H_2^+}(\vec{r}) = \frac{u_{1s,A}(\vec{r}) \mp u_{1s,B}(\vec{r})}{\sqrt{2}}.$$

$$(7.13)$$

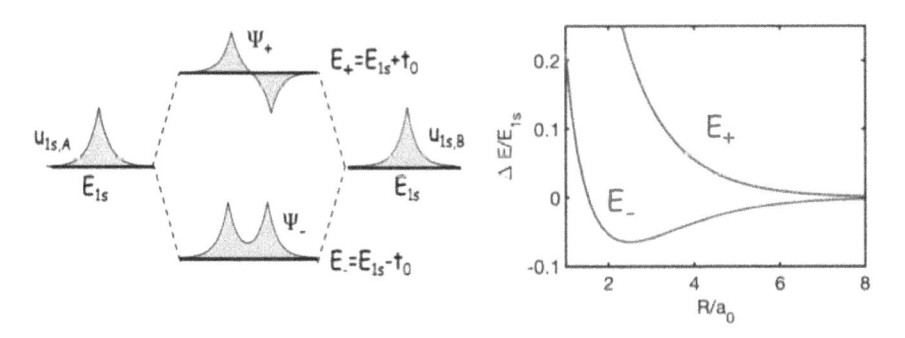

Fig. 7.2 Pooling electrons in the middle through orbital overlap (left bottom) lowers curvature of wavefunction, thus kinetic energy and ultimately bonding energy E_- compared to antibonding E_+ (right).

We can readily see that the lowest energy bonding state Ψ_- corresponds to the symmetric sum of the orbitals, which is the lowest curvature we can get in our LCAO-MO model. Spreading the electron into this bonding state, driven by the overlap $-t_0$, lowers the kinetic energy of the electron wavefunction. In contrast, the higher energy antibonding state Ψ_+ has the highest curvature and thus kinetic energy, and amounts to zero probability in between the atoms — the electrons in the two atoms are sequestered.

We will later see how bands appear in a crystalline solid — a molecular network. Degenerate copies of atomic states split combinatorially into bonding–antibonding pairs (Figs. 8.1 and 8.2), creating valence and conduction bands.

7.8　Bond Hybridization

Remember that covalent bonds are driven by overlap of electron cloud, which are often not along the x, y, z axes. For instance, CH_4 has a tetrahedral structure, with carbon in the middle (as we shall see, same for Si, Ge, and III–V semiconductors like GaAs). The vertices of the tetrahedron make an angle of $109^0 28'$ at the center. To get this combination, we need to hybridize orbitals into so-called sp^3 bonding, by mixing s and p orbitals. The energy lost by promoting an s electron to one of the unoccupied p sites is then gained through enhanced bonding along the tetrahedral axes $\boxed{\text{P7.6}}$.

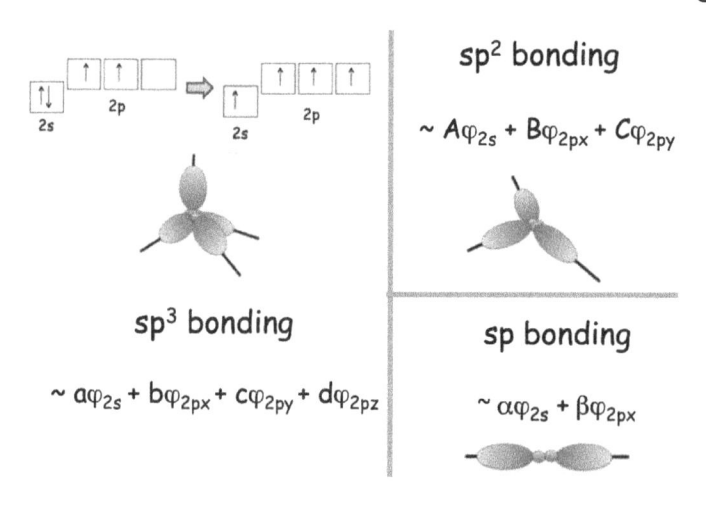

Not just CH_4, but ammonia NH_3 and water H_2O have sp^3 bonding. For ammonia, three of the four tetrahedral axes are bonded, the fourth accommodating a pair of unbonded electrons which repel these bonds and reduce the bond angle to $\sim 106^0$. For water, two of the axes are bonded, the other two each sport a pair of unbonded electrons (four in all), reducing the bond angle to $\sim 104^0$. The bent nature of H_2O makes water polar, which ultimately helps it act as a universal solvent — the basis for life.

For 2D structures, we expect a similar sp^2 bonding, as in double-bonded ethylene C_2H_4, graphene and benzene. Finally, in 1D, we get linear $s-p$ bonding in triple-bonded acetylene C_2H_2.

Speaking of water, the bent nature of water allows water molecules to interlink with other pairs through hydrogen bonds, weakly ionic bonds between the electropositive hydrogen atoms in one molecule and electronegative oxygen atoms in another, to form ice crystals. There are many phases of ice, but the hydrogen bonds keep the molecules further apart than the unbonded jostling configurations in liquid phase that can gyrate right next to each other. This makes ice a unique solid — it is lighter than water and floats on it. This is why lakes freeze at the top, and the floating ice floes keep the rest of the lake unfrozen and allow fishes to survive harsh winters.

Homework

P7.1 **Radial wavefunctions.** For the orbitals guessed in problem P6.6, use the code to calculate and plot the radial wavefunctions and show that they have the same shape as in the cartoon figure.

P7.2 **Ionization potentials.** Work out the energy level diagrams for Na, Cl and Cl^- as shown in Fig. 7.1.

P7.3 **Madelung Constant.** The 3D series sum for $\mathcal{M}$ is poorly convergent and needs special tricks. Instead, work out $\mathcal{M}$ for 1D NaCl.

P7.4 **Crystal field splitting.**

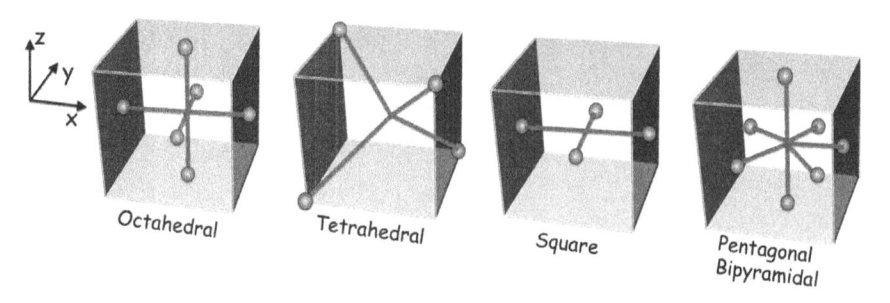

Octahedral   Tetrahedral   Square   Pentagonal Bipyramidal

Let us look at coordination complexes involving ligands (ions/molecules attached to a central metallic atom), shown as red spheres. In a crystal environment, electrons try to stay away from these ligands, splitting the degeneracy of the metal d orbitals. Specifically, we make a distinction between t_{2g} orbitals like d_{xy}, d_{yz}, d_{xz} that thread between crystal x–y–z axes and flip sign upon axial inversion, and e_g orbitals like $d_{z^2}, d_{x^2-y^2}$, which follow the axes and retain sign.

By looking at the distance of various orbitals from the ligand locations (orbitals nearer ligands have higher energy), arrange the 5d orbitals in energy schematically for the four geometries above. In each case, assume a d^8 configuration, and a large energy split compared to double electron pairing, and find the number of unpaired electrons.

P7.5 **Elliptical coordinates.** Let $\vec{r}_{A,B}$ be the two coordinates of a H_2 molecular electron relative to the two nuclei, with separation R. The

Hamitonian elements (Eq. (7.8)) are

$$H_{AA} = \underbrace{\langle u_{1s,A}|T + U_A|u_{1s,A}\rangle}_{E_{1s}} + \underbrace{\langle u_{1s,A}|U_R + U_B|u_{1s,A}\rangle}_{\text{Coulomb } J},$$

$$H_{AB} = \underbrace{\langle u_{1s,A}|T + U_A|u_{1s,B}\rangle}_{E_{1s}S} + \underbrace{\langle u_{1s,A}|U_R + U_B|u_{1s,B}\rangle}_{\text{Exchange } K}. \qquad (7.14)$$

Use elliptical coordinates $\vec{\lambda} = (\vec{r}_A + \vec{r}_B)/R$, $\vec{\mu} = (\vec{r}_A - \vec{r}_B)/R$, $\lambda \in (1, \infty)$, $\mu \in (-1, 1)$, $\phi \in (0, 2\pi)$. This means $r_{A,B} = R(\lambda \pm \mu)/2$. Show that the volume element $d^3\vec{r}$, normalization N, overlap S, electron–nuclear Coulomb and exchange integrals J and K in units of $U_0 = q^2/4\pi\epsilon_0 R$ are

$$d^3\vec{r} = R^3(\lambda^2 - \mu^2)d\lambda d\mu d\phi/8, \quad N = \sqrt{R^3/8\pi},$$

$$S(R) = \int u^*_{1s,A} u_{1s,B} d^3\vec{r} = e^{-\alpha}(\alpha^2/3 + \alpha + 1), \quad \alpha = R/a_0,$$

$$J(R)/U_0 = \int u^*_{1s,A} u_{1s,A} d^3\vec{r}\left(-\frac{R}{r_B} + 1\right) = e^{-2\alpha}(1 + \alpha),$$

$$K(R)/U_0 = \int u^*_{1s,A} u_{1s,B} d^3\vec{r}\left(-\frac{R}{r_A} + 1\right) = S(\alpha) - e^{-\alpha}\alpha(1 + \alpha).$$

$$(7.15)$$

Resulting in $\Delta E_\pm(R) = E_\pm - E_{1s} = \dfrac{J \mp K}{1 \mp S}$. Plot $\Delta E_\pm(R)$ vs R and show how the bonding and antibonding states vary with atom separation R.

P7.6 | **Orbital hybridization.** Let us see how we can take linear combinations of the s, p_x and p_y orbitals to create new orbital basis sets, for instance in benzene. We will choose combinations (sp^2 bases) so that the new superpositions are along the C–C bonds, i.e., at 120° to each other. Write down the expressions for sp_1^2, sp_2^2 and sp_3^2 for a C-atom in terms of its $2s$, $2p_x$ and $2p_y$ orbitals which are assumed to be orthogonalized and normalized. Keep in mind that the new orbitals should also be orthogonal among themselves and normalized. Find the coefficients $a, b, c, d, A, B, C, \alpha, \beta$.

The eigenstates in these new bases are called σ states. Explain why these states would sit far from the Fermi energy and contribute poorly to transport. Also explain why the states created by the remaining p_z orbitals, called π states, sitting perpendicular to the benzene plane, would contribute more to transport.

Chapter 8

Solids: From Bonds to Bands

8.1 Crystal Electronics: States in Fourier Space

A crystalline solid is a network of bonds periodically extended, so we expect its $E-k$ to be simply a Fourier transform of the electronic potential. We will derive the following statements shortly, but they are actually fairly intuitive. Let me first show the painting before I reveal the brushstrokes. We will discover that the $E-k$ is the eigenspectrum of the *Fourier transform* of the Hamiltonian blocks coupling any element of a periodic unit (say an atom labeled 'm') with its neighboring blocks 'n' P8.1–P8.6

$$\left[H(\vec{k})\right] = \sum_n \left[\hat{\mathcal{H}}_{mn}\right] e^{i\vec{k} \cdot (\vec{R}_m - \vec{R}_n)}, \tag{8.1}$$

where each matrix $[\ldots]$ is $b \times b$ in size where b is the number of orbital basis sets in the unit cell, m labels any unit cell in the periodic structure, and n runs over all neighboring unit cells including itself that bear non-zero Hamiltonian matrix elements (i.e., bonds) with unit cell m. The b eigenvalues $E_n(\vec{k})$, $n = 1, 2, \ldots, b$ of $[H(\vec{k})]$ for each $\vec{k}$ connect along the k-axes to give rise to b bands. From our knowledge of periodic structures and Fourier series, we know that there is also a periodicity of the bands in $\vec{k}$ space, related to the ones in real space

$$\vec{r} \to \vec{r} + \vec{R} \ (\vec{R}\text{: lattice periodicity vector}),$$

$$\vec{k} \to \vec{k} + \vec{K} \ (\vec{K}\text{: reciprocal lattice periodicity vector}), \tag{8.2}$$

such that $\vec{K} \cdot \vec{R} = 2\pi$. The smallest periodic unit cell in $\vec{k}$ space that also satisfies the space group symmetry of the crystal is called the *Brillouin zone*.

We calculate all $E(\vec{k})$s within this Brillouin zone to avoid redundancy and facilitate state counting.

Note that the bands are ultimately composed of discrete states, yielding a finite number of eigenvalues. If we assume the entire solid consists of N periodic units, then we must have $b \times N$ eigenvalues. The b comes from the band multiplicity, while N comes from the discreteness of the $\vec{k}$ states. In fact, we will have N equally spaced $\vec{k}$ points within the Brillouin zone.

8.2 Emergence of Bands and k Points in 1D

We can find the bandstructure for a 1D chain of atoms by writing down the periodic Hamiltonian. Given a set of identical basis elements (say $1s$ orbitals) at each atom, $u_n(x) = \psi_{1s}(x - Na)$, we can write down the eigenfunction as $\psi(x) = \sum_n \psi_n u_n(x)$, and then write down the coefficients ψ_n

$$
\begin{pmatrix}
\epsilon_0 & -t_0 & \cdots & \cdots & 0 \\
-t_0 & \epsilon_0 & -t_0 & \cdots & 0 \\
\vdots & \vdots & \vdots & \vdots & \vdots \\
\cdots & \cdots & \cdots & -t_0 & \epsilon_0
\end{pmatrix}
\begin{pmatrix}
\psi_1 \\ \psi_2 \\ \vdots \\ \psi_N
\end{pmatrix}
= E
\begin{pmatrix}
\psi_1 \\ \psi_2 \\ \vdots \\ \psi_N
\end{pmatrix},
\tag{8.3}
$$

where the onsite term $\epsilon_0 = U + 2t_0 = \int u_n^*(x)\hat{\mathcal{H}}u_n(x)dx$ and the hopping term $-t_0 = \int u_n^*(x)\hat{\mathcal{H}}u_{n+1}(x)dx$. This looks a lot like the finite difference Hamiltonian (Eq. (5.4)) described earlier, except instead of grid points we are using localized atom-like basis sets (more specifically, the finite difference arises from rectangular basis sets) $\boxed{P8.7}$.

We can solve the eigenvalue equation numerically for an $N \times N$ Hamiltonian (Fig. 8.1), using the finite difference code described earlier (Section 5.1). As N increases, we see the emergence of a band of eigenvalues with bandwidth $4t_0$. We note that the states get denser at the two band-edges, while the energy levels plotted against index (effectively, k, since the ks are equally spaced by index) upto the Brillouin zone π/a gives us a cosine. There are N eigenvalues capable of holding $2N$ electrons in the band. If each atom donates one electron, the band gets half-filled and the Fermi energy runs right through the middle, giving us a 1D metal (in reality, the 1D chain is prone to becoming mechanically unstable, but we are ignoring that here).

We can in fact get these results analytically. Expanding the nth row

$$
E\psi_n = \epsilon_0\psi_n - t_0\psi_{n+1} - t_0\psi_{n-1}.
\tag{8.4}
$$

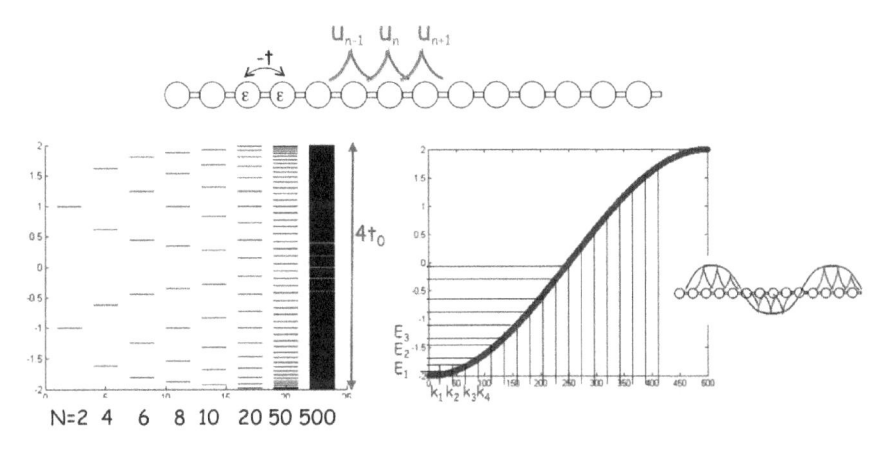

Fig. 8.1 Evolution of energy levels in a 1s orbital-based 1D chain of varying length, ending with a continuous metallic band.

Assuming the ends of the chain are unimportant (i.e., chain is long enough), we can expect the periodic structure to yield a periodic result, i.e., a plane wave. Indeed, a plane wave describing the coefficients, $\psi_n = \psi_0 \exp(inka)$, satisfies the above equation, giving us an energy dispersion and the overall eigenfunction in the 1s basis sets as

$$E = \epsilon_0 - 2 \cos ka$$

$$\psi(x) = \sum_n \psi_n u_n(x) = \sum_n \psi_0 u_n(x) e^{ikx}, \quad x = na \text{ (Bloch's theorem)},$$

$$\text{satisfying } \psi(x + a) = \psi(x) e^{ika} \tag{8.5}$$

The last equation arises because $u_n(x) = \psi_{1s}(x-na)$, so that shifting x by a can be easily neutralized by changing the dummy index $n \to n-1$, but with the phase e^{ika} left behind as a result. The wavefunction is thus the series of periodically arranged basis sets modulated by a plane wave of wavevector k (inset on right in Fig. 8.1). Furthermore, if we assume the chain consists of N atoms in a periodic unit (i.e., on a ring), then $\psi(x = Na) = \psi(x = 0)$ (the wavefunction must revert after a full circle), which applied to the Bloch functions implies that $k = 2\pi n/Na = 2\pi n/L$, $n = 0, \pm 1, \pm 2, \ldots)$, N equally spaced k points within the Brillouin zone.

While the band is in k-space, it secretes in it all information about the electron dynamics in real space. For instance, if we took a wavepacket around a k point and explored the average of two of its components, we get

$$\psi_k = \frac{1}{N} \sum_n u_n(x) e^{i[(k+n\Delta k)x - (\omega + n\Delta\omega)t]}$$

$$\approx \frac{u(x)}{2}[e^{i[(k-\Delta k)x - (\omega - \Delta\omega)t]} + e^{i[(k+\Delta k)x - (\omega + \Delta\omega)t]}]$$

$$= u(x) \underbrace{e^{i(kx - \omega t/\hbar)}}_{\text{Fast}} \underbrace{\cos\left(\frac{\Delta kx - \Delta\omega t}{2}\right)}_{\text{Slow}}. \tag{8.6}$$

We see beats arising from a superposition of nearby frequency modes, creating a fast moving term at a phase velocity $v_{ph} = \omega/k$ from the first pre-factor, modulated by the second factor, an envelope that is slowly moving at group velocity $v_g = \Delta\omega/\Delta k$. Taking $\Delta k, \Delta\omega \to 0$ and $E = \hbar\omega$, we find that the *slope of the band describes its real space group velocity* that behaves intuitively as expected $\boxed{\text{P8.8}}$

$$\boxed{v_g = \frac{1}{\hbar}\frac{\partial E}{\partial k}}. \tag{8.7}$$

However, this is a quantum solution for an electron in a solid, so there must be differences! Indeed there are. For starters, the band is parabolic only at the bottom but deviates pretty soon, in fact, becoming an inverted parabola near the Brillouin zone $\pm\pi/a$, as is easily seen with the exact 1D bandstructure $E = \epsilon_0 - 2t\cos ka$. The velocity $v_g = \partial E/\hbar\partial k = 2at\sin ka/\hbar$ actually vanishes at the Brillouin zone, because the counterpropagating waves $\psi_\pm$ at the equivalent k points $\pm\pi/a$ set up standing waves that are at rest, with their crests and troughs separated by the atomic lattice constant (the half-wavelength $\lambda/2 = \pi/k = a$). If we take superpositions of these waves $\psi_+ \pm \psi_-$, one of them peaks at the atomic locations where the added atomic attraction pulls the band down relative to the parabolic free electron band, while the other peaks mid-way between the atoms and is relatively unaffected, creating a bandgap at the Brillouin zone.

Even at the bottom of the band, the parabolicity is deceptive, because the mass extracted by fitting a parabola $\sim\hbar^2 k^2/2m^*$ is different from the free electron mass m_0. From the kinematic definition of momentum

$$m^* = \frac{p}{v_g} = \frac{\hbar k}{\partial E/\hbar\partial k}.$$

At parabolic band bottoms

$$\boxed{m^* = \lim_{k \to 0} \frac{\hbar k}{\partial E / \hbar \partial k} = \frac{1}{\hbar^2} \frac{\partial^2 E}{\partial k^2}}. \tag{8.8}$$

This makes sense — the effective mass is inversely proportional to the curvature at the band edge. A steeper E–k means a localized wavefunction in k space, which implies a delocalized (and thus more mobile) wavepacket in real space due to the reciprocity between Fourier pairs x and k. In fact, a steeper band with larger bandwidth $4t_0$ within a fixed Brillouin zone $\pm\pi/a$ corresponds to a shallower set of quantum wells in real space with larger hop $-t_0 = \int u_n^*(x)\hat{\mathcal{H}}u_{n+1}(x)dx$ between neighboring wells, and the electron as a result is more mobile and lighter.

For most semiconductors the effective mass is lower than the actual electron mass (e.g., GaAs is at 6%, while silicon is 91% along the longitudinal direction and 19% along the transverse axis). This is because of the net attraction from each nucleus that tries to haul in an electron, speeding it up toward each well whereupon the next well picks it up, and the electron slingshots across the solid like pebbles skipping across water. You can think of it as a quasielectron with a mass that is renormalized by its interactions with the surrounding lattice, in the same way a galloping horse gathering a dust cloud can be thought of as a 'quasihorse = horse + dust cloud' (albeit a larger effective mass in this particular example — maybe the buoyant quasihouse = house + balloons in the movie 'UP' is a more apt case of reduced effective mass).

If we constructed the Hamiltonian out of p_x orbitals, then the overlap $-t_0$ would change sign due to the overlap between the positive half of one p and the negative half of its neighbor, and the band would run downwards, like an inverted parabola. The effective mass m^* is now negative, indicating that the electrons are prone to move in the same direction as an electric field $\mathcal{E}$. Note, however, that such a valence band is usually nearly full of electrons, so we can look at the motion of the small empty top of the band. As the electrons move in one direction, the holes (like the empty chair in a game of musical chairs) moves in the opposite direction, so it is expedient now to think of a positively charged hole with positive effective mass occupying the top of a valence band. In fact, you can think of it as a 'quasipositron', again, with a mass renormalized by lattice interactions.

How can we get multiple bands out of the same Hamiltonian? One way is to think of a set of N dimers, with a strong intradimer hopping $-t_1$ and a weak interdimer hopping $-t_2$. The corresponding Hamiltonian is now

$$
\begin{pmatrix}
\epsilon_0 & -t_1 & \cdots & \cdots & 0 \\
-t_1 & \epsilon_0 & -t_2 & \cdots & 0 \\
0 & -t_2 & \epsilon_0 & -t_1 & \cdots \\
\vdots & \vdots & \vdots & \vdots & \vdots \\
\cdots & \cdots & \cdots & -t_1 & \epsilon_0
\end{pmatrix}
\begin{pmatrix}
\psi_1 \\ \psi_2 \\ \vdots \\ \vdots \\ \psi_{2N}
\end{pmatrix}
= E
\begin{pmatrix}
\psi_1 \\ \psi_2 \\ \vdots \\ \vdots \\ \psi_{2N}
\end{pmatrix}.
\tag{8.9}
$$

We can solve it numerically as before (Fig. 8.2), seeing now that for different ts (here $t_1 = 1$, $t_2 = 0.5$), we get a bandgap. The two bands can be folded into a common Brillouin zone, using the translation and rotational invariance of the bands, $E(\vec{k}) = E(-\vec{k}) = E(\vec{k} + \vec{K})$.

To solve this analytically, we take two of these dimer atoms at one unit, and rewrite the matrix as

$$
\begin{pmatrix}
\alpha & -\beta & \cdots & \cdots & 0 \\
-\beta^\dagger & \alpha & -\beta & \cdots & 0 \\
\vdots & \vdots & \vdots & \vdots & \vdots \\
\cdots & \cdots & \cdots & -\beta^\dagger & \alpha
\end{pmatrix}
\begin{pmatrix}
\varphi_1 \\ \varphi_2 \\ \vdots \\ \varphi_N
\end{pmatrix}
= E
\begin{pmatrix}
\varphi_1 \\ \varphi_2 \\ \vdots \\ \varphi_N
\end{pmatrix},
$$

where

$$
(\varphi_n) = \begin{pmatrix} \psi_{2n-1} \\ \psi_{2n} \end{pmatrix}, \quad
[\alpha] = \begin{pmatrix} \epsilon_0 & -t_1 \\ -t_1 & \epsilon_0 \end{pmatrix}, \quad
[\beta] = \begin{pmatrix} 0 & 0 \\ t_2 & 0 \end{pmatrix}.
\tag{8.10}
$$

This has the same structure as the 1D Hamiltonian, albeit with matrix components. We can thus write out the nth row once again and invoke

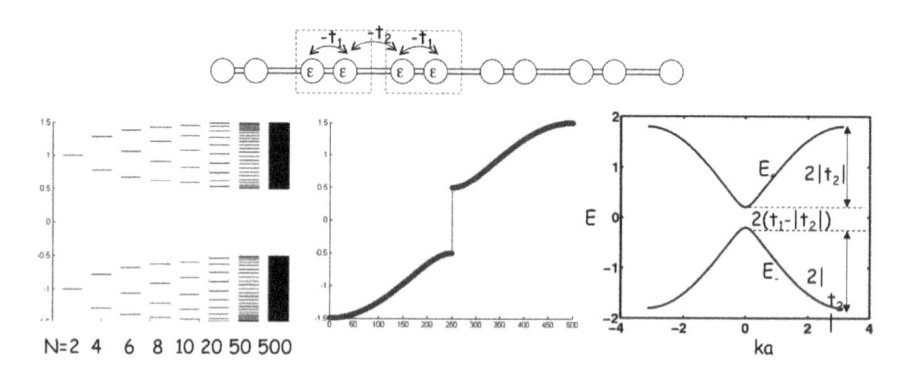

Fig. 8.2 1D dimerized chain yields two bands with a gap proportional to the degree of mismatch in dimer hopping.

Bloch's theorem

$$\alpha\varphi_n - \beta\varphi_{n+1} - \beta^\dagger\varphi_{n-1} = E\varphi_n,$$

$$\varphi_n = \varphi_0 e^{inka} \implies \begin{pmatrix} \epsilon_0 & -t_1 \\ -t_1 & \epsilon_0 \end{pmatrix} - \begin{pmatrix} 0 & 0 \\ t_2 e^{ika} & 0 \end{pmatrix} - \begin{pmatrix} 0 & t_2 e^{-ika} \\ 0 & 0 \end{pmatrix}$$

$$= E \begin{pmatrix} 1 & 0 \\ 0 & 1 \end{pmatrix} \implies E_\pm = \epsilon_0 \pm \sqrt{t_1^2 + t_2^2 + 2t_1 t_2 \cos ka}. \tag{8.11}$$

By comparing values of $E_\pm$ at $ka = 0$ and $ka = \pi$, i.e., across the Brillouin zone, we see a bandgap proportional to $t_1 - t_2$, the extent of dimerization. N dimers ($2N$ atoms) will provide $4N$ states including spin, $2N$ in each band. If each atom again donates one electron, then the lower band is full, the upper empty, and the Fermi energy runs through the bandgap, making this a 1D semiconductor.

8.3 A Toy Model: Kronig–Penney

A simple model for a solid is a set of quantum wells and barriers — a set of step functions. Instead of having an oscillating bond energy, we are using an oscillating onsite energy to create the band doubling needed to initiate a bandgap. In our Matlab code developed in Chapter 5, we can simply use

$$U = U0/2 * (\text{sign}(\sin(n/(N/(2 * \text{pi} * \text{periods})))) + 1), \tag{8.12}$$

with n varying over a loop of spatial gridpoint indices, while periods and $U0$ are self explanatory. N is the number of spatial grid points. For the algebra inclined, we write down the wavefunctions in the wells and the tunneling tails inside the barrier. Let's imagine a barrier of width L_b to the left of $x = 0$ and a well of width L_w to the right. We can write for this first barrier-well pair

$$\Psi_1 = \begin{cases} Ae^{iKx} + Be^{-iKx}, & 0 < x < L_w \text{ (in well)}, \\ Ce^{\kappa x} + De^{-\kappa x}, & -L_b < x < 0 \text{ (in barrier)}, \end{cases} \tag{8.13}$$

where $K = \sqrt{2mE/\hbar^2}$ and $\kappa = \sqrt{2m(U_0 - E)/\hbar^2}$. $L = L_w + L_b$ is the period, equal to the length of the well plus length of the barrier. Now, instead of proliferating coefficients into the next well, we invoke Bloch's theorem at the right side of the barrier, by demanding that

$$\Psi_2 = \Psi_1 e^{ik(L_w + L_b)} = \Psi_1 e^{ikL},\qquad(8.14)$$

where k is different from K and κ and is the wavevector for the periodic lattice, and not the free electron or the tunneling electron. Our aim is to relate k with E.

Matching wavefunctions and derivatives at the left end of the well and the right, we get

$$A + B = C + D,$$

$$iK(A - B) = \kappa(C - D),$$

$$Ae^{iKL_w} + Be^{-iKL_w} = (Ce^{-\kappa L_b} + De^{\kappa L_b})e^{ik(L_w + L_b)}$$

$$iK(Ae^{iKL_w} + Be^{-iKL_w}) = \kappa(Ce^{-\kappa L_b} - De^{\kappa L_b})e^{ik(L_w + L_b)}.$$

$$(8.15)$$

We leave it as an exercise to show (Problem P8.9) that the quasianalytical solution to the above equations leads to a transcendental equation relating k, K and k' (i.e., energy) that can be solved graphically. Only certain energies satisfy this equation — leading to bands and bandgap. The problem asks you to verify that these results are in agreement with the numerical solution (Fig. 8.3) of the potential in Eq. (8.12).

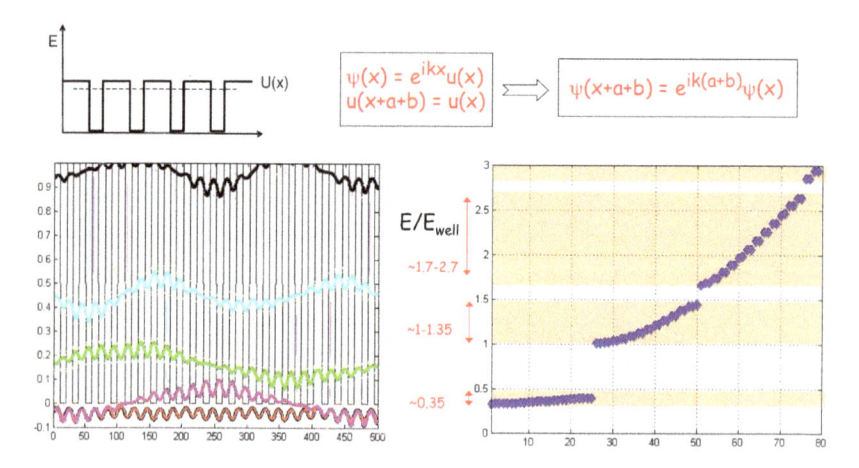

Fig. 8.3 Kronig–Penney model of square wells and barriers showing (left) Bloch states that are box eigenstates modulated by plane waves of varying wavelengths, leading to (right) bands with gaps.

8.4 Working out the Brillouin Zone

Figure 8.4 shows the recipe for finding the first Brillouin zone, following the process in 1D. In 1D, the period in real space was a, and that in reciprocal space was $2\pi/a$, so that the Brillouin zone was obtained by bisecting the nearest neighbor points in reciprocal space separated by $2\pi/a$, giving us a zone between $\pm\pi/a$. Extending beyond the BZ will replicate the eigenvalues $E - k$. The eigenvectors of course oscillate faster due to the longer wavevector, but they would produce the same exact values at the lattice sites which are the basis for our calculated band-structure.

Extending to 3D, we need three vectors $\vec{R}_{1,2,3}$ to define the periodic unit cell in real space. In other words, *every* crystal lattice unit cell (no more no less!) should be reachable from every other with a linear combination $m\vec{R}_1 + n\vec{R}_2 + p\vec{R}_3$, $m, n, p = 0, \pm 1, \ldots$. To get to reciprocal space, we need the vector equivalent of $2\pi/a$. We get it by first defining the direction of each reciprocal lattice vector, say, $\vec{K}_1$ as being perpendicular to the other two, $\vec{R}_{2,3}$, to make it similar to 1D. Then, we enforce the $2\pi/a$ in that direction by ensuring the projection is 2π, i.e., $\vec{K}_1 \cdot \vec{R}_1 = 2\pi$. One way to ensure that is by using

$$\vec{K}_1 = \frac{2\pi \overbrace{\left(\vec{R}_2 \times \vec{R}_3\right)}^{\text{Area}}}{\underbrace{\left|\vec{R}_1 \cdot \left(\vec{R}_2 \times \vec{R}_3\right)\right|}_{\text{Volume}}} \quad \text{and cyclic permutations.} \quad (8.16)$$

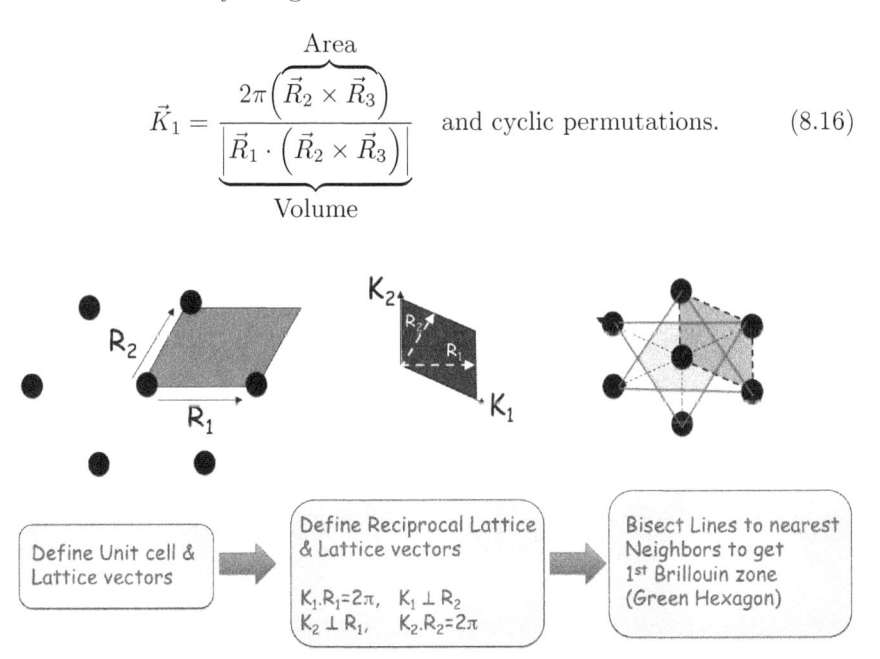

Fig. 8.4 The process for extracting Brillouin zones.

Once again, this means the reciprocal lattice (points we see in a diffraction pattern of the original crystal) can be constructed by a similar linear combination, $M\vec{K}_1 + N\vec{K}_2 + P\vec{K}_3$, $M, N, P = 0, \pm 1, \ldots$.

Once we create the reciprocal lattice, we find the first Brillouin zone (the 3D equivalent of the space between $\pm \pi/a$) by connecting an arbitrary point on the reciprocal lattice with all its nearest neighbors, perpendicularly bisecting each joining line, and identifying the volume enclosed by all these bisectors, such as the green hexagon for the 2D triangular lattice in Fig. 8.4.

Graphene is an interesting example — it would look like a triangular lattice except it has a missing atom at the center of the hexagon, and thus consists of two interweaving triangular sublattices, one created by reinserting the missing atom, and one created by the vacancy. To make it into a periodic lattice, we need to consider each bonded dimer as a unit cell, whereupon graphene becomes a triangular lattice of dimers. The hexagonal Brillouin zone shown above becomes that of graphene. Because the two dimer elements are transferrable into each other with a simple transformation (rotation by 60° about the perpendicular axis), the two bands generated by the dimer p_z orbitals end up gapless, and we get metallic bands. The metallicity is protected by spatial inversion and time reversal symmetry $\boxed{\text{P9.4}}$. More details are in NEMV, Section 5.4.

8.5 The Complexity of 3D Bands

When going through complicated materials with multiple orbitals, such as silicon, we need to look at various ranges of interactions as well as various orbitals participating in those interactions $\boxed{\text{P8.10–P8.11}}$. The silicon unit cell is *diamond cubic*, which means we create two separate face-centered cubes (each with an atom at the eight corners and six face centers), and then stagger these two FCC relative to each other along a quarter of the body diagonal. As a result, each corner silicon atom occupies the center of a tetrahedral unit with four nearest neighbors (see the following figure).

The silicon atom near the bottom left corner at $1/4$ of the body diagonal has four nearest neighbors $\vec{d}_{1,2,3,4}$ connecting along the tetrahedron (shaded in green). These coordinates, expressed as an (x,y,z) triad, relative to the orange silicon atom, are

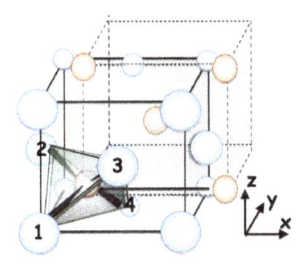

$$\vec{d}_1 = \frac{a}{4}(-1,-1,-1), \quad \vec{d}_2 = \frac{a}{4}(-1,+1,+1),$$

$$\vec{d}_3 = \frac{a}{4}(+1,-1,+1), \quad \vec{d}_4 = \frac{a}{4}(+1,+1,-1). \tag{8.17}$$

When connecting s orbitals with s orbitals and various $p_{x,y,z}$, we can look at the relative sign flips of the corresponding coordinates in the sequence of $\vec{d}$s and record the signs as a *Parity vector* $\vec{P}$.

$\vec{P}$	$s, p_x p_x$	$p_x, sp_x, p_y p_z$	$p_y, sp_y, p_x p_z$	$p_z, sp_z, p_x p_y$	
Si_1	+1	+1	+1	+1	(8.18)
Si_2	+1	+1	−1	−1	
Si_3	+1	−1	+1	−1	
Si_4	+1	−1	−1	+1	

It is easy to understand the signs. For instance, p_x will flip sign whenever an x flips sign in the four atomic coordinates, in this case for $\vec{d}_{3,4}$. For combinations like sp_x, we just multiply the parities from the s and p_x columns. In fact, what we have recorded here is analogous to the multiplication table for a set of symmetry operations relating to the crystallographic point group of the silicon diamond cubic structure.

Accordingly, we get four possible combinations of signs and vectors in the $[H]_{mn}$ Fourier transforms

$$(s_1 s_2 s_3 s_4) \begin{pmatrix} e^{i\vec{k}\cdot\vec{d}_1} \\ e^{i\vec{k}\cdot\vec{d}_2} \\ e^{i\vec{k}\cdot\vec{d}_3} \\ e^{i\vec{k}\cdot\vec{d}_4} \end{pmatrix} = \begin{cases} g_0 \text{ if signs from Column 1} \\ g_1 \text{ if signs from Column 2} \\ g_2 \text{ if signs from Column 3} \\ g_3 \text{ if signs from Column 4.} \end{cases} \tag{8.19}$$

We can thus write down the form of the Hamiltonian in the $sp^3 s*$ basis set for nearest neighbors B of atom A (only the upper triangular part is written here, we generate the lower by adding $H + H^\dagger$)

$$
\begin{array}{c}
\begin{array}{cccccccccc}
s_A & s_B & p_{Ax} & p_{Ay} & p_{Az} & p_{Bx} & p_{By} & p_{Bz} & s_A^* & s_B^*
\end{array} \\
\begin{array}{c}
s_A \\ s_B \\ p_{Ax} \\ p_{Ay} \\ p_{Az} \\ p_{Bx} \\ p_{By} \\ p_{Bz} \\ s_A^* \\ s_B^*
\end{array}
\left(
\begin{array}{cccccccccc}
E_s & V_{ss}g_0 & 0 & 0 & 0 & -V_{sp}g_1 & -V_{sp}g_2 & -V_{sp}g_3 & 0 & 0 \\
 & E_s & -V_{sp}g_1^* & -V_{sp}g_2^* & -V_{sp}g_3^* & 0 & 0 & 0 & 0 & 0 \\
 & & E_p & 0 & 0 & V_{xx}g_0 & V_{xy}g_3 & V_{xz}g_2 & 0 & -V_{s^*p}g_1 \\
 & & & E_p & 0 & V_{xy}g_3 & V_{yy}g_0 & V_{yz}g_1 & 0 & -V_{s^*p}g_2 \\
 & & & & E_p & V_{xz}g_2 & V_{yz}g_1 & V_{zz}g_0 & 0 & -V_{s^*p}g_3 \\
 & & & & & E_p & 0 & 0 & V_{s^*p}g_1 & 0 \\
 & & & & & & E_p & 0 & V_{s^*p}g_2 & 0 \\
 & & & & & & & E_p & V_{s^*p}g_3 & 0 \\
 & & & & & & & & E_s^* & 0 \\
 & & & & & & & & & E_s^*
\end{array}
\right).
\end{array}
$$

$$(8.20)$$

The entries $g_{0,1,2,3}$ are obtained by cross-checking the row–column index (e.g., $s_A p_{Bx}$) with the corresponding column sp_x on the parity matrix. The entries are calibrated to individual bond couplings V_{ss}, $V_{s\sigma}$, $V_{\pi\pi}$, etc. $\boxed{\text{P8.11}}$ using simple trigonometry. One can then check that if we only have s orbitals (i.e., only the first 2×2 block of $[H(\vec{k})]$ exists), we don't have a bandgap along the (100) direction $\boxed{\text{P8.11}}$ where the projected s orbitals do not dimerize (Fig. 8.5). In the (111) direction however, they

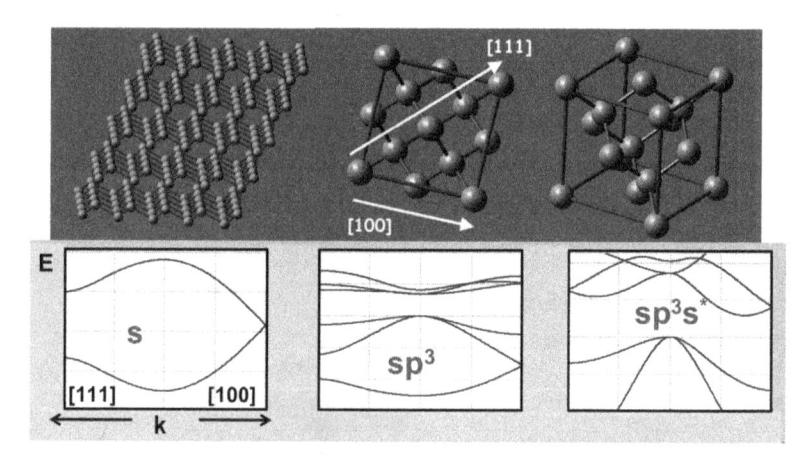

Fig. 8.5 (Left) Si diamond cubic crystal shows no dimerization along [100] ($\pm x, \pm y, \pm z$) axes, only along the cube diagonal [111] direction ($\pm x \pm y \pm z$ axes), so that in an s orbital basis set, a tight binding calculation shows a gap only in [111] but not [100] direction. Adding p orbitals (center) opens a direct bandgap as in the ionic solid GaAs, allowing dipolar coupling leading to optical activity. In an sp^3s^* basis set (right) with virtual excited $3s$ orbitals in addition to $2s$ and $2p$, an indirect bandgap opens at 85% along the Brillouin zone. Fixing the transverse masses requires more bells and whistles, such as $sp^3d^5s^*$ basis sets.

are unequally spaced (big gap between atoms at corners and quarter body diagonals, with nothing at body center), so there is a bandgap. Now, if we bring in p orbitals (first 8×8 block), the p orbitals project unequally along each direction and we get a direct bandgap. To get an indirect bandgap of silicon, we need to include virtual (unoccupied) $3s$ orbitals (we call them s^*) whereupon the V_{ps^*} terms create an indirect bandgap.

Bandstructure is a rich field with a lot of complexity, see the following homework.

Homework

P8.1 **Toy bandstructures.** Consider a 1D dimerized chain, but instead of the couplings 't' oscillating, let the onsite energies oscillate between ϵ_1 and ϵ_2. Following the same principle as a 1D semiconductor, i.e., taking two atoms as one block and solving the nth block row, work out the E–k relation for the two bands for this semiconductor. Find the bandgap in terms of the parameters given.

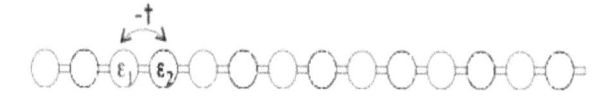

Now repeat this for the following chain shown: Here all atoms have the same energy ϵ and each nearest neighbor bond has coupling $-t$. Take each vertical pair as one 'dimer' and write down α, β, β^+, etc., like in the chapter. Work out and plot the E–k.

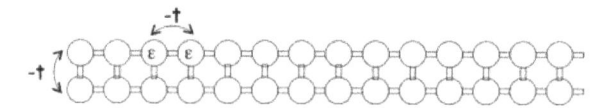

P8.2 **Solitons.** Let us consider a polymer with repeated dimer units.

Note how the double bonds lie to the right of each lower atom on the left section, but switch over so they lie to the left of each lower atom on the right section. At the domain wall shown with a dashed oval, we have a 'defect' known as a soliton, which has a single bond on both sides (as opposed to one single and one double bond). The soliton defect is needed to connect the two dimerized domains that differ in their bond layout.

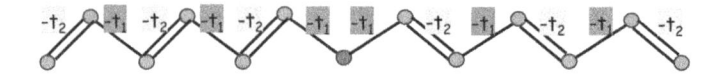

Assume there are 150 double bonds and 150 single bonds on the left domain and similarly 150 double bonds and 150 single bonds on the right domain (so that you have a total of $4 \times 150 + 1 = 601$ atoms). Let $t_1 = -0.8\,\text{eV}$ and $t_2 = -1\,\text{eV}$. Set each onsite energy equal to the negative sum of the bond energies of the two sides.

(a) Find the eigenvalues of the 601×601 matrix, plot them as lines set at the eigenenergies and stack them vertically like in $\boxed{\text{P5.4}}$. Do you notice anything interesting about the bands you get?

(b) If each atom donates one electron, how would electrons distribute themselves at zero temperature. With 601 electrons and 601 protons, what are the net charge and spin of the polymer?

(c) If we now dope the soliton with an extra electron, which energy would the electron reside in? What are the net charge and spin?

$\boxed{\text{P8.3}}$ **Phonons.** Phonons are the quantization of atomic movement in crystals. They are very important to describe heat flow on semiconductor and insulator materials. The displacement from equilibrium x_n of atoms in a crystal can be modeled by a 1D chain of masses m joined by springs with spring constant k. Newton's equation for the nth mass with discrete coordinate x_n gives us $md^2 x_n/dt^2 = -K\Delta x_n$, using Hooke's law for the springs. The net stretch of the spring Δx is given by the stretch $x_n - x_{n+1}$ on one side, plus the stretch $x_n - x_{n-1}$ on the other side.

Now we want to get the dispersion ω vs q (ω: phonon frequency, q: phonon wave vector). Just like we did for a 1D periodic crystal, assume a plane wave solution $x_n = x_q e^{i(qna - \omega t)}$. Replace in the equation of motion you just found and solve for the dispersion $\omega = \omega(q)$. Note that the dispersion is linear at low frequencies. That slope gives us the speed of sound in the crystal and thus we call those phonons acoustic.

Besides acoustic phonons, there are also optical phonons, which are important to recover Fourier's law of heat (an analogue of Ohm's law). Let's try to get the dispersion for the optical phonons. Getting two branches is very similar to what we needed for a semiconductor — i.e., a set of dimers. Consider a 1D crystal with alternating springs with spring constants K_1 and K_2 (Fig. 8.6). Write down Newton's equations for x_n and x_{n+1}. Assume that x_n is proportional to $\exp(-i\omega t)$ and simplify the equations. Then combine your equations in a matrix equation for the whole system that should have the form

$$\omega^2 m\mathbf{x} = K\mathbf{x}, \tag{8.21}$$

with K a matrix and $\mathbf{x}$ the vector of all the x_n.

At this point you should see a similarity between the K matrix and the Hamiltonian for a crystal with two atoms per unit cell. We will solve it analogously by taking each two-atom dimer as a unit cell. Identify the

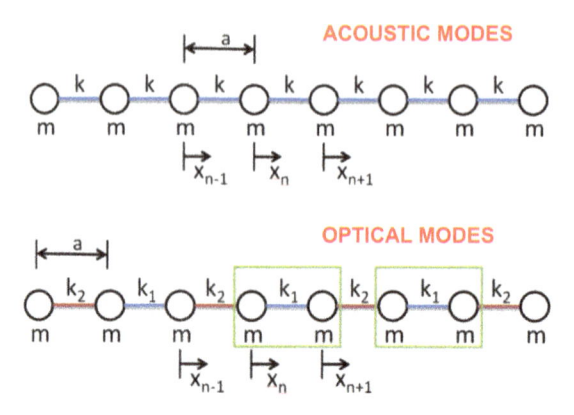

Fig. 8.6 Monomer and dimer atomic chains give rise to acoustic and optical phonon modes.

2×2 on site and off site blocks and find the dispersion. You should get two bands, one that goes through $\omega = 0$, which is the acoustic band, and one that doesn't, which is the optical band.

P8.4 **C60: The death star in the carbon-verse.** We don't have carbonite, but we do have C_{60}. Find its nearest neighbor tight binding energies, numbered as follows.

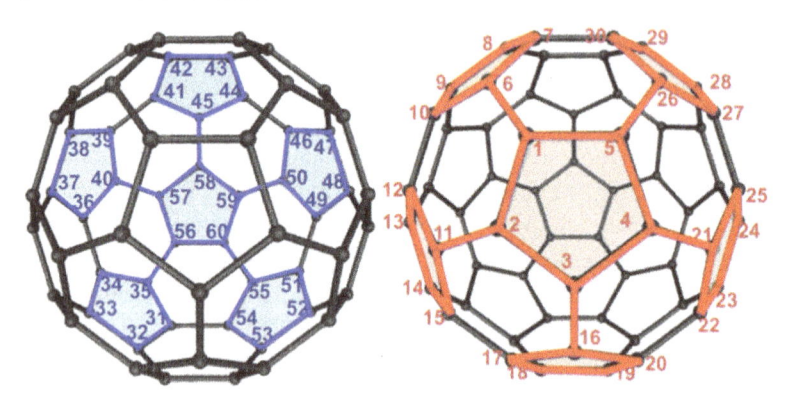

P8.5 **Benzene.** Find the bandstructures and eigenstates of benzene, with a nearest neighbor $V_{\pi\pi} \approx 2.7$ eV interaction for carbon. Treat it as a 1D chain of periodicity six and find the eigenspectrum analytically.

P8.6

Graphene. Work out the bandstructures and eigenvalues for graphene and Bernally stacked bilayer graphene. In the p_z basis sets of two dimer atoms for monolayer graphene and crosslinked Bernal stacked bases for bilayer graphene, the Hamiltonians (NEMV, Sections 5.4 and 5.5) are

$$
\mathcal{H} = \begin{cases}
\begin{pmatrix} 0 & \hbar v_F k_- \\ \hbar v_F k_+ & 0 \end{pmatrix} & \text{(monolayer graphene)} \\[2em]
\dfrac{\hbar^2}{2m^*} \begin{pmatrix} 0 & k_-^2 \\ k_+^2 & 0 \end{pmatrix} & \text{(bilayer graphene)},
\end{cases}
\tag{8.22}
$$

where $k_\pm = k_x \pm i k_y$, $m^* \approx 0.03\, m_0$. Show that monolayer graphene has a linear dispersion and bilayer quadratic. Also show the eigenfunctions

$$
\psi = \begin{cases}
\dfrac{1}{\sqrt{2}} \begin{pmatrix} 1 \\ \pm e^{i\theta} \end{pmatrix} e^{i(k_x x + k_y y)} & \text{(monolayer graphene)} \\[2em]
\dfrac{1}{\sqrt{2}} \begin{pmatrix} 1 \\ \pm e^{2i\theta} \end{pmatrix} e^{i(k_x x + k_y y)} & \text{(bilayer graphene)},
\end{cases}
\tag{8.23}
$$

where $\theta = \tan^{-1}(k_y/k_x)$ is the wavevector angle, and the $+$ $(-)$ sign is for conduction (valence) band. Consult NEMV, Sections 5.4 and 5.5 for the way to derive these Hamiltonians, and the interpretation of the effective masses.

P8.7

Finite element. Assume a net of triangular *half-overlapping* 'tent function' basis sets, $u_n(x) = f(x - na/2)$, where

$$
f = \begin{cases} x, & (0 \leq x \leq a/2), \\ (a - x), & (a/2 < x \leq a), \end{cases}
$$

and zero outside. Evaluate the kinetic energy matrix elements $T_{mn} = -(\hbar^2/2m) \int dx u_m(x) \partial^2 u_n(x)/\partial x^2$. Repeat with a finite difference basis set $u_n(x) = [\Theta(x - (n-1)a) - \Theta(x - na)]/\sqrt{a}$ and show that $[T]$ has the finite difference tridiagonal form (Eq. (5.5)).

P8.8

Bloch oscillations. The periodicity of a crystal can lead to interesting physics! For instance, we will see that a DC field applied to a crystal can actually give an AC output, known as Bloch Oscillation. This has now been observed many times in many systems from superlattices to cold atomic gases. We will work out the physics in what follows, based on our 'semiclassical' understanding of crystal physics.

(a) Newton's law says that $F = dp/dt$. For a constant DC electric field $\mathcal{E}_0$, $F = q\mathcal{E}_0$. The momentum is $p = \hbar k$. Solve this equation to show that the momentum k drifts in time at a constant rate. In other words, find $k(t)$, assuming at t=0 we are at the bottom of the band, $k = 0$.

(b) As k drifts toward and eventually past the Brillouin zone, the slope of the $E - k$ tracks the real space velocity v of the electron. Since the slope switches sign after the Brillouin zone, the electron must turn around, and thus execute an AC oscillation!

Assume a 1D tight binding channel, with $E = 2t_0(1 - \cos ka)$. What is the real space group velocity $v_g(k)$? Using the expression for $k(t)$ from above, write this as $v_g(t)$.

(c) Integrate over time to get $x(t)$, and show that x oscillates with time. Find the Bloch oscillation frequency ω_B, and show that it increases linearly with the electric field. Also show that the oscillation amplitude (maximum value of x) decreases inversely with field $\mathcal{E}_0$.

P8.9 **Kronig–Penney.** The Kronig–Penney model boiled down to a set of four equations in A, B, C and D. To solve them simultaneously, we have to make sure that the determinant of the 4×4 matrix of coefficients of those four quantities must vanish (if not, since we have zero on the other side of the equation, inverting would make A, B, C and D zero. The vanishing determinant prevents that). Show that the determinant equation is

$$\left(\frac{\kappa^2 - K^2}{2K\kappa}\right) \sinh \kappa L_b \sin K L_w + \cosh \kappa L_b \cos K L_w = \cos kL. \qquad (8.24)$$

Plot the left-hand and right-hand sides vs k, and show that the resulting energies agree with the numerical solution. Simplify by assuming the barriers are a series of delta functions (at this point Kronig–Penney becomes a *Dirac comb*), in other words, $L_b \to 0$ and $U_0 \to \infty$ such that $\kappa^2 L_w L_b/2 = P = constant$. In that case, $\kappa L_b \ll 1$ and $\kappa \gg K$. Show that

$$\underbrace{\frac{P}{KL_w} \sin K L_w + \cos K L_w}_{f(KL_w)} = \cos kL \qquad (8.25)$$

The right-hand side ensures that $-1 \leq f \leq 1$, which immediately puts restrictions on E and opens bandgaps.

P8.10 **Bandstructure of gold.** Gold has an FCC crystal structure, while its valence electrons are primarily derived from $6s$ orbitals. The spherical symmetry of 's' orbitals means that the interaction between any two nearest neighbor atoms is identical. Assume a nearest neighbor Au–Au interaction in an FCC lattice given by $-t$, and onsite energy ϵ_0.

(a) Construct the k space Hamiltonian H_k by summing over nearest neighbors. Get a compact expression for $E(k_x, k_y, k_z)$.
(b) Plot along the 100 and 111 directions, upto the relevant Brillouin zones (assume interatomic separation 'a', so that the k is in units of $1/a$. Plot E in units of 't', so the actual value of 't' does not matter).

P8.11 **Tight binding elements.**

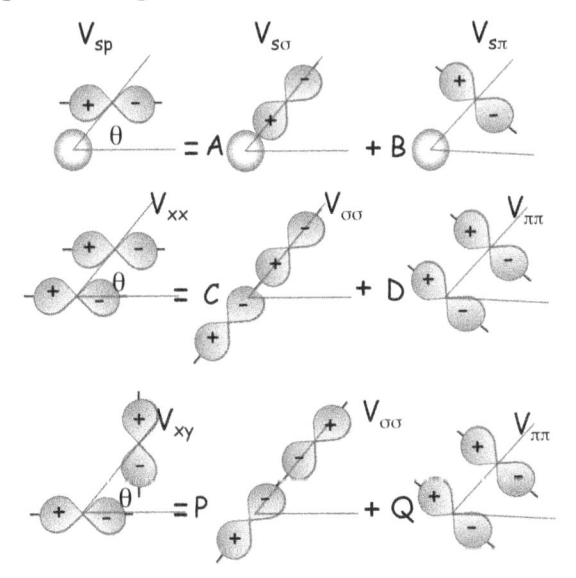

The matrix elements V_{ss}, V_{sp}, V_{xx}, V_{xy} are related to separately calibrated σ and π bonding strengths through direction cosines. Write down the relations between the set V_{xx}, V_{yy}, V_{zz}, V_{xy}, V_{yz}, V_{xz}, V_{sp}, V_{s^*p} and the set $V_{\sigma\sigma}$, $V_{\sigma\pi}$, $V_{s\sigma}$, $V_{s^*\sigma}$ in terms of the angle Θ between the orbital orientations x, y, z and the bond orientation. In other words, the coefficients A, B, C, D, P, Q above. Note that some bonds — e.g., $V_{s\pi}$ shown, and $V_{\sigma\pi}$ (not shown) will be zero from symmetry.

Also work out the angle Θ between the bond and one of the crystal axes in the diamond cubic lattice (i.e., between the 111 and 100 directions).

The nearest neighbor sp^3s^* Vogl parameters (in eV) (a: anion, c: cation) are shown in the adjoining table. The two V_{sp} rows for GaAs are for V_{sapc}, V_{scpa}, V_{s^*apc} and V_{s^*cpa}.

	E_{ss}	E_p	E_{s^*}	V_{ss}	V_{xx}	V_{xy}	V_{sp}	V_{s^*p}
Si	−4.2	1.715	6.69	−8.32	1.72	4.56	5.72	5.36
GaAs(a)	−8.34	1.04	8.59	−6.44	1.96	5.08	4.48	4.84
GaAs(c)	−2.66	3.67	6.74	−6.44	1.96	5.08	−5.8	−4.8

Chapter 9

Spins and Magnetism

9.1 The Discovery of Spins

In addition to the three indices n, l, m, an electron in an atom has a fourth index denoting its spin state. Electrons are spin-half particles, as was demonstrated in the classic experiment by Stern and Gerlach. The idea was to deflect a stream of silver atoms whose magnetic moments were supposed to couple with an external spatially varying magnetic field, so that by looking where the atoms ended up we can infer its angular momentum states. The surprising result was that no matter how the measurements were taken, the result always created two lines (let's call them up and down) meaning there were only two possible magnetic moments that the atom registered.

The situation becomes complicated when we try to filter the atoms by axis. Suppose we use a z-directed field to split the electrons into up and down. Next, we take the up z electrons and put them through an x-directed field, and now separate out the up x electrons into say front and back. If the electrons acted classically, we would expect their state at this stage to be $\uparrow_z \times \uparrow_x$. However, taking these electrons and resending them through the z-directed field once again created two lines, meaning the original up-z electrons have now turned back into a mixture of up and down. The x measurement robbed the state of its z-memory. The spin operators along the x and z directions do not commute.

By analogy with angular momentum, we can then postulate a similar commutation relation for the spin operators, as described in Eq. (7.1). We expect $2s + 1$ lines from the z-component of the spin angular momentum. Empirically, we see two lines, so by setting $2s + 1 = 2$, we get $s = 1/2$. The electron is a spin-half particle with two z-directed eigenstates, $\uparrow_z$ and $\downarrow_z$.

9.2 Spin Matrix Operators

We can write down any spin in the basis set of $\uparrow_z$ and $\downarrow_z$ spins as

$$S = \alpha \uparrow_z + \beta \downarrow_z = \begin{pmatrix} \alpha \\ \beta \end{pmatrix}. \tag{9.1}$$

By definition then,

$$\uparrow_z = \begin{pmatrix} 1 \\ 0 \end{pmatrix} \quad \downarrow_z = \begin{pmatrix} 0 \\ 1 \end{pmatrix}. \tag{9.2}$$

Since the spin z operator has these two states as eigenstates with eigenvalues $\pm\hbar/2$, it must be diagonal with its eigenvalues as diagonal elements

$$\hat{S}_z = \frac{\hbar}{2} \begin{pmatrix} 1 & 0 \\ 0 & -1 \end{pmatrix}. \tag{9.3}$$

This is easily verified by checking that $\hat{S}_z \uparrow_z = (\hbar/2) \uparrow_z$ and $\hat{S}_z \downarrow_z = -(\hbar/2) \downarrow_z$. In fact, it is the only matrix that will accomplish it.

The other two matrices can also be uniquely obtained by using the commutation relations (Eq. (7.1)). We get

$$\hat{S}_x = \frac{\hbar}{2} \begin{pmatrix} 0 & 1 \\ 1 & 0 \end{pmatrix} \quad \hat{S}_y = \frac{\hbar}{2} \begin{pmatrix} 0 & -i \\ i & 0 \end{pmatrix}. \tag{9.4}$$

As expected, these are off-diagonal in the z spin basis set, and they flip a down z spin into up and vice versa, as the Stern–Gerlach experiment demonstrated. Collectively, we can write them as

$$\vec{S} = \frac{\hbar}{2}\vec{\sigma}, \tag{9.5}$$

where $\sigma_{x,y,z}$ are called Pauli matrices (Eq. (3.18)). Together with the 2×2 identity matrix, they completely span the 2D matrix space. We can derive them using spin ladder operators $\boxed{P9.1\text{–}9.2}$, like in problem $\boxed{P6.2}$.

We can now find the eigenstates of these matrices and designate them as up-x, down-x, up-y and down-y

$$\uparrow_x = \frac{1}{\sqrt{2}} \begin{pmatrix} 1 \\ 1 \end{pmatrix} \quad \downarrow_x = \frac{1}{\sqrt{2}} \begin{pmatrix} 1 \\ -1 \end{pmatrix} \quad \uparrow_y = \frac{1}{\sqrt{2}} \begin{pmatrix} 1 \\ i \end{pmatrix} \quad \downarrow_y = \frac{1}{\sqrt{2}} \begin{pmatrix} 1 \\ -i \end{pmatrix}. \tag{9.6}$$

We have a nice analogy between electron spins and photon polarization, with the $\uparrow_z, \downarrow_z$ spin bases replaced by $\hat{x}, \hat{y}$ unit vectors in 3D space.

$$\underbrace{\begin{pmatrix} 1 \\ 0 \end{pmatrix}}_{\uparrow_z}, \underbrace{\begin{pmatrix} 0 \\ 1 \end{pmatrix}}_{\downarrow_z} \implies \begin{cases} E_+ = E_0 \hat{x} \cos(kz - \omega t) \\ E_- = E_0 \hat{y} \cos(kz - \omega t) \end{cases} \quad \text{(linearly polarized light)},$$

$$\underbrace{\frac{1}{\sqrt{2}} \begin{pmatrix} 1 \\ \pm 1 \end{pmatrix}}_{\uparrow_x, \downarrow_x} \implies \begin{cases} E_0 \dfrac{\hat{x} + \hat{y}}{\sqrt{2}} \cos(kz - \omega t) \\ E_0 \dfrac{\hat{x} - \hat{y}}{\sqrt{2}} \cos(kz - \omega t) \end{cases} \quad \text{(45° rotated polarization)},$$

$$\underbrace{\frac{1}{\sqrt{2}} \begin{pmatrix} 1 \\ \pm i \end{pmatrix}}_{\uparrow_y, \downarrow_y} \implies \begin{cases} E_0[\hat{x}\cos(kz - \omega t) + \hat{y}\sin(kz - \omega t)] \\ E_0[\hat{x}\cos(kz - \omega t) - \hat{y}\sin(kz - \omega t)] \end{cases}$$

$$= E_0 Re\underbrace{\left[\frac{\hat{x} \pm i\hat{y}}{\sqrt{2}} e^{i(kz - \omega t)}\right]}_{\text{LCP, RCP}}, \tag{9.7}$$

where in the last case we have left and right circularly polarized light.

The difference between Cartesian basis sets $\hat{x} = (100), \hat{\bar{x}} = (-100), \hat{y} = (010), \ldots$ and spinor bases $\uparrow_z = (10)^T, \downarrow_z = (01)^T, \uparrow_x = (11)^T/\sqrt{2}, \ldots$ (T: transpose) is easy to see since each Cartesian axis, for instance, $\hat{y}$, is orthogonal to the other two ($\hat{x}$ and $\hat{z}$) except its opposite $\hat{\bar{y}}$, while each spinor basis, for instance $\uparrow_y$, is a mixture of the others ($\uparrow_z, \downarrow_z$ or $\uparrow_x, \downarrow_x$) except its opposite $\downarrow_y$.

9.3 Spin Rotation

Finally, we can construct the $\hat{S}$ matrix for an arbitrary rotation (θ, ϕ) and find its eigenstates, by projecting the spin matrices (Eqs. (9.3) and (9.4)) onto the various Cartesian components (see Fig. 9.1 for a refresher on the transformations)

$$\hat{S}(\theta, \phi) = \hat{S}_x \sin\theta \cos\phi + \hat{S}_y \sin\theta \sin\phi + \hat{S}_z \cos\theta$$

$$= \frac{\hbar}{2} \begin{pmatrix} \cos\theta & \sin\theta e^{-i\phi} \\ \sin\theta e^{i\phi} & -\cos\theta \end{pmatrix}. \tag{9.8}$$

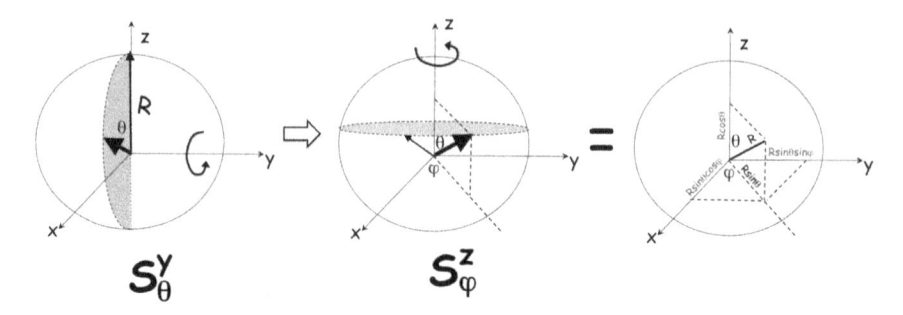

$$S^y_\theta \qquad\qquad S^z_\varphi$$

Fig. 9.1 The sequence of rotations on a sphere that leads to the final resting point of (R, θ, ϕ) — altitude, latitude, longitude. In this case, they involve an anticlockwise rotation (looking in) about the y-axis followed by one about the z-axis to get to that point.

The up and down eigenstates can then be calculated as

$$\uparrow_{\theta,\phi}=\begin{pmatrix}\cos\frac{\theta}{2}e^{-i\phi/2}\\ \sin\frac{\theta}{2}e^{i\phi/2}\end{pmatrix} \qquad \downarrow_{\theta,\phi}=\begin{pmatrix}-\sin\frac{\theta}{2}e^{-i\phi/2}\\ \cos\frac{\theta}{2}e^{i\phi/2}\end{pmatrix}. \tag{9.9}$$

Note the half-angles! They distinguish the state of a spinor from a normal vector (the former are orthogonal at 180° while the latter are orthogonal at 90°). We can see that by rotating the spin from the north pole along a line of longitude through 360°,

$$\uparrow_{0,0}=\begin{pmatrix}1\\0\end{pmatrix}=\uparrow_z, \qquad \uparrow_{2\pi,0}=\begin{pmatrix}-1\\0\end{pmatrix}=-\uparrow_z. \tag{9.10}$$

As we will see later (Section 12.2), the 2π rotation amounts to an exchange operation that yields a negative sign for the electron. We need to go through 4π rotation to recover the original state. This can be better appreciated by thinking of a 180° rotation as a reflection operation in a plane parallel to the spin direction. A regular (polar) vector, like an arrow, flips its direction upon reflection when pointing toward the mirror, and stays unchanged if it lies in the mirror plane. In contrast, a rotation (axial) vector stays unchanged when pointing toward the mirror (a wheel moving in the mirror plane stays unchanged upon reflection), but flips its direction when lying in the plane (a wheel rotating toward the mirror flips direction upon reflection). These symmetry operations are enforced by the (anti)commutation algebra of the Pauli matrices.

9.4 Relation with 3D Spatial Rotation

It is worthwhile drawing a distinction between spinors (which are orthogonal when they are at $180°$) and vectors (orthogonal at $90°$). The rotation of the spinors above is akin to rotation of a vector, except for the minus sign we pick up — we need the former to rotate through 4π before it fully reverts to the original. The reason we can describe a 3D rotation with a 2D Pauli matrix is because out of the three components, only two are independent. The rotation of a vector (x, y, z) preserves the norm $x^2 + y^2 + z^2 = r^2$. For rotation matrices that take us from one configuration to another, say a rotation around the z-axis by θ,

$$\begin{pmatrix} x' \\ y' \\ z' \end{pmatrix} = \underbrace{\begin{pmatrix} \cos\theta & \sin\theta & 0 \\ -\sin\theta & \cos\theta & 0 \\ 0 & 0 & 1 \end{pmatrix}}_{\mathcal{R}_\theta^z} \begin{pmatrix} x \\ y \\ z \end{pmatrix}, \tag{9.11}$$

where the rotation matrix satisfies $(\mathcal{R}_\theta^z)^T \mathcal{R}_\theta^z = I$ to preserve the norm.

Because we preserve the norm, we can equivalently look at the evolution of a 2×2 matrix whose determinant $-(x^2 + y^2 + z^2)$ is already fixed. For the z-rotation, we can write this as a unitary transformation, similar to the way we transform matrices between two basis sets

$$\begin{pmatrix} z' & x' - iy' \\ x' + iy' & -z' \end{pmatrix} = U^\dagger \begin{pmatrix} z & x - iy \\ x + iy & -z \end{pmatrix} U. \tag{9.12}$$

We expect that the spinors rotate the way an ordinary 2D vector does

$$\begin{pmatrix} a' \\ b' \end{pmatrix} = U \begin{pmatrix} a \\ b \end{pmatrix}. \tag{9.13}$$

We will revisit the expression for U shortly when we discuss the spin statistics theorem. In this case, however, we can simply find it by construction. The expression for a spin rotation by angle $\Delta\phi$ about the axis $\hat{n}$ is given by (note $(\vec{\sigma} \cdot \hat{n})^2 = I$)

$$U = S_{\Delta\phi}^n = e^{i\vec{\sigma} \cdot \hat{n}\Delta\phi/2} = I \cos \Delta\phi/2 + i\vec{\sigma} \cdot \hat{n} \sin \Delta\phi/2. \tag{9.14}$$

To get us to the angle (θ, ϕ), we need to do a sequence of rotations (Fig. 9.1). First, we rotate about the y-axis by θ to get to the right latitude, and then we rotate about the z-axis by ϕ to get to the right longitude, so we can write down $U = S_\phi^z \times S_\theta^y$. It is then straightforward to verify that

$$\uparrow_{\theta,\phi} = \begin{pmatrix} e^{-i\phi/2}\cos\theta/2 \\ e^{i\phi/2}\sin\theta/2 \end{pmatrix} = U \begin{pmatrix} 1 \\ 0 \end{pmatrix}, \quad U = S_\phi^z \times S_\theta^y. \tag{9.15}$$

It is straightforward to substitute and verify that the transformed results in Eq. (9.12) are the same as Eq. (9.11). We will explain the physics behind our choice for U, especially the half angle, shortly when we discuss the spin statistics theorem.

In the physics literature, we usually refer to the above equivalence as the homomorphism of SU(2) onto SO(3) representations, i.e., Special Unitary matrices in 2D onto Special Orthogonal matrices in 3D. Note that for the SU(2), we have $0 \le \theta \le 4\pi$, so there is a two-to-one mapping between SU(2) and SO(3) because of the extra minus sign upon a 2π rotation.

9.5 How Do We Combine Spatial and Spintronic Descriptions?

Recall that we used atomic orbital basis sets at each site for a 1D chain to describe a simple metal. With spin in the mix, we now double the size of our Hamiltonian, so that each basis set is now two orbitals with opposite spins, $\{u_{1s\uparrow_z}(1), u_{1s\downarrow_z}(1), u_{1s\uparrow_z}(2), u_{1s\downarrow_z}(2), \dots, u_{1s\uparrow_z}(N), u_{1s\downarrow_z}(N)\}$. If the potential is spin-independent, then it stays diagonal in the spin space, and the tight binding Hamiltonian looks like

$$
\begin{array}{cccc}
\uparrow_{z1} & \downarrow_{z1} & \uparrow_{z2} & \downarrow_{z2} \cdots \cdots \cdots \\
\end{array}
$$

$$
\begin{pmatrix}
\epsilon_0 & 0 & -t_0 & 0 & \cdots & \cdots & \cdots & \cdots & 0 \\
0 & \epsilon_0 & 0 & -t_0 & \cdots & \cdots & \cdots & \cdots & 0 \\
-t_0 & 0 & \epsilon_0 & 0 & -t_0 & 0 & \cdots & \cdots & 0 \\
0 & -t_0 & 0 & \epsilon_0 & 0 & -t_0 & \cdots & \cdots & 0 \\
\vdots & \vdots & \vdots & \vdots & \vdots & \vdots & \vdots & \vdots & \vdots \\
0 & \cdots & \cdots & \cdots & \cdots & \cdots & \cdots & 0 & \epsilon_0
\end{pmatrix} .
$$

$$\tag{9.16}$$

In other words, each element got augmented by a 2×2 identity matrix, $\epsilon_0 \to \epsilon_0 I_{2\times2}$, $-t_0 \to -t_0 I_{2\times2}$, etc. If the Hamiltonian were spin-dependent, say a spin term σ_z, then we will get $\epsilon_0 \to \epsilon_0\sigma_z = \begin{pmatrix} \epsilon_0 & 0 \\ 0 & -\epsilon_0 \end{pmatrix}$, etc. We can write this schematically by the notation $[H] \to [H] \otimes \sigma_z$, where $\otimes\sigma_z$

indicates a direct product with σ_z. The direct product between two entities A and B can be thought of as taking every element of A and tagging on the entire B ensemble. For instance,

$$\underbrace{(a\ b)}_{[A]} \otimes \underbrace{(\alpha\ \beta)}_{[B]} = (a[B]\ b[B]) = (a\alpha\ a\beta\ b\alpha\ b\beta),\ (a\ b) \otimes \begin{pmatrix} \alpha \\ \beta \end{pmatrix} = \begin{pmatrix} a\alpha & b\alpha \\ a\beta & b\beta \end{pmatrix}.$$

$$(9.17)$$

Ultimately, the spin (or pseudospin) structure of the wavefunction is established by its overall symmetry $\boxed{\text{P9.3 and 9.4}}$. Very often, this symmetry manifests itself by evolving the electron phase around the Fermi circle and can be quantified by its overall winding number $\boxed{\text{P9.5}}$.

Homework

P9.1 **Spin ladder.** The spin ladder operator $\hat{S}_+$ (see $\hat{L}_\pm$ earlier) takes a down spin up but annihilates an up spin (as it can't go any further up).

$$\hat{S}_+ \downarrow_z = \hbar \uparrow_z, \quad \hat{S}_+ \uparrow_z = 0. \tag{9.18}$$

Work out $\hat{S}_+$ operator. Similarly, the $\hat{S}_-$ operator (should be its Hermitian conjugate). From these, and the relations $\hat{S}_\pm = \hat{S}_x \pm i\hat{S}_y$, work out $\hat{S}_{x,y}$.

P9.2 **Spin 1.** Work out the 3×3 spin matrices $\hat{S}_{x,y,z}$ for a spin 1 particle, and find the three eigenvectors for each in the $\uparrow_z, 0_z, \downarrow_z$ basis set.

P9.3 **Symmetries of graphene.** (a) Start with the 2D bandstructure of graphene (Eq. (8.22)) and add a diagonal term due to unequal A and B atom onsite energies, and a second nearest neighbor (chiral) hopping

$$\mathcal{H} = \underbrace{\hbar v_F (\tau_z \sigma_x k_x + \sigma_y k_y)}_{\mathcal{H}_{\text{graphene}}} + \underbrace{mv_F^2 \sigma_z}_{\mathcal{H}_{\text{mass}}} + \underbrace{\Delta_2 \sigma_y}_{\mathcal{H}_{\text{chiral}}}, \tag{9.19}$$

where σ describes the pseudospin (orbital/sub-lattice) and $\tau_z = \pm 1$ represents the two valleys. The z-products are meant to be direct products, e.g., $\tau \otimes \sigma$. Write down the new $E - k$ and show that we get a gap (resembling a famous equation!) and that m is the effective mass and Δ_2 gives a shift in Dirac point along the k_y axis away from the Brillouin zone.

(b) We now show that these gapping perturbations are eliminated by the symmetries of graphene. We start with *Time reversal symmetry* (show that $T = -i\sigma_y K$, K: complex conjugation, turns up spin into down), i.e., when you flip time $t \to -t$, you expect $\vec{k} \to -\vec{k}$. Write down Schrödinger equation with these transformations. Also write down the complex conjugated version of Schrödinger equation. For periodic structures satisfying Bloch's theorem, the k dependence is mainly a plane wave modulating a periodic comb, so $\psi(-k) = \psi^*(k)$. Using these relations, prove

$$\boxed{\mathcal{H}^*(k) = \mathcal{H}(-k)}. \tag{9.20}$$

Show that the perturbation term $\mathcal{H}_{\text{chiral}}$ is eliminated by this symmetry. Also show that the mass term $mv_F \sigma_z$ and a spin–orbit term $\lambda \tau_z \sigma_z s_z$ coupling $\vec{\tau}$ with real spin $\vec{s}$ are allowed by time-reversal symmetry.

(c) For *spatial inversion* (*orbital or sub-lattice*) *symmetry*, switching $(x, y) \rightarrow (-x, -y)$ should turn atom A into B and vice versa, along opposite ks. Show that this means $\psi(\vec{k}) = \sigma_x \psi(-\vec{k})$ in the A–B atom basis set. Substituting in Schrödinger equation, show that this implies

$$\boxed{\sigma_x \mathcal{H}(k)\sigma_x = \mathcal{H}(-k)} \tag{9.21}$$

Show that the perturbation term $\mathcal{H}_{\text{mass}}$ does not satisfy this symmetry because of its chiral nature ($\sigma_z = -i\sigma_x\sigma_y$). *In other words, the gaplessness of graphene is protected by these two dual symmetries in 2-D.*

P9.4 Weyl Semi-metals

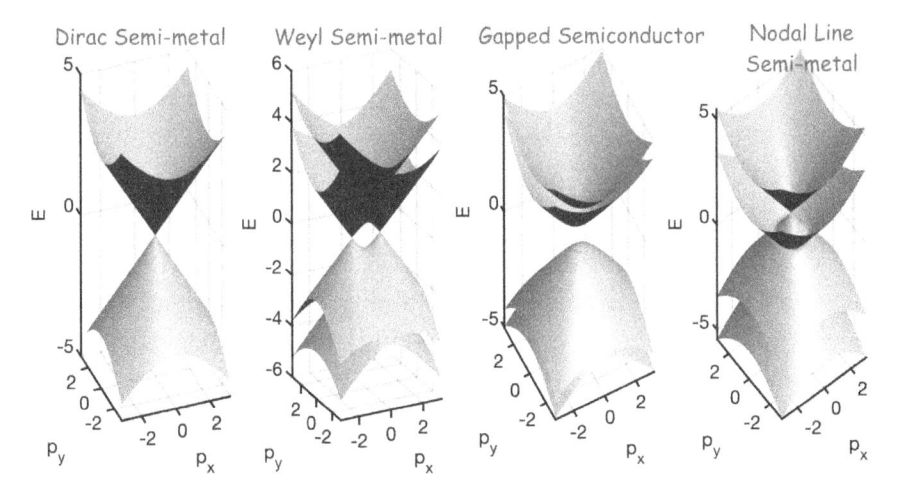

(a) A *Dirac semi-metal* (DSM) is a 3D analog of graphene, with a Hamiltonian that can be classified in terms of pseudospin $\vec{\sigma}$ and real spin $\vec{\tau}$.

$$\mathcal{H}_1 = \tau_x \otimes \hbar v_F (\vec{\sigma} \cdot \vec{k}) = \hbar v_F \begin{pmatrix} 0 & 0 & k_z & k_- \\ 0 & 0 & k_+ & -k_z \\ k_z & k_- & 0 & 0 \\ k_+ & -k_z & 0 & 0 \end{pmatrix}. \tag{9.22}$$

(In Matlab, we can do the $\otimes$ with the *kron* argument for Kronecker tensor product. Also note that the order matters — *the outer indices* — *e.g.,* τ *in the matrix above, come first*). Show that this gives the 3D analogue of graphene bands. Note that since the added terms in P9.3 that act as perturbations in 2D are part of the natural 3D Dirac Hamiltonian, the gaplessness is automatically protected without needing added symmetries. Any small perturbation $\alpha I + \vec{\beta} \cdot \vec{\sigma}$ linear in $\vec{k}$ shifts the E–k vertically or laterally but does not open a gap. Squeezing to 2D graphene eliminates the $\sigma_z k_z$ term, making the E–k sensitive to perturbations $\propto \sigma_z$ that sublattice symmetry ($[\mathcal{H}, \sigma_z] = 0$) eliminates.

(b) Next, for DSMs like Na$_3$Bi and Cd$_3$As$_2$, we add a term that breaks spatial inversion symmetry

$$\mathcal{H}_{\mathrm{DSM}} = \mathcal{H}_1 + M(\vec{k}) I \otimes \sigma_z, \tag{9.23}$$

with $M(\vec{k}) = M_0 - M_1 k_z^2 - M_2(k_x^2 + k_y^2)$. For $M_{0,1,2} < 0$, show that the Dirac points are at $k_z = \pm\sqrt{M_0/M_1}$.

(c) We now add a magnetic field m that separates the up–down spins τ for a *Weyl semi-metal* (WSM) like TaAs

$$\mathcal{H}_{\mathrm{WSM}} = \mathcal{H}_1 + b I \otimes \sigma_z + m \tau_z \otimes I. \tag{9.24}$$

Show that the solutions are $\epsilon_{sp} = s\sqrt{m^2 + b^2 + v_F^2 p^2 + 2\mu b\sqrt{m^2 + v_F^2 p_z^2}}$, where $s, \mu = \pm 1$. Also show that by setting $k_x = 0$, we get for $E(k_y, k_z)$ (i) a DSM for $m = b = 0$, (ii) a gapped semiconductor for $|m| > |b|$ (e.g., $m = 1$, $b = 0.5$), and finally (iii) two inverted overlapping bands with two Weyl points for $|b| > |m|$ (e.g., $m = 0.5$, $b = 1$), touching at $v_F \vec{p} = (0, 0, \pm\sqrt{b^2 - m^2})$. Plot $E(0, k_y, k_z)$ in each case.

(d) We can get a *line-node semi-metal* by adding a different perturbation breaking term representing a fast spatially varying magnetic field along the σ_x direction, switching signs between $\tau_z = \pm 1$ sectors.

$$\mathcal{H}_{\mathrm{LNS}} = \mathcal{H}_1 + b' \tau_z \otimes \sigma_x \tag{9.25}$$

Show that the solutions are

$$E = s\sqrt{p_x^2 v_F^2 + \left[v_F\sqrt{p_y^2 + p_z^2} + b'\mu\right]^2}, \quad s, \mu = \pm 1.$$

Show that for $p_x = 0$, we get zero energy along a circle $p_y^2 + p_z^2 = (b'/v_F)^2$. Plot $E(0, k_y, k_z)$ for $b' = 1$.

Later we will show how the Fermi surfaces of a WSM are fractured into pieces along its surfaces connected by bulk tunneling.

P9.5 **Berry phase, Chern number and Skyrmions.**

(a) The celebrated *Aharonov–Bohm* effect showed that an electron carried around a region of localized magnetic flux (e.g., a solenoid) can pick up a non-trivial phase that shows up in interference experiments. We account for it with a modification to the momentum $\vec{p} \to \vec{p} - q\vec{A}$, where the magnetic vector potential $\vec{A}$ is related to the magnetic field through a curl, $\vec{B} = \vec{\nabla} \times \vec{A}$. Show that this modification generates the correct Lorenz force $\vec{F} = q\vec{v} \times \vec{B}$. Note that for flow equations of a vector field that is both position- and time-dependent, as in hydrodynamics, we must use $d/dt = \partial/\partial t + \vec{v} \cdot \vec{\nabla}$.

(b) Show that a free electron as a result picks up an extra phase γ proportional to the areal flux of the magnetic field $\Phi_B = \int \vec{B} \cdot d\vec{S}$, in addition to the trivial phase $\exp\left[-i \oint \vec{p}(t).d\vec{r}/\hbar\right]$ around the path.

(c) We can now generalize this for any Hamiltonian by tracking the extra phase $\gamma(t)$ from motion in $\vec{k}$-space, by substituting

$$\psi[\vec{k}(t)] = e^{i\gamma(t)} e^{-i/\hbar \int_0^t dt' \epsilon[\vec{k}(t')]} \phi[\vec{k}(t)]. \qquad (9.26)$$

Substitute in Schrödinger equation and find the Berry phase as a time integral of a time derivative. Since time enters explicitly through $\vec{k}(t)$, you can switch to integral and derivative of $\vec{k}$ to generalize the equation above. Show that we get the Berry phase as a flux of a magnetic field in k-space

$$\gamma(t) \propto \int \vec{B}_{\vec{k}} \cdot d\vec{S}_{\vec{k}}, \quad \vec{B}_{\vec{k}} = \vec{\nabla}_{\vec{k}} \times \vec{A}_{\vec{k}} \text{ (Berry curvature)},$$

$$\vec{A}_{\vec{k}} = i\phi^*[\vec{k}] \vec{\nabla}_{\vec{k}} \phi[\vec{k}] = i u_{\vec{k}}^* \vec{\nabla}_{\vec{k}} u_{\vec{k}}, \quad u \text{ being the Bloch part.} \quad (9.27)$$

The vector potential is set by Ampere's law $\oint \vec{B} \cdot d\vec{l} = \mu_0 I$ with current density (Eq. (3.5)), albeit in k-space, $J_k \propto i u_k^* \nabla_k u_k$.

(d) The Chern number C_n refers to the quantum of Berry flux, $\gamma = 2\pi C_n$. For graphene, use the spinor wavefunction in the two component basis (Eq. (8.23)) and show that $C_n = \pm 1/2$. The corresponding Berry phase γ gives an overall minus sign upon rotation, as expected for fermions.

(e) Set up a numerical code for graphene plus an out-of-plane magnetic field. On a 2D k-grid, find the numerical eigenvectors at each point, and then take the normalized dot product of two neighboring points $\Psi_{k_x,k_y}^{\dagger} \cdot \Psi_{k_x+dk,k_y}$ and $\Psi_{k_x,k_y}^{\dagger} \cdot \Psi_{k_x,k_y+dk}$ for each band. Extract the phase by taking imaginary part of the log of each dot product, then plot the local phases as Berry curvature, and the sum to get the Chern number of each band.

(f) For a generic spinor (Eq. (9.15)), show that the magnetic field/Berry curvature is $\vec{B}_{\vec{k}} = i\hat{k}/2k^2$, the field due to a magnetic monopole, and the Berry phase γ is given by negative half of the solid angle (Gauss's Law).

(g) A *skyrmion* is a magnetic vortex with a non-zero Berry phase, created by inversion symmetry breaking such as at the interface between a thin magnetic film and a heavy metal underlayer. Use the same spinor description, assuming the magnetization azimuthal angle $\theta(r)$ depends only on distance r from the core, while the angle $\phi = \phi(\varphi)$ changes with φ around the vortex periphery by 2π. Substitute and show that in 2D polar coordinates (r, φ), $A_r = 0$, $A_\varphi = 1$. Assuming the skyrmion magnetization unit vector $\hat{m}$ flips from down in the core to up at the boundary, show that

$$C_n = [m_z(r = \infty) - m_z(r = 0)]/2 = 1, \quad \text{where } m_z(r) = \cos\theta(r).$$
$$(9.28)$$

Chapter 10

How Do Spins Interact with Their Surroundings?

10.1 Zeeman Effect: Aligning with Magnetic Fields

As the Stern Gerlach experiment showed, spins interact with magnetic fields by trying to align with them like miniature tops. The Zeeman Hamiltonian describing this part of the interaction, assuming the magnetic field $\vec{B} = B_0 \hat{z}$, is $\boxed{\text{P10.1}}$

$$\hat{\mathcal{H}}_{\text{Zeeman}} = -\vec{\mu} \cdot \vec{B} = -\frac{g\mu_B \vec{S}}{\hbar} \cdot \vec{B} = -\frac{g\mu_B \vec{\sigma}}{2} \cdot \vec{B} = -\frac{g\mu_B B_0}{2} \begin{pmatrix} 1 & 0 \\ 0 & -1 \end{pmatrix},$$

$$(10.1)$$

which suggests that the lower energy eigenstate is the electron $\uparrow_z$ oriented up along the B field. Here $g \approx 2$ is the Lande g-factor, the Bohr magneton $\mu_B = q\hbar/2m \approx 60 \ \mu\text{eV/T}$, and $\gamma = g\mu_B/\hbar \approx 1.76 \times 10^{11} \text{rad}/s - T$ is the gyromagnetic ratio.

The Zeeman splitting between the two energies $\Delta = g\mu_B B_0$. To create some dynamics in the spins, we need to create off-diagonal terms coupling the $\uparrow_z$ and $\downarrow_z$ terms, in other words a field in the $x-y$ plane to couple to $\vec{S}$ (remember the $S_{x,y}$ matrices are off-diagonal in the $\uparrow_z, \downarrow_z$ basis). However, this will simply create new spin split states in the $x-y-z$ plane. To create dynamics, we also need a time-dependent AC field whose frequency is resonant with the Zeeman splitting. The corresponding electron spin resonance (ESR) Hamiltonian

$$\hat{\mathcal{H}}_{\text{ESR}} = -g\frac{\mu_B \vec{S}}{\hbar} \cdot (B_0 \hat{z} + B_1 \hat{x} \cos \omega t) = -\frac{g\mu_B}{2} \begin{pmatrix} B_0 & B_1 \cos \omega t \\ B_1 \cos \omega t & -B_0 \end{pmatrix}.$$

$$(10.2)$$

We can solve this analytically by assuming $B_1 \ll B_0$ and Taylor expanding $\boxed{P10.2}$, and making some approximations (e.g., near resonance, i.e., $\hbar\omega \approx \Delta$). Alternately, we can solve this numerically using the codes developed earlier (Section 5.2) for time-dependent problems. We can see in what follows how in a DC field B_0 responsible for a Zeeman split of $10\,\mu\mathrm{eV}$ and an AC field with $B_1 = B_0/5$ at resonance, we get a periodic transfer of spins from up to down. The colored numerical results are overlayed with the black lines that give us the quasianalytical near-resonance expression $P_\downarrow(t) = \sin^2\omega_R t$ (red) and $P_\uparrow(t) = \cos^2\omega_R t$ (blue), where the transfer frequency $\omega_R = g\mu_B B_1/2\hbar$.

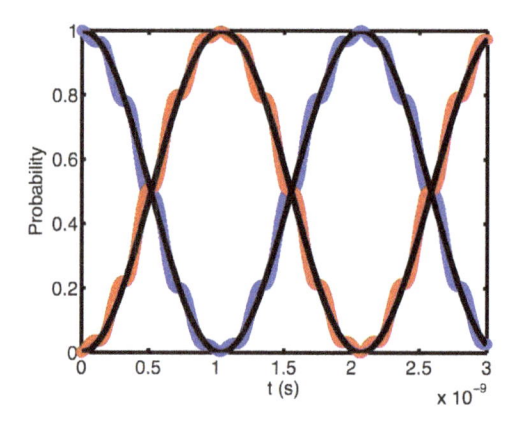

10.2   Internal Fields from Exchange: Ferromagnetism

Ferromagnetism is caused by a strong Hund's rule coupling (Section 7.1) — putting spins in parallel keeps them apart through exchange (Pauli exclusion) and cuts down the sizeable Coulomb repulsion for localized electrons in partially filled d orbitals of the transition metal series. In this case, we can approximately treat the impact of parallel spins as an effective magnetic field to which the spins in turn respond. The Zeeman split acts on the entire metallic band, splitting it by an exchange energy (Fig. 10.1) so that the majority spins are more numerous at the Fermi energy than the minority spins. The Hamiltonian for a collection of spins can be simplified if we pretend each spin is experiencing an average field due to all other

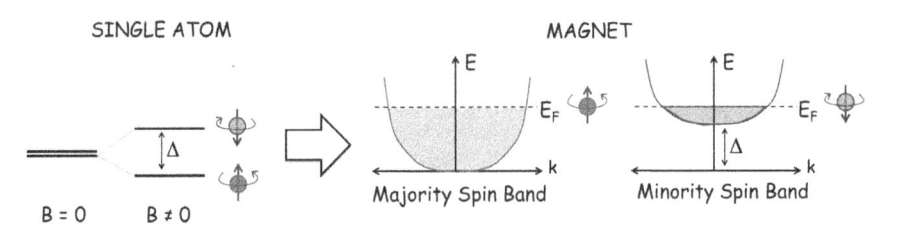

Fig. 10.1   Zeeman split between atomic spins expands into internal exchange fields in a ferromagnet to create a split between up and down spin bands.

spins, and this field is the same for all spins $\boxed{\text{P10.3 and 10.4}}$

$$\hat{\mathcal{H}}_{\text{FM}} = -\sum_{ij} J_{ij}\vec{S}_i \cdot \vec{S}_j - g\mu_B \vec{H}_{\text{ext}} \cdot \sum_i \vec{S}_i$$

$$= -g\mu_B \sum_i \vec{S}_i \cdot \vec{B}_i/\hbar \approx -g\mu_B \sum_i \vec{S}_i \cdot \langle \vec{B} \rangle/\hbar$$

$$\vec{B}_i = \frac{\hbar}{g\mu_B}\sum_j J_{ij}\vec{S}_j + \mu_0\vec{H}_{\text{ext}} \underbrace{\approx \langle\vec{B}\rangle}_{\text{Mean field approximation}} = \mu_0(\vec{H}_{\text{ext}} + \lambda\vec{M}).$$

(10.3)

The internal field comes from the collective average magnetization set up by the other spins with an adjustable factor λ, so that their average dipole moment, in units of gyromagnetic ratio $\gamma \approx 17.6$ GHz/KOe, is

$$\langle\vec{\mu}\rangle = \underbrace{(g\mu_B/\hbar)}_{\substack{\text{Gyromagnetic ratio } \gamma}} \times \underbrace{N\langle\vec{S}\rangle}_{\vec{M}}$$

(10.4)

where N is the number of dipoles per unit volume. The average is found by a Boltzmann weighted sum over the discrete values of $S_z - m\hbar$ for m running from $-S$ to S

$$\langle S_z \rangle = \frac{\sum_{m=-S}^{m=S} m\hbar e^{-\beta g\mu_B mB}}{\sum_{m=-S}^{m=S} e^{-\beta g\mu_B mB}} = \hbar S B_S(S\Delta), \quad \beta = 1/k_B T, \quad (10.5)$$

where $\Delta = \beta g\mu_B B$, and the Brillouin function $\boxed{\text{P10.5}}$

$$B_J(x) = \frac{2J+1}{2J}\coth\frac{(2J+1)x}{2J} - \frac{1}{2J}\coth\frac{x}{2J},$$

(10.6)

varies between $x(J+1)/3$ for $x \ll 1$ to 1 for $x \gg 1$, so that

$$M = N g \mu_B S B_S(\beta g \mu_B \mu_0 (H_{\text{ext}} + \lambda M)). \tag{10.7}$$

For ferromagnets, we need to see *spontaneous magnetization* at zero external fields $H_{\text{ext}} = 0$. The left-hand side of Eq. (10.7) is a $45°$ angled straight line when plotted vs M. The right-hand side starts with a linear approximation $\sim N(g\mu_B)^2 \lambda \mu_0 M \beta S(S+1)/3$ for small M, and then saturates for large M to $N g \mu_B S$. This second curve already intersects the first curve at $M = 0$, but it will also intersect at a higher finite M provided it has an initial slope higher than $45°$ that eventually bends to zero at saturation and crosses the straight line, in other words

$$N(g\mu_B)^2 \mu_0 \lambda \beta S(S+1)/3 > 1 \Longrightarrow T < \underbrace{\frac{(g\mu_B)^2 N \lambda \mu_0 S(S+1)}{3k_B}}_{T_C}. \tag{10.8}$$

In other words, at low temperature thermal fluctuations are small enough that we get a spontaneous magnetization driven by exchange. Note that all we establish is consistency, i.e., at low temperature the ferromagnetic state is energetically possible.

The feasibility of ferromagnetism is ultimately suggested by the Stoner criterion $\boxed{\text{P12.4}}$, when the Coulomb repulsion responsible for exchange is large enough, $U_0 D(E_F) \gg 1$, and the spins align to cut down the Coulomb cost through Pauli exclusion. On the other hand, antiferromagnetism arises when we try to reduce the kinetic energy by spreading the electrons around (remember that kinetic energy is set by the curvature of the wavefunctions). Such a spreading is allowed if the spins are oppositely aligned, especially near half-filling, blocked otherwise once again by Pauli exclusion.

The final sign of the exchange term J for elemental transition metals is given by the Bethe–Slater curve, which starts negative at small interatomic separations (Mn, etc.), then goes to high positive at higher separations (Fe, Co, Ni), then reduces to small positive for rare earths like Gd.

10.3 Internal Fields from Nuclear Attraction: Spin–Orbit

We already saw the emergence of a spin–orbit term $\propto \vec{\sigma} \cdot \vec{\nabla} \times \vec{p}$ in the Hamiltonian arising from what could be called 'relativistic' effects $\boxed{\text{P3.9}}$. The term came as a natural consequence of low energy expansion of the Dirac equation. Let us discuss the physics now. An electron orbiting around a heavy nucleus sees in turn a large charge current arising from the cluster of protons in the nucleus orbiting relative to itself. The nuclear current creates an internal magnetic field that orients the electrons along itself like spinning tops. The potential arising from this spin–orbital coupling is what we are after in this section. Relativity theory tells us the effective magnetic field arising from a rotating charge/electric field, and the corresponding Zeeman term, are (details in NEMV, Section 5.7.)

$$\vec{B} = \vec{v} \times \vec{\mathcal{E}} \propto \vec{k} \times \vec{\mathcal{E}}$$

$$\mathcal{H}_{SO} = \frac{g\mu_B}{\hbar} \vec{S} \cdot \vec{B} \propto \begin{cases} \vec{L} \cdot \vec{S} & \text{for atoms } U = U(r), \ \vec{\mathcal{E}} = \dfrac{\vec{r}}{qr}\dfrac{dU}{dr}, \\ (S_x k_y - S_y k_x) & \text{for gated structure, } \vec{\mathcal{E}} = \hat{z}\mathcal{E}, \end{cases}$$

$$(10.9)$$

where the factor of 2 arises from Thomas precession, an adjustment to the relativistic equation valid for two inertial frames (constant velocity separation) to account for rotation. Such gateable spin–orbit effects can in fact be used to generate spin switching device concepts ⟨P10.6⟩.

10.4 Macrospin Dynamics: The LLG Equation

In presence of a magnetic field $\vec{H}$ with potential $\hat{\mathcal{U}}_{\text{ext}} = -\gamma \vec{H} \cdot \vec{M}$, we can work out the dynamics of the spin (dot represents time derivative). Separating out the magnitude and direction of the magnetization vector $\vec{M} = M_s \hat{m}(\theta, \phi)$, we get ⟨P10.7 and 10.8⟩ (Eq. (10.22))

$$\dot{\hat{m}} = \underbrace{-\gamma \left(\hat{m} \times \vec{H} \right)}_{\text{Conservative } \vec{T}_{\text{cons}}} + \vec{T}_{\text{damping}} + \cdots \text{(LLG equation)}. \qquad (10.10)$$

This magnetic field can include internal forces coming from the potential landscape, such as shape anisotropy and magnetocrystalline anisotropy. We can evaluate them by the same procedure by which we can extract the field from the Hamiltonian above, namely,

$$\vec{H}_{\text{eff}} = -\frac{\vec{\nabla}_{\hat{M}} U}{\gamma} = -\left(\hat{\theta} \frac{\partial U}{\partial \theta} + \frac{\hat{\phi}}{\sin\theta} \frac{\partial U}{\partial \phi} \right) \frac{1}{\gamma M_s}. \qquad (10.11)$$

In addition, we have some *non-conservative* forces that cannot be readily written as a gradient of a potential. One term is phenomenological Gilbert

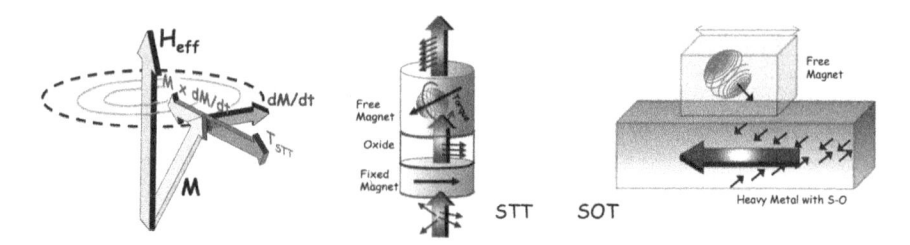

Fig. 10.2 (Left) Vector components of spins, field and torques. (Center) Spin transfer torque (STT) due to spin injected from a hard magnet absorbed by a soft magnet. (Right) Spin–orbit torque (SOT) due to spin separation and subsequent injection from a heavy metal.

damping, which tries to damp out the dynamics (often originating from the loss of angular momentum to a surrounding lattice)

$$\vec{T}_{\text{damping}} = \alpha \hat{m} \times \frac{d\hat{m}}{dt}. \tag{10.12}$$

This can be countered by antidamping torques (Fig. 10.2), the most common being adiabatic spin transfer torques that arise when conduction electrons try to follow the local texture of the spatially varying magnetization (effectively the *inertial term* of a hydrodynamic equation), with the number density n of torquing electrons set by angular momentum conservation

$$\vec{T}_{\text{STT}} = \left(\vec{v} \cdot \vec{\nabla} \right) \hat{m}$$

$$\vec{v} = \frac{\vec{J}}{qn}, \qquad \underbrace{n\frac{P\hbar}{2} = \frac{M_S}{\gamma}}_{\text{ang. mom. conserv.}} \implies v = |\vec{v}| = \frac{Jg\mu_B P}{2qM_s}, \tag{10.13}$$

where $P = (N_\uparrow - N_\downarrow)/(N_\uparrow + N_\downarrow)$ is the polarization. Note that much like the other torques, this one is perpendicular to the instantaneous magnetization ($\hat{m} \cdot \vec{T} = 0$) so that the magnitude is preserved and only the direction changes. The adiabatic STT is also coplanar with $\hat{m}$ and the current density $\vec{J}$. In a magnetic tunnel junction structure where we have a fixed magnet with magnetization $\vec{M}_1$ and a free magnet with magnetization $\vec{M}_2$, this gradient term can be simplified as $v\hat{m}_2 \times (\hat{m}_1 \times \hat{m}_2)$. There could be another perpendicular component orthogonal to both $\hat{m}$ and $\vec{T}_{\text{STT}}$, along $\hat{m} \times \vec{T}_{\text{STT}}$, arising from other considerations — for instance, a *non-adiabaticity* term that arises due to the inability of conduction electrons to instantaneously follow the spatial variations of $\hat{m}$, and a *field-like* term.

Then there are torques applied by spin–orbit effects, such as from current flowing in a heavy non-magnetic metal underlying the magnet. Their torques can be understood in terms of the local electric field (direction set by the symmetry axes of the device) and the corresponding spin–orbit torque. For current flowing along $\hat{x}$ (i.e., $k_y = 0$, $k_x > 0$) and a magnet/non-magnet stack with interface in the $x-y$ plane, the potential (Eq. (10.9)) looks like $\sim - S_y k_x$, and the corresponding spin–orbit torque would look like

$$\vec{T}_{\text{SOT}} = c_J \theta_{\text{SHE}} \hat{m} \times (\hat{m} \times \hat{y}), \tag{10.14}$$

where θ_{SHE} is the spin Hall angle. Finally, there is a thermal torque acting through a fluctuating field, in effect a white noise

$$\vec{T}_{\text{thermal}} = -\gamma \hat{m} \times \vec{h}_{th}, \ \vec{h}_{th} = h \underbrace{\hat{n}}_{\text{random orientation}},$$

$$\langle h_i(t) \rangle = 0, \ \langle h_i(t) h_j(t') \rangle = D \delta_{ij} \delta(t - t'), \quad D = \frac{2\alpha k_B T}{\gamma M_S V}. \tag{10.15}$$

The correlation function can be obtained by solving the equivalent Fokker–Planck (drift diffusion) equation, as outlined in NEMV, Sections 9.3 and 9.4.

Homework

P10.1 **Spin dynamics.** An electron is at rest in an oscillating magnetic field: $\mathbf{B} = B_0 \cos(\omega t)\hat{k}$, where B_0 and ω are constant. So we can construct the Hamiltonian matrix for this system by using S_z:

$$[H] = -\frac{g\mu_B B_0 \cos\omega t}{2}\begin{bmatrix} 1 & 0 \\ 0 & -1 \end{bmatrix} \tag{10.16}$$

(a) Solve for the two eigenvectors of $S_x = \hbar\sigma_x/2$, labeled as $\uparrow_x$, $\downarrow_x$.

(b) Assuming the electron starts in the spin-up state along the x-axis at $t = 0$, $\psi(t = 0) = \uparrow_x$, find the wavefunction $\psi(t)$ at any subsequent time. *Hint*: Start with the time-dependent Schrödinger equation (Eq. (3.6)), where $\psi(t) = \begin{bmatrix} a(t) \\ b(t) \end{bmatrix}$. Substituting above, you get two equations for derivatives of a and b with the initial state specified above. This should give $\psi(t)$.

(c) Find the probability of getting $-\hbar/2$ (spin-down state $\psi_{\downarrow_x}$) if you measure S_x. (*Hint*: find the overlap $\psi_{\uparrow_x}^\dagger \psi(t)$ and then take modulus squared to get the probability. The † means turn rows into columns and do complex conjugate, so $\psi_{\uparrow_x}^\dagger$ will be a 1×2 vector while the original $\psi_{\uparrow_x}$ was 2×1).

(d) Find the expectation values $\langle \hat{S}_{x,y,z} \rangle$.

P10.2 **Spin resonance.** Let us first work out the ESR oscillations analytically. Writing Ψ as in Problem P10.1, write down the differential equations for a and b. When solving, use the *Rotating Wave Approximation* meaning that near resonance $\omega \approx \Delta/\hbar$, terms like $e^{i(\Delta/\hbar+\omega)t}$ oscillate quickly and are dropped, while terms like $e^{i(\Delta/\hbar-\omega)t}$ are retained. Show that the probabilities are as described in the section on ESR, i.e., squares of sines and cosines with a transfer frequency $\omega_R = g\mu_B B_1/2\hbar$.

Now, solve numerically as well using the Crank–Nicholson scheme (Section 5.2), for the given initial condition. You should reproduce the ESR figure earlier in the chapter.

P10.3 **Ferromagnetic resonance.** Let us work out the lineshape of a ferromagnetic resonance (FMR) measurement. where a small RF field resonant with the Zeeman split in a magnet causes electrons to transition between the two (similar to ESR for single spins). Start with the Landau–Lifshitz–Gilbert equation for the dynamics of the overall magnetization, $\partial \vec{M}/\partial t = -\gamma(\vec{M} \times \vec{B}_{\text{eff}})$, where $\vec{B}_{\text{eff}} = -\partial U/\partial \vec{M}$ is the effective magnetic field including demagnetization, anisotropy and external fields.

(a) Let us first get an idea of the ferromagnetic resonance frequency. Assuming no damping, we still need to include the internal magnetic field $\vec{B}_{\text{eff}} = \vec{B} - \mu_0 \overset{\leftrightarrow}{N} \cdot \vec{M}$, where $\overset{\leftrightarrow}{N}$ is the ellipticity tensor (thus, the vector $\overset{\leftrightarrow}{N} \cdot \vec{M} = N_{xx}M_x\hat{x} + N_{yy}M_y\hat{y} + N_{zz}M_z\hat{z}$ with x, y, z along the principal axes of the ellipsoidal magnet). Note that $\text{Tr}(\overset{\leftrightarrow}{N}) = 1$, and also the diagonals are equal for a sphere, all zeros except one for a disk, and one zero two equal non-zeros for a cylinder). We assume $M_z \approx M_s \gg M_{x,y}$, and simplify to get

$$\frac{\partial \vec{M}_x}{\partial t} = -\gamma M_y[B - \mu_0 M_s(N_{zz} - N_{yy})],$$
$$\frac{\partial \vec{M}_y}{\partial t} = \gamma M_x[B - \mu_0 M_s(N_{zz} - N_{xx})]. \tag{10.17}$$

From this equation, show that the FMR frequency

$$\omega = \gamma\sqrt{[B - \mu_0 M_s(N_{zz} - N_{yy})][B - \mu_0 M_s(N_{zz} - N_{xx})]}$$
$$\approx \gamma\sqrt{B(B + \mu_0 M_s)}, \tag{10.18}$$

where we assumed the oscillating field is applied perpendicular to the easy axis of a disk, so that $N_{xx} = 1, N_{yy} = N_{zz} = 0$. Clearly, at zero field there is no resonance, unless we explore it along a different direction.

(b) We can add a Gilbert damping term $\alpha\gamma[\vec{M} \times \partial\vec{M}/\partial t]/M_s^2$ to the LLG equation, and get the lineshape

$$\Delta H = (2\pi\alpha/M_s)f + \Delta H_0, \tag{10.19}$$

where the last term is inhomogeneous broadening, that is frequency-independent. We can plot the FMR absorption linewidth vs frequency and extract α from the slope.

P10.4 **Magnon spectrum.** Let us now add anisotropy. A tight binding like equation can be derived for excitations around the ground state of a ferromagnet, where all spins are pointing up along the z-axis. The Heisenberg Hamiltonian in presence of uniaxial anisotropy is

$$\hat{H} = -2J \sum_n \vec{S}_n \cdot (\vec{S}_{n+1} + \vec{S}_{n-1}) + D \sum_n S_{nz}^2. \qquad (10.20)$$

The effective magnetic field $\gamma \vec{B}_{\text{eff}}^n = -\partial E / \partial \vec{S}_n$. Show that

$$\gamma \vec{B}_{\text{eff}}^n = 2J(\vec{S}_{n+1} + \vec{S}_{n-1}) + 2D\hat{z}S_{nz}. \qquad (10.21)$$

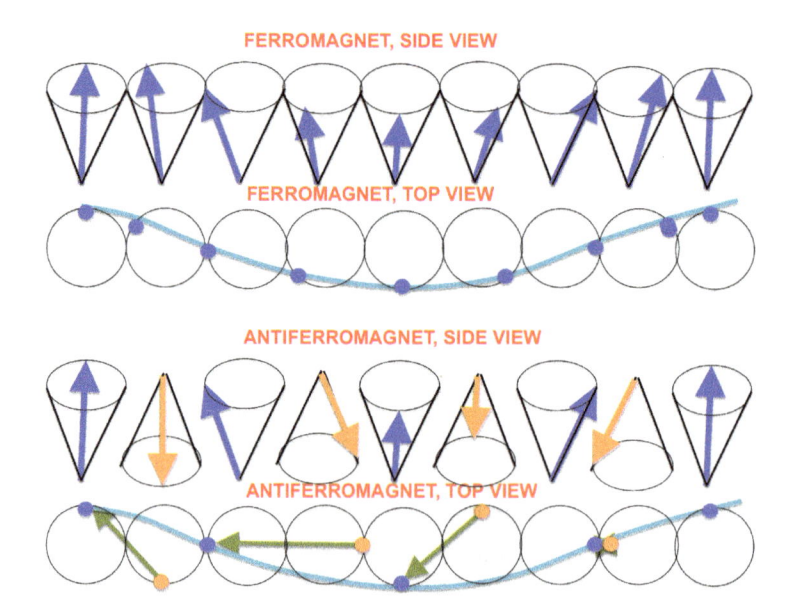

Using the commutation relations for spin $[S_n^x, S_n^y] = i\hbar S_n^z$ and cyclic permutations, and spin evolution $i\hbar d\vec{S}_n/dt = [\vec{S}_n, \hat{\mathcal{H}}]$, show once again, that the equation for motion of the spins is

$$\frac{d\vec{S}}{dt} = -\gamma \vec{S} \times \vec{B}_{\text{eff}}. \qquad (10.22)$$

Write out the x and y component evolutions of S_n, assume small perturbations around the easy axis (i.e., $S_n^z \approx S \gg S_n^{x,y}$, and drop binary products of $S_n^{x,y}$). This should look like the tridiagonal tight binding equation. Now, assume a plane wave solution $S_n^{x,y} = u^{x,y}e^{i(nka-\omega t)}$ and find the dispersions of the magnetic excitations. Show that $\omega \approx \omega_{ex}(1 - \cos ka) + \omega_A$, where $\omega_{ex} = 4JS/\hbar$ and $\omega_A = 2DS$.

Repeat for an antiferromagnet, consisting of two sublattices. The even ones satisfy $S_{2p}^z \approx S$ while the odd ones satisfy $S_{2p+1}^z \approx -S$. Show that the magnon dispersion $\omega = \sqrt{(\omega_{ex} + \omega_A)^2 - \omega_{ex}^2 \cos^2 ka}$.

For typical ferromagnets $\hbar\omega_{ex} \approx 10\,\text{meV}$ while $\hbar\omega_A \approx 1\,\mu\text{eV}$. This means the long wavelength ($k \to 0$) ferromagnetic magnons have a frequency of $\sim 1 - 10\,\text{GHz}$, while the antiferromagnetic magnons have a much higher frequency, $\sim 100\,\text{GHz} - 1\,\text{THz}$.

P10.5 **Ferromagnetic Curie temperature.** Using the identity $\sum_{n=0}^{N} x^n = \dfrac{x^{N+1} - 1}{x - 1}$ for $|x| < 1$, show that $\sum_{m=-S}^{m=S} e^{m\Delta} = \left(\dfrac{\sinh{(S+1/2)\Delta}}{\sinh{(\Delta/2)}} \right)$. Going one more step, show that $\langle S_z \rangle = \hbar S B_J(\beta g \mu_B B S)$. Plot the transcendental equation for M for various temperatures T to show that there is a critical temperature for a finite spontaneous magnetization.

P10.6 **Spin modulator.**

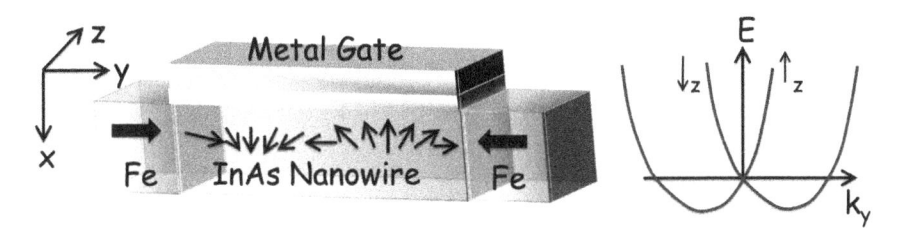

The Datta-Das spin 'transistor' consists of a 1D channel with a spin–orbit coupled Rashba Hamiltonian. The idea is to inject electrons from a ferro magnetic source, apply a vertical electric field with a gate to rotate the spin in transit (using spin–orbit coupling), vary the alignment of that spin with a ferromagnetic drain, and thus modulate the current transmitted through a ferromagnetic drain (we assume contacts are 100% polarized, meaning all spins injected at the source removed by the drain are along $+y$).

We start with electrons in a 2D inversion layer in the y–z plane, injected along the y-axis, so that $k_z = 0$ and $k_y = 0$. In the basis set of spin up and down electrons along the z-direction, we can write down a 2×2 Hamiltonian:

$$\begin{bmatrix} H_0 & 0 \\ 0 & H_0 \end{bmatrix} + \alpha_R \vec{\mathcal{E}} \cdot (\vec{\sigma} \times \vec{k}), \tag{10.23}$$

where $H_0 = \dfrac{\hbar^2 k_y^2}{2m^*}$, α_R is the spin–orbit coupling strength, k is the electron wavevector, and the components of σ are the three Pauli spin matrices worked out in class.

(a) Simplify H_{so} for $\mathcal{E}$ along the x-axis, by using the expressions for σ, and write down the total Hamiltonian as 2×2 matrix for a given k_y (assume $k_z = 0$).

(b) Solve for the two eigenvalues to give you two bands. Plot the 1D $E\text{–}k_y$ diagrams schematically, showing both bands on the same plot. (The band splitting means that the $z_{\uparrow,\downarrow}$ states have different energies for the same k_y, as if a magnetic field acted along z and created a Zeeman split).

(c) At a given Fermi energy E_F, you will now get two Fermi wavevector values $k_\pm$ for the positive k_y branches for the $z_\uparrow$ and a $z_\downarrow$ states, meaning that the two states evolve at different rates along the $+y$ directed channel. Find $k_\pm$ and their difference, $\Delta k = k_+ - k_-$.

(d) Imagine we inject an electron using a ferromagnetic source whose magnetization is polarized entirely along the $+y$ axis. Write down the eigenstate as follows, and find A and B.

$$\uparrow_y = A \underbrace{\begin{bmatrix} 1 \\ 0 \end{bmatrix}}_{z_\uparrow} + B \underbrace{\begin{bmatrix} 0 \\ 1 \end{bmatrix}}_{z_\downarrow} \tag{10.24}$$

(e) The electron then propagates across the channel from source to drain along the y-direction, with the $\pm z$ states evolving with different $k_\pm$ values as above, giving us at the end of the channel a new evolved state:

$$\Psi_1 = A \begin{bmatrix} 1 \\ 0 \end{bmatrix} \exp(ik_+ L) + B \begin{bmatrix} 0 \\ 1 \end{bmatrix} \exp(ik_- L), \tag{10.25}$$

current is given by the transmission probability between the electron at the drain end (i.e., Ψ_1 above) and the drain state (which we will call Ψ_2). Assuming the drain is also polarized in the same direction as the source, the wavefunction in the drain $\Psi_2 = \uparrow_y$. The transmission probability now is $P = \left| \Psi_2 \Psi_1^+ \right|^2$ (remember that dagger is conjugate transpose of a matrix).

    Work through the algebra to simplify P, and show that it depends on the Δk that you worked out earlier — in fact, only on $\Delta k L$.

(f) Plot schematically how P (and thus the current) varies with gate voltage (the gate voltage determines $E_x = V_G/t_{ox}$ with t_{ox} being the oxide thickness, and this field determines Δk), based on the expressions you just obtained. This gate-dependent modulation of the current describes the action of a spin 'transistor' (strictly speaking, a modulator).

(g) Describe in words the physical origin of the behavior you see, i.e., the current modulation you just plotted.

P10.7 **Domain walls.** The Hamiltonian for a magnet with uniaxial anisotropy is

$$\mathcal{H} = -J \sum_{ij} \vec{S}_i \cdot \vec{S}_j - K \sum_i S_{iz}^2. \tag{10.26}$$

Let us imagine a domain wall between up states and down states in the ferromagnet and estimate its width. If we ignore K and only J exists, argue that the domain wall will be infinitely wide. Conversely, if we ignore J, argue why the domain wall is infinitely abrupt. In presence of J and K then, the width is $\propto \sqrt{J/K}$. Consider N spins ($N \gg 1$) across the domain wall switching between angles of 0 and π, so that $\Delta\theta = \pi/N$. Assuming only nearest neighbor interactions, show that the energy

$$E(N) = JS^2\pi^2/2N + KS^2\pi N/2. \tag{10.27}$$

Find the optimal number of spins in the domain wall, and thus the domain wall width (assume lattice constant a).

For a continuous set of spins in 1D, we can write the nearest neighbor spins through Taylor expansion as $\theta \pm ad\theta/dx$. In 1D, each spin is defined by an angle θ and its spatial derivative $\dot{\theta} = d\theta/dx$. Show that

$$E = \int dx \underbrace{\left[A(\dot{\theta})^2 - K\cos^2\theta\right]}_{\mathcal{H}}. \tag{10.28}$$

Find A, and the momentum $p = \partial\mathcal{H}/\partial\dot{\theta}$, and use Hamilton's equations, $\dot{\theta} = \partial\mathcal{H}/\partial p$, $\dot{p} = -\partial\mathcal{H}/\partial\theta$, to write down the differential equation for θ and solve it in 1D with boundary conditions $\dot{\theta}_{x=\infty} = 0$. This will give us the equation for the domain wall. Show that the equation is

$$\theta(x) = 2\tan^{-1}\left(e^{(x - x_0)/\Delta}\right), \quad \Delta = \sqrt{A/K}. \tag{10.29}$$

This equation is easily extended to 2D polar coordinates to describe a Neel skyrmion $\boxed{\text{P9.6}}$, $\theta(r) = 2\tan^{-1}[\sinh(r/\Delta)/\sinh(R/\Delta)]$. Plot it!

$\boxed{\text{P10.8}}$ **Walker breakdown.**

(a) Assume $x_0(t)$ in Eq. (10.29) moves rigidly in a domain wall. Converting to polar θ, ϕ coordinates around the x-axis, use simplifications (e.g., $\partial\hat{m}/\partial\theta = \hat{\theta}$, similar for $\partial\hat{m}/\partial x$, $\partial\hat{m}/\partial\phi$), to show

$$d\hat{m}(\theta(x_0(t)), \phi(t))/dt = \sin\theta(-\hat{\theta}\dot{x}_0/\Delta + \hat{\phi}\dot{\phi}). \qquad (10.30)$$

(b) Include in the LLG equation and simplify the torques $\vec{T}_{\text{damping}}$, $\vec{T}_{\text{STT}}$, $\vec{T}_{\text{cons}}$, with the anisotropy potential $U = (K_1 m_z^2 + K_2 m_x^2)$. You will get two coupled dynamical equations in (x_0, ϕ). Simplify by using the *Thiele approximation*, by integrating all sides across the domain wall expanse over $\int_{-\infty}^{\infty} dx_0$. Eliminate $\dot{x}_0$ to get an equation for $\dot{\phi}$ alone.

(c) Walker breakdown happens where energy moves into azimuthal motion $\dot{\phi}$ and the domain wall motion $\dot{x}_0$ slows down. To avoid it, we need to be able to satisfy $\dot{\phi} = 0$. Show that this equality cannot be satisfied for $v > v_{cr}$. Find the critical velocity.

(d) Repeat including $\vec{T}_{\text{SOT}}$ and show and explain that at high spin–orbit coupling, Walker breakdown is postponed.

Chapter 11

Counting States

Energy band diagrams in real space show the allowed energy levels ϵ_i at a given point. However, we can go ahead and do some 'bean counting' — identify the degeneracy of the levels, their multiplicity — how many states lie in a given energy range. We will see later that the conductivity of an object depends on how many propagating states ('modes') lie near the Fermi energy. The density of states can thus be written as

$$D(E) = \sum_i \delta(E - \epsilon_i). \qquad (11.1)$$

In practice, we cycle over all eigenstates ϵ_i and maintain a counter that tallies where each sits on the energy spectrum, accounting for duplications. The histogram of these counts is the density of states. For some simple geometries, this is easy to estimate. For instance, for a 1D parabolic band (Fig. 11.1), $E_k = E_c + \hbar^2 k_x^2 / 2m^*$, we can count the number of states (each k_x has two states for two spins) in an energy range dE that corresponds to a momentum range dk_x, related by the band dispersion. The count is easy if you remember that the k_x states are equally spaced with spacing $2\pi/L$ for a channel of length L. Also remember that there are positive and negative k_x states for a given energy E. We thus get

$$\underbrace{dN_S = D_{1d}(E)dE}_{\text{definition}} = \underbrace{4}_{\text{spins, direction}} \times \underbrace{\frac{dk_x}{2\pi/L}}_{\text{No. of } k \text{ pts.}}$$

$$= \frac{2LdE}{\pi(dE/dk_x)} = \frac{2LdE}{\pi\hbar\sqrt{2(E - E_c)/m^*}} \sim (E - E_c)^{-1/2}. \qquad (11.2)$$

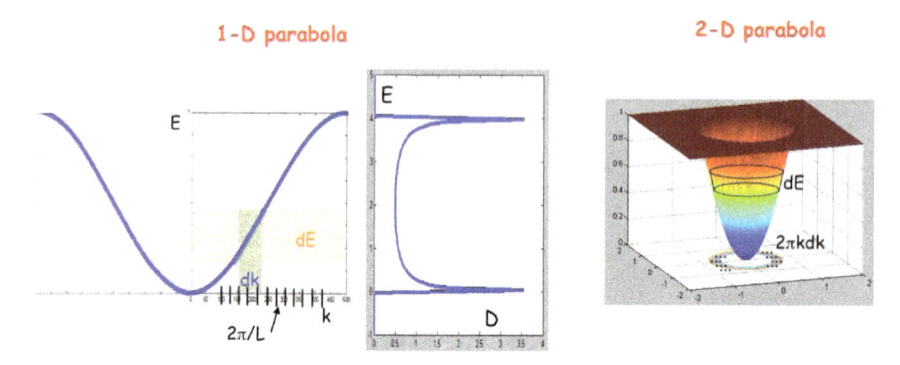

Fig. 11.1 DOS arises from counting states (*k points*) over an energy slice dE. The k points are spread out equally within the Brillouin zone.

This means the states are denser at the bottom of the conduction band, near the bandedge — or at the top of the valence band where it goes as $\sim (E_v - E)^{-1/2}$. We see this clearly in Figs. 8.1 and 8.2. When we go to 2D, the bands become spherically symmetric, with $E_k = E_c + \hbar^2 k^2 / 2m^*$, which is the equation to a 2D sphere in phase space, $k_x^2 + k_y^2 = 2m^*(E - E_c)/\hbar^2$. Accordingly, we go to spherical coordinates. Along the radial axis we get the same 1D counter giving us a density of states $\sim (E - E_C)^{-1/2}$. However, we also have an angular integral which gives us more points for higher energies since we have a bigger spherical circumference. In fact, the angular sum gives us an added factor of $\sim k \sim (E - E_C)^{1/2}$, which cancels the original radial count. If we do the algebra carefully P11.1–P11.3 , we find a step function leading to a constant density of states.

$$D_{2d}(E) = (m^* S / \pi \hbar^2) \Theta(E - E_C) \sim \underbrace{(E - E_C)^{-1/2}}_{\text{radial integral}} \times \underbrace{(E - E_C)^{1/2}}_{\text{angular integral}} .$$

(11.3)

Predictably, 3D gives a second angular integral over ϕ, and $D_{3d} \sim (E - E_C)^{-1/2} \times (E - E_C)^{1/2} \times (E - E_C)^{1/2} \sim (E - E_C)^{1/2}$.

For symmetric bands, the energy-dependence of the density of states can be estimated just by keeping track of units. The density of states in d dimensions is proportional to the d-dimensional phase–space volume $\Omega_d \sim k^d$, and inversely proportional to energy $\sim 1/E$ (since DOS is states per unit energy), so that the DOS per unit real space volume

$$D_d(E) \sim \frac{k^d}{E} \sim E^{(d/n)} - 1, \quad \text{for dispersion } E \propto k^n. \tag{11.4}$$

11.1 Confinement Creates Subbands

Since the energies are additive in the various dimensions, $E \sim k_x^2 + k_y^2 + \cdots$, it is straightforward to show that we can get higher dimensional density of states by a convolution of lower dimensional ones. For instance,

$$D_{3d}(E) = \int dE' D_{2d}(E') D_{1d}(E - E') = D_{2d}(E) \otimes D_{1d}(E), \tag{11.5}$$

which follows since the delta functions that sum into the DOS satisfy the same convolution rule, with higher dimensional delta functions decomposable into convolutions of lower dimensional ones, using the trivial integral of delta function as the underlying property $\boxed{\text{P11.4}}$.

Let us work through this decomposition (Fig. 11.2). We confine a 3D block (energy bands are 3D paraboloids) along the z-direction to a height W, whereupon the k_z values get quantized in units of $(p\pi/W)$, creating a set of 2D paraboloidal subbands with quantized band bottoms

$$
\begin{aligned}
E_{3d} &= \frac{\hbar^2(k_x^2 + k_y^2 + k_z^2)}{2m^*} \to E_{2d}^p \\
&= \frac{\hbar^2(k_x^2 + k_y^2)}{2m^*} + \underbrace{\frac{\hbar^2(p\pi/W)^2}{2m^*}}_{p^2 \epsilon_z}, \quad p = 1, 2, \dots.
\end{aligned} \tag{11.6}
$$

At this stage, the original 3D square root density of states decomposes into a set of 2D step functions that each represents a 2D density of states.

We can continue this process further, going all the way to a 'purely 2D' structure (i.e., where only one 2D subband exists *in the energy range of interest* and no inter-subband transitions occur — this happens when the subband separation $\gg k_B T$, and potential variations due to applied voltages or defects are small). Next, we can confine the box along y-axis, creating a quasi-1D nanowire. The original 2D band now breaks into a set of separated 1D subbands, and the single step function DOS breaks into a sequence of 1D horn-like densities of states. Once again, we can proceed till we reach 1D, then confine along the x-axis as well to get quantum dots with discrete molecular densities of states.

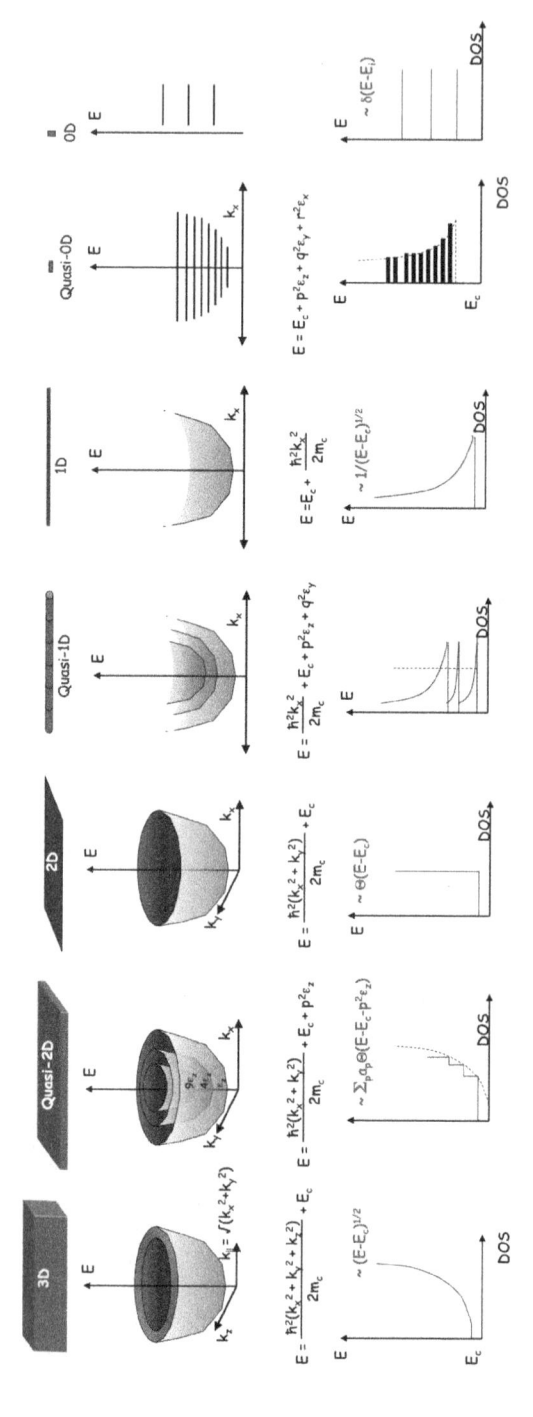

Fig. 11.2 (Left to right) Evolution of structure, dispersion E–k and density of states from bulk to quantized lower dimension.

11.2 Numerical Counting of the Density of States

Numerically, we need to cycle through all k_x, k_y and k_z entries within the Brillouin zone. For a simple cubic, this is fairly easy ($\pm\pi/a$ for each) $\boxed{\text{P11.5}}$. For harder structures like body or face-centered cubic, we can still use the same cubic unit cell, but that will typically involve multiple Brillouin zones combined together. We will then scale down the DOS by a factor of 2 or 4 — the number of atoms in the real space unit cell. We now maintain a histogram counter that tallies each energy registered by the k points, and plot them $\boxed{\text{P11.6}}$. An example is shown in what follows for a 3D Matlab code. The three dimensions are easily combined using a cubic storage matrix, each slice of which is a regular square matrix. We use the repmat, permute, reshape and histc Matlab commands to do this in a fairly compact way.

Matlab code for 3D simple cubic lattice Density of States:

```
Nx=551;kx=linspace(-pi,pi,Nx);ky=kx;kz=kx; %%k grid

t=1; f=-2*t*cos(kx); %% 1D tight binding dispersion

%% Create 3D matrix with E(kx,ky)
Ex=-repmat(f,[Nx,1,Nx]);
Ey=permute(Ex,[2 1 3]);
Ez=permute(Ex,[1 3 2]);
E=4*t+Ex+Ey+Ez;
Emin=min(min(min(E)));Emax=max(max(max(E)));

%% Create bin for DOS and count histogram of above E's into Ebin
NNE=151;Ebin=linspace(Emin-1e-9,Emax+1e-9,NNE);
dE=Ebin(2)-Ebin(1);
DOS=histc(reshape(E,1,Nx*Nx*Nx),Ebin);
DOS=DOS/(sum(DOS)*dE)
axis([Emin-5*dE Emax+5*dE 0 1.2*max(DOS)])

plot(Ebin,DOS)
```

Note that we assumed the k points are equally distributed over the Brillouin zone. That is not always necessary. The key features of the DOS arise primarily from select symmetry points of the reciprocal lattice — so it maybe more efficient to take a dense grid of k points from those segments as long as we know how to weight the results appropriately to account for the unequitable selection of k points.

More generally, the density of states can be rewritten as a sum of spikes, each a Lorentzian with vanishing width and adjusted height so that the area under it is unity, and then expanding each Lorentzian

$$D(E) = \sum_{n=1}^{\infty} \delta(E - \epsilon_n) = \sum_n \lim_{\eta \to 0^+} \underbrace{\frac{\eta/\pi}{(E - \epsilon_n)^2 + \eta^2}}_{\text{Vanishing Lorentzian}}$$

$$= \sum_n \frac{i}{2\pi} \Big(\underbrace{\frac{1}{E + i\eta - \epsilon_n}}_{G} - \underbrace{\frac{1}{E - i\eta - \epsilon_n}}_{G^\dagger} \Big). \tag{11.7}$$

Green's function G, which we will explain later in the book, can easily be matricized, leading to the following easily executable equation for the density of states

$$D(E) = Tr[\delta(EI - H)] = \frac{i}{2\pi} Tr(G - G^\dagger), \quad G(E) = [(E + i\eta)I - H]^{-1},$$
$$\tag{11.8}$$

where Tr refers to the trace (sum of diagonals). This allows us to do the count for any DOS given a Hamiltonian matrix $[H]$ $\boxed{\text{P11.7 and P11.8}}$. The η we need to choose should be small, but large enough compared to the separation between the eigenvalues of the entire system to create a continuum plot. For an $N \times N$ Hamiltonian with N eigenenergies distributed over a bandwidth $\sim 2dt_0$, d being the dimension (the bandwidth equation strictly works for cubic structures), the energy separation $\delta\epsilon \sim 2dt_0/N$, so we need to make sure our chosen $\eta > \delta\epsilon$. We discuss more details in NEMV toward the end of Section 8.3, because the introduction of the complex term is subtle and relates to the concept of irreversibility.

11.3 Density of States Effective Mass

Most materials have multiple bands with different effective masses, each describable by its own constant energy contours. For instance, a band with dispersion $E = \hbar^2/2(k_x^2/m_x + k_y^2/m_y + k_z^2/m_z)$ leads to constant energy ellipsoids with a standard ellipsoidal equation $k_x^2/a^2 + k_y^2/b^2 + k_z^2/c^2 = 1$, where the lengths of the *k-space* principal axes $(a, b, c) = \sqrt{2E/\hbar^2}(\sqrt{m_x}, \sqrt{m_y}, \sqrt{m_z})$. If we're counting states to tally the total charge, we need to add the volumes $4\pi abc/3$ of each of these 'cigars' — six equivalent ones for silicon conduction bands along the [100] directions, eight for Ge along the [111] directions, one for GaAs at the Γ point (valence bands are warped but can be broken into rough parabolas). Once we add the k-space volumes up to Ω_k, we can pretend the sum originated from a single sphere of k-space radius R with the same exact volume Ω_k at the Γ point. Setting $a = b = c = R$ in the above expression for principal axes, we see that this size can be associated with a *density of states effective mass* m_{DOS} using $R = \sqrt{2m_{\mathrm{DOS}}E/\hbar^2}$. This single parameter gives us an idea for charge content — higher mass equals higher charge.

11.4 From States to Modes

While DOS gives us an idea of the energy level distribution, we can get a bit more detailed information by looking at the local density of states (LDOS), which weights the DOS with the occupation probability, in other words,

$$\mathrm{LDOS}(x_n, E) = \sum_i \mathrm{DOS}(E)|\psi_i(x_n)|^2, \tag{11.9}$$

in effect telling us the individual atomistic contributions to the overall DOS. For transport calculations such as conductance, for instance, we also need to weight it by the velocity, to distinguish propagating vs localized modes, such as elecrons pinned at defects. This added weighting leads to the concept of 'modes' $M(E)$, in principal spatially resolvable with wavefunctions once again. The Landauer theory that we will invoke later tells us in essence that the low temperature low bias conductance in a ballistic conductor is the mode count at a special point called the Fermi energy, times a universal constant.

Specifically, for a 1D system, we can make the connection

$$M(E) = 2\pi\gamma_{\mathrm{eff}}D, \tag{11.10}$$

where γ_{eff} can be identified as the effective escape rate, ultimately connected to the broadening of the levels by coupling to the surrounding contact continuum through the energy–time uncertainty principle (Eq. (4.14)),

$\gamma_{\text{eff}} = \hbar/\tau$. For a simple homogenous system, we can identify $\tau = L/v$, so that $M = (hvD/L)$. Once again for a 1D system with a single parabolic band, we already worked out the density of states $D = nL/2\pi\hbar v$ n being the degeneracy, in which case $M = n$; *mode count equals degeneracy*.

Note that the charge density increases with increasing DOS effective mass m_{DOS} while transmission, velocity and current will decrease with increasing transport effective mass — typically the longitudinal mass m_l along the transport direction (tunneling current decreases exponentially). For a 1D wire, there is only one mass, and the velocity and charge having opposite mass dependences cause them to cancel, giving us a universal ballistic conductance in the absence of internal scattering. In 3D, however, an optimal solution would be to have materials with highly elongated cigar-like asymmetric bands with $m_l \ll m_t$, so that we have large transverse (and thus DOS) effective mass m_t for high charge density, and low transport effective mass m_l for high speed (except when tunneling sets in). See NEMV, Section 18.2, for more details on the rich proliferation of effective masses in oriented solids.

Homework

P11.1 **DOS counting.** The E–k for a 1D chain is given in (Eq. (8.5)). Find it's density of states analytically, and show that it looks like Fig. 11.3 top left, in particular, peaking at both ends of the band. *Note that we need a factor of 2 to account for both $\pm k$ states* (in higher dimensions, the angular integral takes care of this). Integrate over the bandwidth to calculate the total number of states. Can you rationalize this simple expression?

P11.2 **SuperModels.** Calculate the 1D E–k for the toy model in Fig. 8.5, and show that the DOS, obtained using Eq. (11.8), looks like *Batman* if you tilt your head sideways. Repeat using the same E–k, but extended to 2D and 3D assuming spherical symmetry. With a little bit of imagination with your head tilted sideways, you can see the 2D and 3D DOS resembling the *Incredible Hulk* and *Ironman* (you may want to shade them black, green and red-yellow to bring home the resemblance).

P11.3 **Graphene DOS.** Given the E–ks for monolayer and bilayer graphene that you worked out in Problem P8.10, calculate and plot their densities of states. Now, open a gap — for instance, by adding unequal diagonal elements $\pm \epsilon_1$. Find the bandgap, effective mass and density of states and plot it.

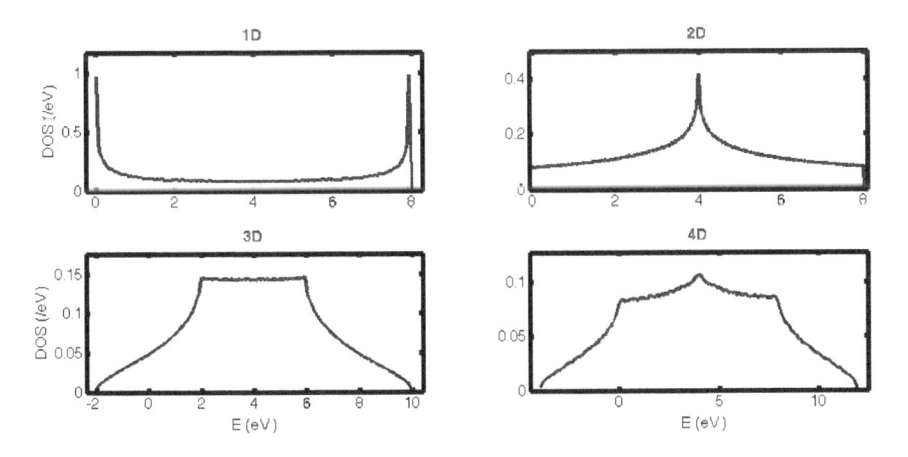

Fig. 11.3 Square lattice DOS in multiple dimensions, heading toward a Gaussian.

P11.4 **High-d DOS.** Show that higher-d DOS is a convolution of lower-d DOS (Eq. (11.3)).

P11.5 **Square lattice in multiple dimensions.** Show the evolution of the square lattice DOS from 1D to 2D to 3D. Best way to do so is to start with 3D, then reduce one of the couplings (say along x-axis) progressively, and then the ones along the y-axis. You can use the codes provided in the shaded regions, then do your own extension to 4D as well.

P11.6 **Valence bands.** In an atom, the s orbitals have lower energy than the p orbitals ($n+1$ rule). Upon bonding, the hybridization splits the bonds, with the upper conduction band primarily of the s-type, while the valence band is primarily p-type. Since s orbitals are spherical, conduction bands near the Γ point are spherically symmetric, and more ellipsoidal away from that point. In contrast, p-type valence bands are highly directional and thus non parabolic. Scientists often use a 3 band $\vec{k} \cdot \vec{p}$ model,

$$\mathcal{H}_{KP} = E_v + \begin{pmatrix} Lk_x^2 \\ +M(k_y^2 + k_z^2) & Nk_xk_y & Nk_xk_z \\ Nk_xk_y & Lk_y^2 \\ & +M(k_z^2 + k_x^2) & Nk_yk_z \\ Nk_xk_z & Nk_yk_z & Lk_z^2 \\ & & +M(k_x^2 + k_y^2) \end{pmatrix}.$$

$$(11.11)$$

Consider $L = -1.9 \times \hbar^2/m_0$, $M = -6.7 \times \hbar^2/m_0$ and $N = -7.5 \times \hbar^2/m_0$. Show that you get three valence bands, one heavy hole with large mass, and two light hole bands, each doubly degenerate. Plot the constant energy contours using the Matlab *isosurface* and *contourf* commands with energy $-0.1\,\mathrm{eV}$ and plot their DOS. Compare with results from 6 band k.p (see literature), and the sp^{3}s* results (Problem P8.11).

P11.7 **Magnetic DOS.** Work out the DOS using Eq. (11.8) for a 2D square with $t_0 = 1\,\mathrm{eV}$, using the Hamiltonian in Eq. (5.7), and also using a magnetic field which requires modifying the second pair of off-diagonal terms, as discussed right below that equation. Assume a magnetic field that gives a ratio $\Phi/\Phi_0 = 1/8$. Repeat by varying the energy and magnetic

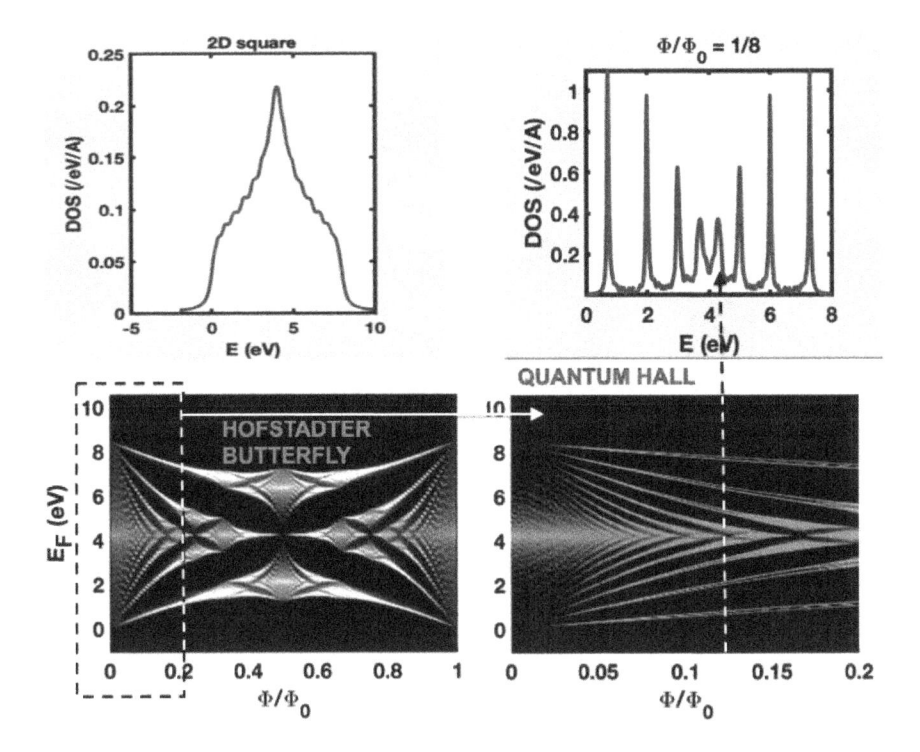

Fig. 11.4 2D DOS in a magnetic field.

field and plot the DOS vs E and Φ/Φ_0. For low fields, this should give us Quantum Hall plateaus coincident with the DOS peaks (Fig. 11.4) at the bottom right. For larger fields, this should expand to give a Hofstadter butterfly. To understand this beautiful piece of physics, refer to NEMV, Section 21.3.

P11.8 **More examples.** Work out the density of states (analytically or numerically) for

- 3D acoustic phonons using the dispersion worked out in P8.3.
- Ferromagnetic magnons, using the dispersion worked out in P10.4.
- Antiferromagnetic magnons, using the dispersion in P10.4.
- Diamond Cubic crystals like Si, using the E–k in Eq. (8.20) and parameters listed in P8.11.
- Photons with polarization 2 and linear E–k, show that $D_{ph}(E) \propto E^2$ in 3D.

Chapter 12

Filling the States: Statistical Physics

12.1 Fermi–Dirac and Bose–Einstein

In the last several chapters, we worked out the solutions to the one-electron Schrödinger equation. A convenient way to lump together the myriad eigenstates is to extract the overall density of states at each energy. This is akin to determining the number of rooms on each floor of a multistoreyed building. To get the hotel's occupancy (charge density), we also need to know how many people can occupy each room. This occupancy gives us an electron distribution function that we will now determine.

Of all possible ways to distribute electrons among various energies ϵ_i subject to a fixed total number N and total energy E, the most likely is the one with largest number of states Ω_i $\boxed{\text{P12.1}}$. For Fermions, we can at most put one electron per state (two if we included spin), while for Bosons, we can put as many as we want. These constraints arise from the spin-statistics theorem that leads to Pauli exclusion for the electrons and none for the bosons. Since the density of states in 3D increases with energy ($\sim \sqrt{E}$ if you recall), we get the largest number of variational possibilities if we put minimum electrons in the higher energy states (placing fewest tenants on the floor with most rooms generates the largest number of combinatorial possibilities). Classically, this generates the Maxwell–Boltzmann distribution, $f_{\mathrm{MB}} \propto e^{-\beta(E-\mu)}$, where $\beta = 1/k_B T$. For electrons however, there is this ceiling of unity that it cannot cross on the occupancy, so the maximum entropy result gives us the Fermi–Dirac distribution $f_{FD} \propto 1/[e^{\beta(E-\mu)} + 1]$ $\boxed{\text{P12.2}}$. For bosons like photons or phonons, there is no restriction, so the maximum entropy gives us the Bose–Einstein distribution $f_{\mathrm{BE}} \propto 1/[e^{\beta(E-\mu)} - 1]$ $\boxed{\text{P12.3}}$.

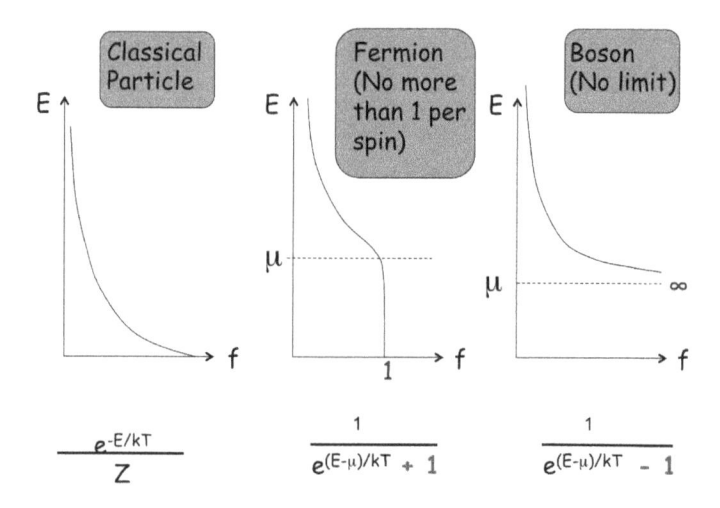

Fig. 12.1 Occupancies for classical, Fermi and Bose gases.

To summarize (Fig. 12.1)

$$f = \begin{cases} \mathcal{Z}^{-1}e^{-\beta(E-\mu)}, & \text{(Maxwell–Boltzmann)}, \\[2mm] \dfrac{1}{e^{\beta(E-\mu)}+1}, & \text{(Fermi–Dirac)}, \\[2mm] \dfrac{1}{e^{\beta(E-\mu)}-1}, & \text{(Bose–Einstein)}. \end{cases} \tag{12.1}$$

12.2 Spin Statistics Theorem

Let us quickly recall the operator for a translation or rotation. The generator of translation was the linear momentum operator $\boxed{\text{P2.4}}$, while that of azimuthal rotation was the angular momentum ($\hat{p}_\phi = \hat{L}_z$) operator, so that the translation and rotation operators were

$$\mathcal{T}(a) = e^{i\hat{p}a/\hbar}, \quad \mathcal{R}(\phi) = e^{i\hat{L}_z\phi/\hbar}, \quad \hat{p} = -i\hbar\partial/\partial x, \quad \hat{L}_z = -i\hbar\partial/\partial\phi. \tag{12.2}$$

Since spin operators have similar commutation rules to angular momentum, we can also write down a spin rotation operator about an axis $\hat{n}$

$$\mathcal{S}_\theta^{\hat{n}} = e^{i\hat{S}_n\theta/\hbar} = e^{i\sigma_n\theta/2}, \quad \text{since } \hat{S}_n = \hbar\sigma_n/2, \tag{12.3}$$

with $\sigma_n = \vec{\sigma} \cdot \vec{n}$. Acting on an eigenstate of the spin operator, i.e., a pure spin state, we can simply replace σ_n with its eigenvalue s_n. Now, when we exchange two spins, this is equivalent to a 2π rotation (each spin rotates π around the other), so that an exchange operator can be written as

$$\underbrace{\mathcal{E}}_{\text{Exchange}} = S_{2\pi}^{\hat{n}} = e^{is_n\pi} = (-1)^{s_n}, \tag{12.4}$$

so that for odd half-integer spins $S_n = \hbar s_n/2$ for which $s_n = 1, 3, 5, \ldots$ we pick up a minus sign under exchange, $\Psi_{12} = -\Psi_{21}$, while for even half-integer spins S_n for which $s_n = 0, 2, 4, \ldots$ we pick up a plus sign, $\Psi_{12} = \Psi_{21}$. This is the *spin-statistics theorem* for odd vs even half-integer spins S_n, which separates fermions from bosons. We saw this example of a negative sign with electrons when we did a 2π rotation of the $\psi_{\theta,\phi}$ state.

12.3 Equilibrium Charge Density

Given the density of states and the occupancy distributions, we can now directly calculate the equilibrium *free electron density* $\boxed{\text{P12.4}}$ in the conduction band and free hole density in the valence band, with the electrochemical potential μ replaced by the Fermi energy E_F. The 3D density of states for a single parabolic band is $D(E) = A\sqrt{E - E_C}$ for electrons, and $A\sqrt{E_V - E}$ for holes, where $A = (1/2\pi^2)(2m^*/\hbar)^{3/2}$ with the appropriate effective mass. The charge densities are given by the integrated product of the density of states and occupancy (like the number of hotel occupants = number of rooms times average occupancy)

$$n = \int_{-\infty}^{\infty} dE D(E) f_{FD}(E - E_F) \approx \int_{E_C}^{\infty} dE A \underbrace{\left(E - E_C\right)}_{E'}^{1/2} e^{-\beta(E - E_F)}$$

$$= A \int_{0}^{\infty} dE' \sqrt{E'} e^{-\beta(E' + E_C - E_F)}$$

$$= e^{-\beta(E_C - E_F)} \underbrace{A \int_{0}^{\infty} dE' \sqrt{E'} e^{-\beta E'}}_{N_C}$$

$$= e^{-\beta(E_C - E_F)} \underbrace{A(k_B T)^{3/2} \overbrace{\int_{0}^{\infty} dx x^{1/2} e^{-x}}^{\Gamma(3/2) = \sqrt{\pi}/2}}_{N_C}, \quad N_C = 2/\lambda_n^3$$

$$p = \int dE D(E)[1 - f_{FD}(E - E_F)]$$

$$\approx \int dE A \sqrt{E_V - E}\left[1 - e^{-\beta(E - E_F)}\right]$$

$$\approx N_V e^{-\beta(E_F - E_V)}, \quad N_V = 2/\lambda_p^3. \tag{12.5}$$

To summarize then,

$$n = N_C e^{-\beta(E_C - E_F)}, \quad N_C = \frac{2}{\lambda_n^3}, \quad \underbrace{\lambda_n = \sqrt{\frac{2\pi m_n^* k_B T}{\hbar^2}}}_{\text{thermal de Broglie wavelength}},$$

$$p = N_V e^{-\beta(E_F - E_V)}, \quad N_V = \frac{2}{\lambda_p^3}, \quad \lambda_p = \sqrt{\frac{2\pi m_p^* k_B T}{\hbar^2}}.$$

$$\tag{12.6}$$

As expected, the lumped densities of states $N_{C,V}$ are given by the 'sizes' of the electrons, which are set by their kinetic energy $\sim k_B T$ around the bandedges, in other words, the thermal de Broglie wavelengths. The equations also say that as we move the Fermi energy E_F toward a bandedge (say conduction band), we increase charges in that band but reduce charges in the other band, in a way that their product is fixed. In fact,

$$np = N_C N_V e^{-\beta\overbrace{(E_C - E_V)}^{E_G}} = n_i^2, \tag{12.7}$$

where $n_i = \sqrt{N_C N_V} \exp\left[-E_G/2k_B T\right]$ is called the *intrinsic* concentration. As we move E_F around, the geometric mean of n and p stays pinned to n_i.

What does n_i signify? If we take a semiconductor at equilibrium, at low temperatures it will have neither free electrons in the conduction band nor holes in the valence band. As we ramp up the temperature, we start exciting charges from the valence to the conduction band, creating free electrons n in the conduction band and leaving behind holes p in the valence band. For an *intrinsic* semiconductor with no impurities, each electron leaves behind a corresponding hole so that $n = p$. Since their product is pinned to n_i^2, we get $n = p = n_i$, the free charge density at equilibrium in an intrinsic, undoped semiconductor. For silicon, $m_n^* = 1.084 m_0$, $m_p^* = 0.81 m_0$, so that $N_C \approx 2.82 \times 10^{19} \text{cm}^{-3}$ and $N_V \approx 1.83 \times 10^{19} \text{cm}^{-3}$ (a bit smaller N_C and larger N_V for Ge). Using $E_G = 1.12\,\text{eV}$ for silicon at room temperature, we get $n_i \sim 10^{10}\text{cm}^{-3}$ at room temperature. This number is much larger $\sim 3 \times 10^{14}$ for Ge (smaller bandgap $\sim 0.66\,\text{eV}$), and much smaller $\sim 6 \times 10^{8}$ for GaAs (larger bandgap $\sim 1.4\,\text{eV}$). Substituting this n_i into the expressions for n and p, we can extract the intrinsic Fermi energy

$$E_i = \frac{E_C + E_V}{2} + \frac{k_B T}{2} \ln\left(\frac{N_V}{N_C}\right). \tag{12.8}$$

We can approximately ignore the second term, because the effective masses and lumped densities of states are not different enough to compensate for the small value of $k_B T$ compared to E_G. In other words, E_i *is approximately midgap, a tad nearer the band with lower lumped density of states* following a 'lever' rule in order to balance the electrons and holes.

If we now find a way to increase the charge content in a band without diminishing the other band, such as by electron doping, then n will deviate exponentially from n_i as E_F deviates from E_i. With a few algebra steps, we can rewrite the equations for n and p in a convenient way

$$\begin{aligned} n &= n_i e^{\beta(E_F - E_i)}, \\ p &= n_i e^{\beta(E_i - E_F)}. \end{aligned} \tag{12.9}$$

As E_F soars above E_i, so does n while p plummets. Makes sense!

12.4 Doping

How do we make n deviate from n_i, in other words, break the equality $n = p$? That equality came from charge conservation, so we need some added sources of charge to break the logjam. One way to do so is through doping. If we mix in trace amounts of phosphorus (N_D of them per cc, D signifying 'donor'), then each pentavalent phosphorus atom sits as a substitutional impurity replacing a tetravalent silicon atom in its tetrahedral environment (recall Fig. 8.5). Once the four covalent bonds with the four tetrahedral partners are restored, phosphorus is left with a fifth electron. This electron has a very weak binding energy to the parent phosphorus atom, as can be seen by comparing the binding energy of a hydrogen atom (Eq. (2.14)), and noting that we have lower effective mass in silicon and much higher dielectric constant (i.e., screening). The energy of this fifth electron must sit inside the silicon bandgap quite near the conduction band edge, separated by this reduced binding energy, so that it is easily thermionized and liberated as a free electron in the conduction band to be shared by the entire matrix of atoms. Assuming all N_D atoms are fully ionized (i.e., positively charged) at room temperature, we now have a different inequality, $n = p + N_D$, which allows us to have many more electrons in the conduction band than holes in the valence band. In fact, we can approximately set

$$n \approx N_D \gg n_i, \quad p \approx n_i^2/N_D \ll n_i \quad (n\text{-type Si}) \tag{12.10}$$

meaning we gained electrons without creating many holes. From Eq. (12.9), we then get

$$E_F - E_i = k_B T \ln(N_D/n_i) \quad (n\text{-type Si}). \tag{12.11}$$

This continues until we reach high doping where E_F crosses E_C, after which we start sensing the flat part of the Fermi–Dirac distribution (degenerate doping) and the movement of E_F with doping subsequently slows down to a crawl.

It is straightforward to get the corresponding result for hole doping. We use boron, a trivalent atom, which by substituting silicon needs to tear away an electron from a neighbor and thus release a hole. This hole again has a weak binding energy and can diffuse through the entire silicon matrix. The equations are

$$p \approx N_A, \quad n \approx n_i^2/N_A, \quad E_i - E_F = k_B T \ln(N_A/n_i) \quad (p\text{-type Si}). \tag{12.12}$$

All these equations arose from two limitations — the geometric mean of p and n was set by n_i, while their difference is set by the neutralizing charge on the ionized donors,

$$\underbrace{n + N_A^-}_{\text{net negative charge}} = \underbrace{p + N_D^+}_{\text{net positive charge}}, \tag{12.13}$$

where $N_D^+ = N_D/[1 + g_D \exp(E_C - E_D)/k_B T] \approx N_D$ gives us the number of ionized donors (almost complete when $E_C - E_D \sim k_B T$), $g_D \approx 2$ being the band degeneracy. Similarly, with $N_A^- \approx N_A$ and a degenerarcy $g_A \approx 4$.

To get a sense of numbers, note that the large bandgap ($\sim 1.1\,\text{eV} \sim 40 k_B T$) for Si means we have very few intrinsic carriers $n_i \sim 10^{10}/\text{cm}^3$ (about one electron out of a trillion atoms is 'free', the atomic density being $\sim 10^{22}/\text{cm}^3$). For a doping $N_D \sim 10^{16}/\text{cm}^3$, we now get one ionized dopant electron out of million atoms, in other words, the sample is now inundated with free electrons.

12.5 Nonequilibrium: How Current Flows

Let us apply a voltage bias between the source and the drain to drive a current through the system (Fig. 12.2). Assuming each battery terminal only influences the metal contact in touch with it, we will ground the source and keep its energy spectrum intact, while we lower the drain density of states by the applied voltage, $U = -qV_D$. Under this condition, we have two quasi-Fermi energies (also called electrochemical potentials) $E_{F1,F2}$ in the two contacts, each locally in equilibrium and charge neutral. The channel states in between try to achieve equilibrium with each contact, but since the latter are kept separate by a battery, the channel is in the process driven out of equilibrium. In particular, for states energetically between $E_{F1} = E_F$ and $E_{F2} = E_F - qV_D$, the left contact tries to fill the states as they lie below E_{F1} while the right contact tries to empty them as they live above E_{F2}. In the process, an electron current moves from source to drain $\boxed{\text{P12.5}}$. It is assumed that this extra electron rapidly thermalizes by relaxing in energy to the local equilibrium in the drain, and in the process an equivalent electron moves from drain to battery and thence to source to fill up the hole left behind in the source and brings the latter into local equilibrium as well.

Since all we need to drive a current is a difference in Fermi functions, we can accomplish this by having a difference in temperature as well between the contacts. The hotter contact will have more electrons above the Fermi energy, driving electrons to the colder contact by diffusion (current

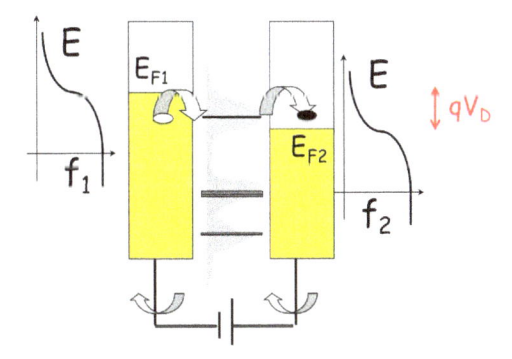

Fig. 12.2 Source and drain Fermi functions $f_{1,2}$ separated by an applied voltage bias create an energy window between E_{F1} and E_{F2} for electrons to flow between contacts and back through the battery without being Pauli blocked by filled contact states in yellow.

$\propto f_1 - f_2$). The colder contact, on the other hand, has more electrons below the Fermi energy since it has promoted fewer electrons above it, and thus the current below the Fermi energy moves from colder to hotter contact. These two counter-currents can cancel. However, for a system with a density of states asymmetrically placed across the Fermi energy, one current is favored over another. For n-type conduction bands, a net electron current will move from hotter to colder contact, while for p-type, electron current moves the other way.

We will work out expressions for the currents later when we do Landauer theory. But we can set up the definitions at the moment. The charge I and heat currents I_Q are related to the applied voltage difference ΔV and temperature difference ΔT by linearizing (Taylor expanding) the Fermi function difference $f_1 - f_2$ above. The response coefficients are ultimately obtained by casting the voltage build-up and thermal current in terms of temperature difference and charge current.

Thus, one set of coefficients requires zero temperature gradient, while the other requires zero charge current (open circuit).

$$I \approx \underbrace{G}_{\text{Conductance}} \Delta V + G_S \Delta T$$

$$\Longrightarrow \boxed{\Delta V = \underbrace{R}_{\text{Resistance}} I - \underbrace{S}_{\text{Seebeck}} \Delta T}$$

$$I_Q \approx G_Q \Delta V + G_K \Delta T$$

$$\Longrightarrow \boxed{I_Q = \underbrace{\Pi}_{\text{Peltier}} I + \underbrace{\kappa}_{\text{Thermal conductance}} \Delta T} . \qquad (12.14)$$

The Seebeck coefficient S measures the conversion from heat into electrocity. S is the *open circuit* ($I = 0$) voltage V_{OC} built up per unit temperature gradient, $S = -V_{\text{OC}}/\Delta T$. The corresponding electromotive field $\vec{\mathcal{E}}_{\text{emf}} = -S\vec{\nabla}T$. The Peltier coefficient Π is the heat flux generated per unit current drive at zero temperature difference, in other words, the efficiency of conversion of electricity into heat. Finally the thermal conductance κ is the *open circuit* thermal current per unit temperature difference.

In later chapters, we will use the Landauer formula to estimate the parameters above, G, G_S, G_Q and G_K, and thence R, S, Π and κ $\boxed{\text{P19.3}}$.

Homework

P12.1 Entropy and quantum statistics. Our aim here is to find the most likely distribution of electrons at low temperature. For a large number of particles this will become the Fermi–Dirac distribution. A material has the following states — $D_3 = 7$ states at energy $E_3 = 4$, $D_2 = 5$ states at energy $E_2 = 2$, and $D_1 = 2$ states at energy $E_1 = 0$, in some energy units. We have a total of $N = 5$ electrons that need to be distributed among these states (think of them as boxes each capable of holding one electron). The total energy we need to get is $E = 12$. This will be accomplished by dividing the five electrons among the states, so we get N_1 electrons at energy E_1, N_2 at energy E_2 and N_3 at energy E_3, with $N_1 + N_2 + N_3 = 5$ and $N_1 E_1 + N_2 E_2 + N_3 E_3 = 12$.

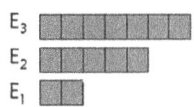

(a) Let us label each configuration of electronic distributions by the triad N_1, N_2, N_3. Identify the complete set of triads that satisfy the two constraints above, on the total number and total energy.

(b) Going back to your permutation/combination knowledge (you can look it up!), write down the number of ways each configuration can be accomplished (e.g., how do you distribute N_1 electrons among D_1 boxes at energy E_1, N_2 among D_2 and N_3 among D_3?). Use the actual values of Ns and Ss and run through the calculation all the way — each answer should be an actual number. List each combination as W_{N_1,N_2,N_3}.

(c) Find the most likely configuration N_1, N_2, N_3 whose W is maximum.

(d) For that most likely configuration, the distribution of electrons should represent Fermi–Dirac (at least in the limit where the number of boxes, energies and particles is large, this will be a crude approximation). Define the fractional occupancy of electrons at each energy, as the number of particles divided by the number of boxes at that energy (i.e., $f_1 = N_1/D_1$, and so on). Plot the calculated f values vs the E values. This will be a crude approximation to the actual Fermi–Dirac distribution for a few particles.

P12.2 **Entropy and Fermi–Dirac.** The Fermi–Dirac distributions can be obtained by maximizing the number of microstructures ('entropy') Ω for a fixed particle number $N = \sum_i n_i$ and energy $E = \sum_i n_i E_i$, the variable being the set of particles $\{n_i\}$ at the ith energy level. We maximize

$$F = \ln \Omega - \alpha \sum_i n_i E_i - \beta \sum_i n_i, \qquad (12.15)$$

with α and β as unknown parameters (known as Lagrange multipliers) used to enforce the constant N and E. Assume we have D_i possible microstructures at the ith level among which to distribute the n_i particles $\boxed{\text{P12.1}}$.

(a) For electrons we can have only 0 or 1 particle in each of the D_i microstructures. How many ways Ω_i can I distribute n_i particles among D_i configurations?

(b) Assuming then Ω is the product of the various Ω_is, write down F in terms of n_i.

(c) Set $dF/dn_i = 0$ and then find the maximizing occupancy n_i/D_i. To do the derivatives, use the Stirling approximation (valid for large N) to simplify the factorials first. $\ln(N\,!) \sim N \ln N - N$.

(d) Show that we get Fermi Dirac if $\beta = -\mu/k_B T$ and $\alpha = 1/k_B T$.

(e) Repeat now for Bosons. The main difference is in step (a), i.e., how to distribute n_i particles among D_i configurations, but this time I can have as many particles as I need within each box instead of just 0 and 1.

(This is a standard problem in permutation theory. Realize that removing the restriction on number of particles in each box is equivalent to removing the partitions between boxes altogether, then distributing the particles, and then reinserting the partitions in various ways. This means we need to treat the $(D_i - 1)$ partitions on a same footing as the n_i particles. The question then is how many ways can we take $n_i + D_i - 1$ objects and choose n_i particles and $D_i - 1$ partitions from them?). Do the same steps as before and show that we now get the Bose–Einstein distribution.

P12.3 **Entropy and Bose–Einstein.** Let us try to write down rate equations for the addition and removal of photons or phonons. We start with the equilibrium problem. The occupation probability P_N of a state with energy E_N and containing N phonons is proportional to $\exp\left(-[E_N - \mu N]/k_B T\right)$, the proportionality determined by normalization. This is similar to the electronic problem worked out in the lecture notes. The main difference is that electrons are Fermions while photons are Bosons. This means that a single level can have at most one electron of a given spin, but can hold an arbitrary number of photons.

(a) Calculate the normalization $Z = \sum_N \exp\left(-[E_N - \mu N]/k_B T\right)$. Assume that the phonons are non-interacting, i.e., $E_N = N\epsilon_0$, and allow the sum to run over all N from zero to infinity.

(b) Substituting Z, you get P_N. Find the average occupancy $\langle N \rangle = \sum_N N P_N$. Show that this result, the Bose–Einstein distribution $N_0 = N(\epsilon_0)$, looks almost identical to the Fermi–Dirac function $f_0 = f(\epsilon_0)$ for electrons, except for a small difference (what?). For that matter, repeat the exercise for electrons N summed over $0, 1$ and show that $\langle N \rangle = f(\epsilon_0)$.

(c) Let us next try to calculate the rates of transition between two electronic states connected by a photon absorption or emission. As before, the ratio of the absorption and removal rates is given by $\exp\left[-(\epsilon_0 - \mu)/k_B T\right]$. For absorption and emission of electrons, we can write this as $f_0/(1 - f_0)$. Show that this same ratio can be realized for photon absorption in terms of Boson occupancies as $N_0/(N_0+1)$ using the expression from the last part.

(d) In fact, we just proved that the absorption rate of a photon can be written as AN_0, while the emission rate is $A(N_0 + 1)$, where N_0 is the number of photons available. The terms proportional to N_0 are called 'stimulated absorption' and 'stimulated emission', while the constant emission rate A independent of N_0 is called 'spontaneous emission'.

P12.4 **Stoner criterion for ferromagnetism.** Recall that strong Coulomb repulsion U_0 in partially filled d or f band materials tends to give ferromagnetism. The right side of Fig. 10.1 shows ferromagnetic metal bands split by exchange energy, Δ, with corresponding shifted densities of states $D_{\uparrow,\downarrow}(E)$. Assume they are constant $\approx D_0$, but shifted by Δ. The total energy and charge density in each band, in absence of an external magnetic field, can be written at zero temperature as

$$E = U_0 n_\uparrow n_\downarrow + \int_0^{E_F} D_\uparrow(E) E \, dE + \int_\Delta^{E_F} D_\uparrow(E) E \, dE,$$

$$n_\uparrow = \int_0^{E_F} D_\uparrow(E) \, dE, \quad n_\downarrow = \int_\Delta^{E_F} D_\downarrow(E) \, dE. \tag{12.16}$$

Rewrite E_F and thus $E(n, m)$ in terms of the total electron density $n = n_\uparrow + n_\downarrow$ and magnetization $m = n_\uparrow - n_\downarrow$. The total electron density n is fixed, but the magnetization m is our variable.

To get ferromagnetism, we need to make sure that a positive $m > 0$ lowers E, in other words, $\partial^2 E(n, m)/\partial m^2 < 0$. Show that this leads to the *Stoner criterion* for ferromagnetism, $U_0 D_0 > 1$. (Note that I am ignoring higher-order terms $\sim m^4$, that prevent m from running away to $\pm\infty$. The actual energy should look like $E \approx A(T - T_C) m^2/2 + B m^4/4$, i.e., the *sombrero* potential $\boxed{\text{P5.4}}$ for $T < T_C$.)

$\boxed{\text{P12.5}}$ **Light emitting diodes (LEDs) and lasers.** Consider a voltage applied across a channel. Electrons can enter the band from the left contact with probability $f(\epsilon_1 - \mu_1)$, emit a photon of energy $h\nu$ with probability $A(N_0 + 1)$ and then exit the band into the right contact with probability $\bar{f}(\epsilon_2 - \mu_2) = 1 - f(\epsilon_2 - \mu_2)$. Conversely, an electron can enter from the right with probability $f(\epsilon_2 - \mu_2)$, absorb a photon with probability AN_0 and exit to the left with probability $\bar{f}(\epsilon_1 - \mu_1) = 1 - f(\epsilon_1 - \mu_1)$.

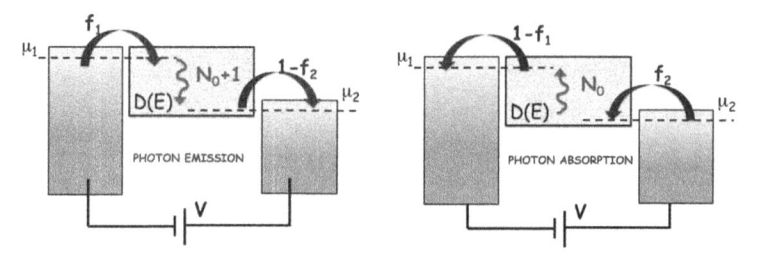

The electron current and photon production rate, using $\bar{f} = 1 - f$, is

$$I \propto \frac{dN_0}{dt} = Af(\epsilon_1 - \mu_1)\bar{f}(\epsilon_2 - \mu_2)(N_0 + 1) - Af(\epsilon_2 - \mu_2)\bar{f}(\epsilon_1 - \mu_1)N_0, \tag{12.17}$$

with $\epsilon_1 - \epsilon_2 = h\nu$ by energy conservation.

(a) Find the voltage threshold for photon emission $dN_0/dt > 0$ at frequency ν at zero temperature, where $N_0 = 0$.

(b) Using the relations $[1 - f(\epsilon)]/f(\epsilon) = \exp[(\epsilon - \mu)/k_BT]$ and $N_0/(N_0 + 1) = \exp[-h\nu/k_BT]$ that we proved earlier, and energy conservation $\epsilon_1 - \epsilon_2 = h\nu$, show that I has a diode characteristic. Also show that if the photons are at a higher temperature $T_0 > T$, we can get a negative short circuit current at zero bias, like in a solar cell $\boxed{\text{P16.2}}$.

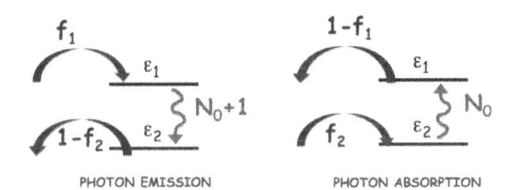

PHOTON EMISSION PHOTON ABSORPTION

(c) For a laser, consider a single contact at μ and two energy levels $\epsilon_{1,2}$ like above. Write down a similar rate equation, except we need the build-up of N_0 to be exponential, meaning the coefficient of the N_0 term on the right-hand side must exceed zero. Show that this requires the occupancy of the higher energy level to exceed the lower. (This is called *population inversion*, and is usually achieved by pumping electrons to a higher energy ϵ_0 with a short lifetime with an intermediate level ϵ_1 with long lifetime, so that electrons relax from ϵ_0 to ϵ_1 faster than they relax from ϵ_1 to ϵ_2).

Part 2

Top–Down: Classical Transport

Chapter 13

Drift–Diffusion

13.1 Basic Definitions

When a device is subjected to non-equilibrium conditions, such as a voltage or a temperature difference across its ends, the response is to drive a current. At each end, charges will flow to try and bring the system into local equilibrium with the reservoir at its boundary, but since the boundaries do not belong to a common equilibrium, the process does not cease and there is a continuous steady-state flow of current. For charge current, the quantity setting the equilibrium level is the local Fermi energy $\mu(x)$, also called *Quasi-Fermi energy* or *Electrochemical potential* — describing the maximum energy upto which the electrons fill the band at low temperature. Thus, the flow of current is directly related to a variation in $\mu(x)$, with current density (amperes per square centimeter)

$$\vec{J} = \frac{\sigma}{q}\vec{\nabla}\mu, \qquad (13.1)$$

where the conductivity tensor σ connects the driven current with the driving potential difference. We can relate this to our familiar version of Ohm's law by assuming a voltage V applied across the ends of a wire of uniform cross-section S and length L, so that $\nabla\mu = qV/L$, and

$$I = JS = (\sigma V/L)S = V/R, \quad \text{where the resistance } R = L/\sigma S. \qquad (13.2)$$

From the current density above we can separate out a part that depends on the local charge density $n(x)$, given by how far this Fermi energy lies above the bandedge $E_C(x) = E_C - q\phi(x)$ shifted by the local electrostatic potential $\phi(x)$ (this difference $\mu(x) - E_C(x)$, which determines charge density, is called the *chemical potential*). We can thus write

$$\vec{J} = \frac{\sigma}{q}\vec{\nabla}[\mu - E_C + q\phi] - \sigma\vec{\nabla}\phi. \tag{13.3}$$

We identify the variation in electrostatic potential as the electric field, $\vec{\mathcal{E}} = -\vec{\nabla}\phi$, and connect the difference between the band bottom and the local electrostatic potential with the electronic charge density per unit volume

$$n = N/\Omega = \int \bar{D}(E)f(E - \mu + E_C - q\phi)dE, \tag{13.4}$$

where $\bar{D}$ is the density of states per unit volume at the Fermi energy, so that $\vec{\nabla}n = \int dE\bar{D}(E)\left(-\partial f/\partial E\right)\vec{\nabla}\left(\mu - E_C + q\phi\right)$. We can then write

$$\vec{J} = \underbrace{q\mathcal{D}\vec{\nabla}n}_{\text{Diffusion}} + \underbrace{\sigma\vec{\mathcal{E}}}_{\text{Drift}}, \tag{13.5}$$

where the conductivity is related to the density of states and the diffusion coefficient as

$$\boxed{\sigma = q^2 \int dE\bar{D}\mathcal{D}\left(-\frac{\partial f}{\partial E}\right) = \langle\langle q^2 \bar{D}\mathcal{D}\rangle\rangle}. \tag{13.6}$$

In other words, the conductivity depends on the density of states/volume and the diffusion constant averaged $\langle\langle\ldots\rangle\rangle$ over the vicinity of the Fermi energy where $-\partial f/\partial E$ is significant, roughly a Lorentzian shape. For a single parabolic band in the non-degenerate limit (where we only sample the exponential tail $\sim e^{-(E - E_F)/k_B T}$ of the Fermi function), we can use $-\partial f/\partial E = f/k_B T$, and the charge density $n = \int dE\bar{D}(E)f(E - E_F)$, so

$$\sigma = \frac{nq^2\mathcal{D}}{k_B T}. \tag{13.7}$$

At this point, we introduce the concept of *mobility*, relating to how fast the carriers are responding to an external field, and connect to conductivity, which also accounts for how many carriers are moving.

$$\sigma = nq\mu. \tag{13.8}$$

This gives us the Einstein relation in the non-degenerate limit

$$\frac{\mathcal{D}}{\mu} = \frac{k_B T}{q} \quad \text{(Einstein relation, non-degenerate limit)}. \tag{13.9}$$

The generalized version for degenerate materials can be inferred from the general equation for conductivity (Eq. (13.6)).

13.2 Understanding Mobility and Diffusion

In the next chapter, we will encounter various restorative processes that show how aggressively an electronic system attempts to revert to equilibrium, assisted by trap assisted recombination–generation in indirect bandgap matterials. Let us now look at how they are driven from equilibrium in the first place. The first is through electrical potential gradients, i.e., electric fields, which give rise to drift. The second is through randomly generated thermal kicks, i.e., diffusion. The former is characterized by a mobility μ while the latter by a diffusion constant $\mathcal{D}$.

In presence of an electric field, the electrons accelerate, but are slowed down by scattering processes that generate microscopic 'friction'. Through the balance between these speed-up and slow-down events, the electrons accomplish the equivalent of terminal velocity that a parachutist encounters during an air drop. The drift velocity is proportional to the electric field and the slope is called the *mobility* (we can see that it has units of cm^2/Vs)

$$v_e - -\mu_e \mathcal{E},$$
$$v_h = \mu_h \mathcal{E}, \tag{13.10}$$

also showing the direction of flow of the charges (electrons and holes) in an electric field. We can estimate this by writing down Newton's equation, say for the electron

$$m_e \frac{dv_e}{dt} = \underbrace{-\gamma_e v_e}_{\text{Damping}} -q\mathcal{E}. \tag{13.11}$$

In the absence of any electric field, we would get $v_e = v_0 e^{-t/\tau_e}$, where $\tau_e = m_e/\gamma_e$ is the momentum scattering time for the electron. Now at steady-state we drop the term on the left-hand side and get $v_e = -q\mathcal{E}/\gamma_e = -q\mathcal{E}\tau_e/m_e$, so that the mobility can be written as

$$\mu_{e,h} = \frac{q\tau_{e,h}}{m_{e,h}}, \qquad (13.12)$$

meaning the mobility is high if either the carrier effective mass is low (light particles), or the scattering time is high (infrequent scattering events).

The electron mobility depends on the effective mass and scattering time and determines the speed of an electron in response to an external electric field, $\mu = v/\mathcal{E}$. This relation follows since

$$\sigma = \frac{L}{R_{cl}A} = \frac{L}{V_D A/I} = \frac{\overbrace{I/A}^{J}}{\underbrace{V_D/L}_{\mathcal{E}}} = \frac{J}{\mathcal{E}}, \qquad (13.13)$$

with $I = Nq/\tau = (nLA)q/(L/v) \implies J = I/A = nqv$. The conductivity thus splits into two separate components — the mobility describing the intrinsic speed of electrons ($\mu \sim 1450$ for n-Si, ~ 450 for p-Si, ~ 50 for metallic Au and $\sim 200{,}000$ for room-T graphene on hBN, in units of cm^2/Vs). The free electron density $n \sim 10^{16}\,\text{cm}^{-3}$ for lightly doped silicon, $\sim 10^{10}\,\text{cm}^{-3}$ in intrinsic silicon, $\sim 2 \times 10^{22}\,\text{cm}^{-3}$ for metallic Au and $\sim 10^{10}\,\text{cm}^{-2}$ for graphene with a moderate doping. The resulting resistivity ρ is roughly $\mu\Omega\text{cm}$ in metals, and about a million times higher for n-Si.

We can then calculate the drift current density

$$J_{e,h}^{\text{drift}} = \mp q n_{e,h} v_{e,h} = \underbrace{q n_{e,h} \mu_{e,h}}_{\sigma_{e,h}} \mathcal{E}, \qquad (13.14)$$

where $\sigma_{e,h}$ are the electron and hole *conductivities*. Their reciprocals $\rho_{e,h}$ are the resistivities and have units of Ω-cm. Since the drift current densities are in the same direction for electrons and holes, the conductivities simply add up ($\sigma_{\text{eff}} = \sigma_e + \sigma_h$).

If we eliminate the net electric field and observe the charges under thermal kicks, we will see a velocity once again, but one that will simply try to homogenize the charges by moving them from a denser region (with more outward kicks than inward) to a rarer region. Across a small region of width l in 1D, the net x-directed current is

$$J_n^{\text{diffusion}} = -q n(x) v + q n(x + l) v = -qvl \frac{dn}{dx}. \qquad (13.15)$$

The local drift terms only exist between two collisions, in other words, $l = v\tau_n$. This gives

$$J_e^{\text{diffusion}} = q \underbrace{\left(\frac{l^2}{\tau_e}\right)}_{\mathcal{D}_n} \frac{dn}{dx}. \tag{13.16}$$

A similar term $J_h^{\text{diffusion}} = -q\mathcal{D}_p \partial p/\partial x$ describes diffusion of holes. Both electrons and hole velocities move downgradient, along direction of decreasing charge density, but the currents for those two components are opposite due to their opposite charge, hence the relative negative sign between the two. The unit of diffusion constant is cm^2/s. We can rationalize them from intuition, since electrons have negative charge and our convention looks at positive charge determining the sign of the current. For instance, an electric field in the $+x$ direction gives a flow of electrons along $-x$ and holes along $+x$, both giving a drift current therefore along $+x$ (opposite velocities, parallel currents), so that $J_{p,n}^{\text{drift}} \propto +q\mathcal{E}$. On the other hand, a positive density gradient along $+x$ causes both charges to go down along $-x$, and their diffusion currents are now opposite, $J_{n,p}^{\text{diff}} \propto \pm d(n,p)/dx$ (parallel velocities, opposite currents).

We can then summarize the drift–diffusion equation as

$$\left.\begin{aligned} J_n &= qn\mu_n\mathcal{E} + q\mathcal{D}_n\vec{\nabla}n \\ J_p &= qp\mu_p\mathcal{E} - q\mathcal{D}_p\vec{\nabla}p \end{aligned}\right\} \quad \text{(drift–diffusion)}. \tag{13.17}$$

Since drift is directed motion, the average of the particle coordinate, $\langle x \rangle$, increases linearly with time, and the slope is given by $\mu\mathcal{E}$. For diffusion, the average $\langle x \rangle - 0$, but the variance $\langle x^2 \rangle$ increases linearly with time, with the proportionality constant being $\mathcal{D}$.

13.3   Charge Extraction and the Equation of Continuity

Drift–diffusion basically provides a mechanism to create a current — through potential gradients established by an electric field, like going down a water slide, and through thermal scattering events, bumping and bouncing off of each other if the slide is packed tight. At equilibrium, $J \propto \nabla\mu = 0$, and we need a way to clear the logjam by taking the charges out. In short ballistic samples, the contacts take the electrons out at their existing energies at rates given by their bonding strength at the channel edges, and the number of states to escape into the contact (hopefully numerous). In large classical samples, we have inelastic scattering where electrons in the conduction band can find holes in the valence band and recombine (this can be tricky as often the holes are static and the electrons drifting, so they need to offload their momentum to other entities — traps, etc. before they can jump down). We will discuss the physics of recombination–generation in two chapters, but for now, we can consider them as imposing an escape rate R or a generation rate G (for instance, if new electrons are injected into the system, optically as in a solar cell).

How do the drift and diffusion currents relate to each other, as well as fields, charges and recombinations? This is where the conservation of charge (formally called *Equation of Continuity*) comes in. Basically, any increase in positive charge over time requires an *influx* (negative divergence of positive current density) or *generation* (e.g., electron–hole production by photoexcitation). Conversely, a decrease in positive charge over time requires a positive divergence or recombination, i.e., charge extraction. For electrons, we switch the sign of current. We can write this succinctly as

$$\left.\begin{aligned}
\frac{\partial n}{\partial t} &= \frac{1}{q}\vec{\nabla}\cdot\vec{J}_n + G_n - R_n \\[2mm]
\frac{\partial p}{\partial t} &= -\frac{1}{q}\vec{\nabla}\cdot\vec{J}_p + G_p - R_p
\end{aligned}\right\} \quad \text{(equation of continuity)}.$$

$$(13.18)$$

We discuss the physics of generation and recombination in two chapters. A good way to characterize them is with a lifetime, which quantifies how quickly a system tries to restore toward equilibrium once it deviates from it. Typically for an n-type system, the rate limiting step is the slower minority p-type carriers p_n, and their recombination rate — analogous for electrons n_p in a p-type system. Accordingly, it depends on the difference between the current charge density and the equilibrium charge density that it is trying to attain. The lifetimes in this case are the minority carrier lifetimes, the time for the minority charges to diffuse and find a trap to absorb momentum and allow recombination.

$$R_n \approx \frac{\Delta n_p}{\tau_n}, \quad R_p \approx \frac{\Delta p_n}{\tau_p}. \tag{13.19}$$

13.4 Electrostatics and Gauss's Law

The band bottoms are set by electrostatics $\boxed{\text{P13.1}}$, which causes currents to move around through drift–diffusion and charges to accumulate by equation of continuity, and these accumulated or depleted charges in turn influence the electrostatics through Coulomb/Gauss's law. Gauss's Law says the total number of electrostatic field lines across (perpendicular to) a closed surface depends on the amount of charge *enclosed* by the surface in Coulombs, $\oint \vec{E} \cdot d\vec{S} = Q_{\text{encl}}/\epsilon$ $\boxed{\text{P13.2 and P13.3}}$. Using some theorems from geometry, we can rewrite it in terms of charge density per unit volume (Coulombs/m^3)

$$\underbrace{\vec{\nabla} \cdot (\epsilon \vec{\mathcal{E}})}_{\text{Field divergence}} = q \underbrace{\left(p - n + N_D^+ - N_A^- \right)}_{\text{Charge density}} \quad \text{(Gauss's law)} . \tag{13.20}$$

In the next chapter, we will discuss the physics of mobility $\mu_{n,p}$ and its dependence on carrier density, temperature and electric field. The other parameter, the diffusion constant $\mathcal{D}$, is in fact related directly to mobility through the Einstein relation, $\mathcal{D}_{n,p} = \mu_{n,p} k_B T / q$ (Eq. (13.9)). Thus, knowing the mobility of a semiconductor in units of cm^2/Vs, we simply multiply it by the thermal voltage $k_B T / q \approx 25\,\text{mV}$ to get the diffusion constant in units of cm^2/s.

In the chapter after that, we will discuss the physics of recombination–generation $R_{n,p}$ and $G_{n,p}$.

13.5 Minority Carrier Diffusion Equation

For bipolar devices driven primarily by diffusion, we can drop the electric fields. We can ignore generation unless there is incident light $\boxed{\text{P13.4–P13.6}}$, and use the relaxation time approximation for the deviation from equilibrium $\Delta n_p = n_p \quad n_{p0}$, $\Delta p_n = p_n \quad p_{n0}$, ultimately giving us in 1D the *Minority Carrier Diffusion Equation*

$$\frac{\partial \Delta n_p}{\partial t} = \mathcal{D}_n \frac{\partial^2 \Delta n_p}{\partial x^2} - \frac{\Delta n_p}{\tau_n},$$

$$\frac{\partial \Delta p_n}{\partial t} = \mathcal{D}_p \frac{\partial^2 \Delta p_n}{\partial x^2} - \frac{\Delta p_n}{\tau_p}. \tag{13.21}$$

At steady state where $\partial/\partial t$ terms drop to zero, we need charge deviations whose second derivatives simply become constant multipliers, implying exponential distributions. Assuming the system goes back to equiibrium for $x = \infty$, we get

$$\Delta n_p(x) = \Delta n_p(0)e^{-x/L_n}, \quad \underbrace{L_n = \sqrt{D_n \tau_n}}_{\text{Diffusion length}}, \tag{13.22}$$

or if it vanishes at $-\infty$, we get a positive exponent. Similarly, for $\Delta p_n(x)$. The electron diffusion current density is then given by

$$J_n(x) = q\mathcal{D}_n \partial \Delta n_p/\partial x = -q \underbrace{(\mathcal{D}_n/L_n)}_{v_n^{diff}} \Delta n_p(x), \tag{13.23}$$

with the diffusion velocity $v_n^{diff} = \mathcal{D}_n/L_n = \sqrt{\mathcal{D}_n/\tau_n} = L_n/\tau_n$ — basically diffusion length divided by minority carrier lifetime.

Homework

P13.1 **Gradient doping field.** Consider a semiconductor that is non-uniformly doped with donor impurity atoms with a concentration of $N_D(x)$. Show that the induced electric field in the semiconductor in thermal equilibrium is given by

$$\mathcal{E}(x) = -\left(\frac{k_B T}{q}\right)\frac{1}{N_D(x)}\frac{dN_D(x)}{dx}.$$

P13.2 **Gauss's law.** Let's use Gauss's law to get the electric field $\mathcal{E}$ for a few simple geometries, where we can guess the direction of the electric fields, shown as follows (note that we are staying away from edges, where there are 'fringing fields' that are hard to account for). Given a charge Q on each metallic geometry, and the Gaussian surface shown in dashed line in each case at a distance $r > R$ for the sphere, $r > R$ for the cylinder, and position $x > 0$ for the plate, find the electric fields in each case, $\mathcal{E}_{\text{sph}}(r)$, $\mathcal{E}_{\text{cyl}}(r)$ and $\mathcal{E}_{\text{plate}}(x)$. In each case, argue that the field is perpendicular to the surface and constant in magnitude, so it can be pulled out of the integral.

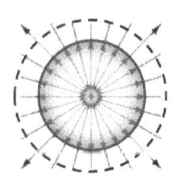 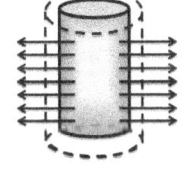 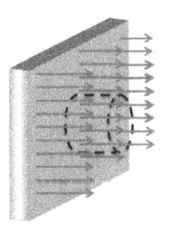

- Sphere of radius R
- Charge Q
- Gaussian surface of radius r

- Long Cylinder of radius R
- Charge Q
- Gaussian surface of radius r, length L

- Large rectangular plate
- Charge Q
- Gaussian surface of CS area A, height x

From the field, calculate the potential $V(r)$ whose derivative is the field. In other words, by integrating $V(r)_{\text{sph,cyl}} = -\int_{r_0}^{r} \mathcal{E}_{\text{sph,cyl}}(r)dr$, $r > R$, $V(x)_{\text{plate}} = -\int_{x_0}^{x} \mathcal{E}_{\text{plate}}(x)dx$, $x > 0$. The lower bound of the integral is the ground where the potential is assumed zero. For the sphere and the cylinder, we assume $r_0 = R$ and for the plate, we assume $x_0 = 0$, i.e., at the metal each time.

P13.3 **Capacitors.** We now apply a voltage V between two concentric spheres of radii R_1 and R_2 with equal and opposite charges $\pm Q$ (so that their field lines don't cancel), concentric cylinders of radii R_1 and R_2 and equal and opposite charge $\pm Q$, and two parallel plates separated by $x_2 - x_1 = d$ with equal and opposite charges $\pm Q$. In each case, the two limits of your voltage integral above will be the two radii/positions. Calculate the voltage drop in terms of Q, and set that drop equal to applied bias V. The ratio gives you the capacitance C for each geometry. Find the capacitances in each case. For the sphere and the cylinder, you can move the outer partner to infinity ($R_2 \to \infty$).

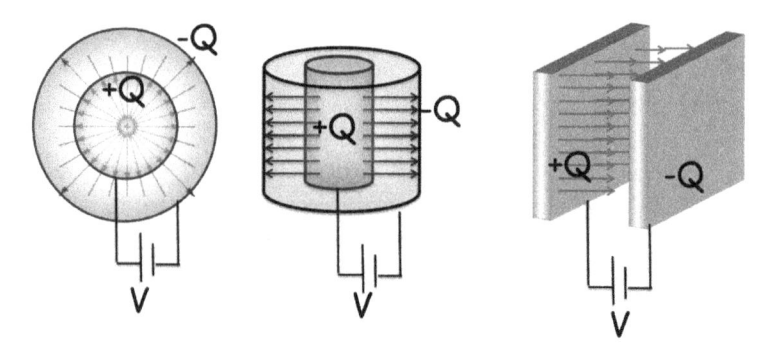

For the parallel plate, insert two dielectrics in parallel (side-by-side) between them, and show that the two parallel capacitors add in series, $C = C_1 + C_2$. Next, put the dielectrics in series, one above the other, then show that the capacitors add in parallel $C = C_1 C_2/(C_1 + C_2)$.

P13.4 **Drift diffusion for spins.** In a material with magnetic properties, we expect to see a spin-dependent conductivity $\sigma_{\uparrow,\downarrow}$ (e.g., P22.7). At hetero-interfaces therefore, we expect to see a splitting of spin electrochemical potentials $\mu_{\uparrow,\downarrow}$ P19.4 and P22.8. A non-magnetic voltage probe like Au will measure a charge potential, while a magnetic voltage probe like Fe would also detect the spin potential, depending on its orientation θ relative to the up spin direction.

$$\mu_{Au} = \mu_{\text{charge}}, \quad \mu_{Fe} = \mu_{\text{charge}} + \mu_{\text{spin}} \cos\theta = \mu_\uparrow \cos^2\theta/2 + \mu_\downarrow \sin^2\theta/2$$

$$\mu_{\text{charge}} = (\mu_\uparrow + \mu_\downarrow)/2, \quad \mu_{\text{spin}} = (\mu_\uparrow - \mu_\downarrow)/2 \tag{13.24}$$

The diffusion equation for the two opposite spins $S, \bar{S} = \uparrow, \downarrow$ in 1D are

$$\frac{\partial n_S}{\partial t} = \frac{1}{q}\frac{\partial J_S}{\partial x} + \left(-\frac{n_S}{\tau_{S\bar{S}}} + \frac{n_{\bar{S}}}{\tau_{\bar{S}S}}\right), \quad S = \uparrow, \downarrow, \quad \bar{S} = \downarrow, \uparrow$$

$$J_S = \underbrace{q n_S \mu_S}_{\sigma_S}\frac{\partial \mu_S}{\partial x} = \sigma_S \mathcal{E} + q D_S \frac{\partial n_S}{\partial x}. \tag{13.25}$$

Let's multiply the $\uparrow$ equation with $\sigma_\downarrow$ and vice versa, and subtract to get rid of the term proportional to $\mathcal{E}$. Write $n_{\uparrow,\downarrow}(x) = (n_\uparrow + n_\downarrow)/2 \pm (n_\uparrow - n_\downarrow)/2 = \bar{n} \pm m(x)/2$, where we ignore the variation of $\bar{n}$ and focus on the variation of $m(x)$. Show that for $\mathcal{E} = 0$, we get a spin diffusion equation for $m(x)$ with solution given a spin diffusion length $\lambda_{\text{eff}} = \sqrt{\mathcal{D}_{\text{eff}}\tau_s}$, where $\mathcal{D}_{\text{eff}} = (\mathcal{D}_\uparrow\sigma_\downarrow + \mathcal{D}_\downarrow\sigma_\uparrow)/(\sigma_\uparrow + \sigma_\downarrow)$ and $1/\tau_s = 1/\tau_{\uparrow\downarrow} + 1/\tau_{\downarrow\uparrow}$.

P13.5 **Spherical diffusion.** Let us look at the diffusion problem in 3D spherical coordinates, for instance, a pellet of charge continuously injected with a syringe into the center of a sphere, that then diffuses radially. The minority carrier diffusion equation is given by

$$\frac{\partial \Delta n_p(r)}{\partial t} = D_n \nabla^2 \Delta n_p(r) - \frac{\Delta n_p(r)}{\tau_n}. \tag{13.26}$$

(a) In radial coordinates, write out ∇^2 assuming Δn_p is only a function of radial distance r, and independent of angle. Use boundary conditions $\Delta n_p(r = \infty) = 0$ to simplify the analytical expression for $n(r)$. You will still have an unspecified constant.

(b) Let us assume the total amount of charge injected is known, i.e., $q \int d\Omega n(r) = Q$. Use spherical volume integral to nail down the coefficient and show that it is given by an areal density set by a specific lengthscale.

(c) Find the diffusion current J_n and total current $I = \int \vec{J} \cdot d\vec{S}$ over the spherical surface area element $d\vec{S}$, in terms of Q, D_n and τ_n.

P13.6 **Photoabsorption.** Radiation on a solar cell creates electron–hole pairs at a rate G_L in units of $cm^{-3}s^{-1}$, per unit volume per second. Once the carriers are photogenerated, they are separated by the built-in field $\mathcal{E}$, flowing down gradient to create a current in absence of a voltage, i.e., a short-circuit current.

(a) Let us look at the minority carrier diffusion equation for holes in an n-type semiconductor in presence of recombination time τ_p and also a generation rate G_L. Write down the minority carrier diffusion equation, and impose steady-state condition ($\partial/\partial t = 0$). Show that far away ($x \to \infty$), there is a non-zero minority charge build-up per unity volume. Find $\Delta p_n(x \to \infty)$ in terms of G_L and τ_p.

(b) From the $x = \infty$ boundary condition, and that at $x = 0$ where the photo-excitation happens, $\Delta p_n(0)$, find $\Delta p_n(x)$ and the hole diffusion current density $J_p(x)$.

Chapter 14

Physics of Mobility: Fermi's Golden Rule

The ability of a material to transport electrons depends on its effective mass and its scattering time, which together determine the mobility. We already know how to calculate the effective mass from the bandstructure. Let us now discuss the scattering time. We can approximately treat each scattering process independently, and add their scattering rates so that the scattering time reduces with each added scattering mode (this additivity is called *Matthiessen's rule*). Two prominent scattering mechanisms for electrons in bulk semiconductors are Coulomb scattering from ionized impurities and acoustic phonon scattering from lattice vibrations. We can estimate the impact of each scattering event between an initial state i and all possible final states f using Fermi's Golden rule, which states that

$$\frac{1}{\tau_i(E)} = \frac{2\pi}{\hbar} \underbrace{\sum_f}_{\text{final states}} \underbrace{\left| \int \psi_f^*(x) U(x) \psi_i(x) \right|^2}_{\text{scattering matrix M}_{\text{if}}} \underbrace{\delta(E_f - E)}_{\text{energy conservation}} . \qquad (14.1)$$

In other words, scattering depends on the available phase space for scattering described by the summation, as well as the strength of the scattering potential described by the integral, subject to the constraint of energy conservation described by the delta function. If the integral is roughly independent of f, then we can pull it out of the summation, whereupon the sum over a set of delta functions yields the density of states for the final electron, $\sum_f \delta(E_f - E) = D(E)$

$$\frac{1}{\tau(E)} \approx \frac{2\pi}{\hbar} \left| \int \psi^*(x) U(x) \psi(x) \right|^2 D(E). \qquad (14.2)$$

Since we are adding probabilities we are throwing away coherence effects and treating them as classical scattering centers.

Fermi's Golden Rule is traditionally derived in a complicated way — we first work out the transition probability between a discrete initial state and a continuum of final states using time-dependent perturbation theory. For an oscillating potential near resonance, we get the square of a *sinc* function (i.e., $\sin^2 x/x^2$) for the transition probability. For long times, this gives a delta function times time, so that we get the equation above for the rate of change of probability, i.e., scattering rate. The derivation is much more elegant once we go into Green's functions and invoke the concept of a scattering cross-section later in this chapter. We direct the interested readers to consult the detailed derivations for NEMV, but we will motivate it shortly with simple examples.

14.1 Ionized Impurity Scattering

Let us first estimate the impact of scattering processes as captured with Fermi's Golden Rule. Ionized impurity scattering has less effect on the electrons if we increase the temperature, because the energized electrons can tear themselves away from the scattering potential (much like it is easier to jet ski over choppy oceanwaves if we go full throttle). We can rationalize this expression from Fermi's Golden Rule. The 3D density of states $D \sim \sqrt{E} \sim \sqrt{k_B T}$. The screened Coulomb potential

$$U(r) = \frac{q^2}{4\pi\epsilon_0 r}e^{-r/L_D},\tag{14.3}$$

where L_D is the Debye screening length that we will revisit later. For the two wave functions in the matrix element let us focus on long wavelength scattering by ignoring the Bloch functions and treating them as plane waves far from the scattering center, $\psi(x) \sim e^{ikx}$. This means the matrix element is just the Fourier transform

$$M_{\text{if}} = \int e^{i(\vec{k} - \vec{k}') \cdot \vec{r}} U(r) d^3\vec{r} = \frac{q^2}{\epsilon_0 \left(|\vec{k} - \vec{k}'|^2 + L_D^{-2} \right)}.\tag{14.4}$$

For elastic scattering (fixed energy), $k^2 = (k')^2 = 2mE/\hbar^2$, and $|\vec{k} - \vec{k}'|^2 = 2k^2(2 - 2\cos\theta) = 4k^2 \sin^2(\theta/2) \propto E \sin^2(\theta/2)$. It is a fun exercise to carry through the integrals, but it is easy to see even without that for weak screening the matrix element $M_{if}^2 \sim 1/k^4 \sim 1/E^2 \sim 1/(k_B T)^2$, meaning that at higher temperature the effect of the Coulomb scattering reduces as the higher electron momenta allow them to escape scattering. From Eq. (14.2) we then get

$$\frac{1}{\tau_{\text{imp}}} = \frac{2\pi}{\hbar} \underbrace{M_{if}^2}_{\sim T^{-2}} \underbrace{D}_{\sim T^{1/2}} N_I \sim N_I T^{-3/2}, \tag{14.5}$$

where N_I is the number of impurities per unit volume. The last term arose because we worked out M_{if} for a single impurity above, whereas the sum over final states must account for all impurities. This additivity of scattering rates for individual uncorrelated impurities comes from Matthiessen's rule.

14.2 Acoustic Phonon Scattering

Phonon scattering, on the other hand, gets stronger with temperature as there are more vibrations in the picture (in the jet ski analogy, our waves are getting choppier as well). In Fermi's Golden rule, the electron–phonon matrix element M_{if} arising from the deformation potential is roughly independent of temperature. However, instead of number of impurities N_I we now have number of acoustic phonons $N_{\text{ph}} = 1/[\exp(\hbar\omega/k_B T) - 1] \sim k_B T/\hbar\omega$ at low energies, meaning the number of phonons and thus their cumulative scattering effect increases linearly with T, i.e., $\sum_f M_{if}^2 \sim k_B T/\hbar\omega$. Together with the density of states $D \sim \sqrt{k_B T}$, we thus get $1/\tau_{ph} \sim T^{3/2}$.

14.3 Scattering Cross-Section

Fermi's Golden Rule tells us that the efficiency of a scattering process depends on the absolute value of the scattering probability between one initial and all final states connected by energy conservation. The matrix element itself could impose added conservation rules, for instance, linear, orbital or spin angular momentum conservation. An intuitive way to visualize this is to define a scattering cross-section $\sigma \approx \pi r^2$ that describes the range r upto which the scattering potential can influence the electron trajectory. To connect the two concepts, we consider a box of area A and length Δx, which carries $A\Delta x N_T$ scatterers, N_T being the scatterer density per meter cubed. Each scatterer has a cross-sectional area σ, so the capture probability for an electron is given by the ratio of the area of all traps to the available area, $\Delta P = (\sigma \times A\Delta x N_T)/A = \sigma \Delta x N_T$. In a given time Δt the length covered by the electrons moving at the thermal velocity is $\Delta x = v_{th}\Delta t$. This gives us the probability of scattering per unit time,

$$\frac{dP}{dt} = \underbrace{1/\tau}_{\text{s}^{-1}} = \underbrace{\sigma}_{\text{cm}^2} \underbrace{v_{th}}_{\text{cm/s}} \underbrace{N_T}_{\text{cm}^{-3}} \tag{14.6}$$

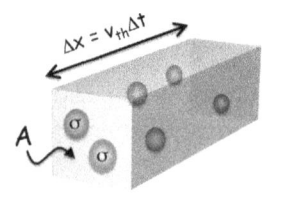

To understand capture cross-section σ, consider Fig. 14.1. We are interested in calculating the *differential* scattering cross-section $d\sigma/d\Omega$, which is the ratio of the number of particles per second scattered into a solid angle $dN/d\Omega$, over the number of incoming particles per second per unit area (current flux) $dN/d\sigma$. Integrating over the solid angle $d\Omega = \sin\theta d\theta \int_0^{2\pi} d\phi = 2\pi \sin\theta d\theta$, we then get the capture cross-section σ that goes into the scattering time above.

As the figure shows, for a classical particle we can relate the angular scattering to the *impact parameter* b using geometry. We see that

$$\frac{d\sigma}{d\Omega} = \frac{2\pi b db}{2\pi \sin\theta d\theta} = \frac{b}{\sin\theta}\left|\frac{db}{d\theta}\right|. \tag{14.7}$$

For a hard sphere of radius R, it is straightforward to show that the cross-section σ upon 3D integration is its projected area πR^2 P14.1 and P14.2.

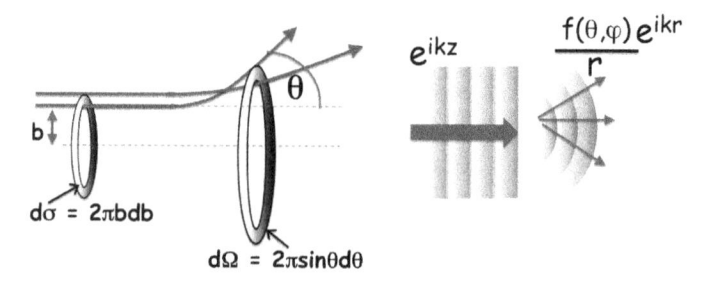

Fig. 14.1 (Left) Differential cross-section given by the annular area ratio before and after scattering between two nearby electrons separated by angle $d\theta$. Corresponding wavefunctions change from plane to spherical at a point scatterer.

For a quantum particle, we need Schrödinger's equation $\boxed{\text{P14.3}}$. In presence of scattering, we can write

$$\psi(\vec{r}) = \underbrace{e^{ikz}}_{\text{incident plane wave } \psi_{\mathrm{p}}} + \underbrace{f(\theta, \phi)\frac{e^{ikr}}{r}}_{\text{outgoing spherical wave } \psi_{\mathrm{s}}}, \tag{14.8}$$

where f is obtained by solving the radial Schrödinger equation in presence of a potential V. By calculating the scattered particle density per unit solid angle per unit time, and the incident current flux, we can easily see that the differential scattering cross-section

$$\frac{d\sigma}{d\Omega} = \frac{(\hbar k/m)|\psi_s|^2/d\Omega}{(\hbar k/m)|\psi_p|^2/\underbrace{r^2 d\Omega}_{d\sigma}} \approx |f(\theta, \phi)|^2. \tag{14.9}$$

14.4 Green's Function and the First Born Approximation

To get the f function, we need to solve Schrödinger equation approximately. A convenient way is to use Green's functions. Basically, we break the equation into an inhomogeneous (Helmholtz) equation with an adjustable source (the scattering potential). Green's function G is the solution when the source is a spike (a delta function). Given that any source can be constructed out of a combination of spikes, it follows by linearity that the solution can be constructed by the exact same combination of Green's functions. For instance, consider the inhomogeneous linear differential equation

$$\hat{L}u(x) = \lambda(x). \tag{14.10}$$

This could represent, for instance, the Schrödinger equation for an electron with a source term. We first find the solution to the equation with an impulse response — Green's function, satisfying

$$\hat{L}G(x, x') = \delta(x - x'), \tag{14.11}$$

with the appropriate boundary conditions. Typically we solve it piecemeal away from the spike and then adjust the coefficients at the boundary, much like we solved the particle on a spike earlier (Section 4.3). The advantage of ignoring λ here is that G depends only on the intrinsic properties of $\hat{L}$ and boundary conditions. It is then easy to show by straightforward substitution that the solution for an arbitrary source $\lambda(x)$ is

$$u(x) = \int dx' G(x, x')\lambda(x'). \tag{14.12}$$

Going back to the scattering theory problem at hand, we can define $k^2 = 2mE/\hbar^2$ and a source term $S(\vec{r}) = 2mU(\vec{r})\psi(\vec{r})/\hbar^2$. We then get

---

Green's Function For Schrödinger Equation

$$\left(\nabla^2 + k^2\right)\psi(\vec{r}) = S(\vec{r})$$

Solved by $\left(\nabla^2 + k^2\right)G(\vec{r},\vec{r}\,') = \delta(\vec{r} - \vec{r}\,')$

Leading to $\psi(\vec{r}) = \underbrace{\psi_0(\vec{r})}_{\text{homogeneous solution}} + \underbrace{\int G(\vec{r},\vec{r}\,')S(\vec{r}\,')\ \overbrace{d^3\vec{r}\,'}^{\text{3D vol element}}}_{\text{particular integral}}$

$$(14.13)$$

Referring back to our H-atom solution, recall that in ∇^2 the quantity R/r acted as a 1D Schrödinger equation. The radial part in 3D would require an extra $1/r$ factor over the solution to Green's function in 1D (clearly a plane wave once we set the delta function to zero), so the solution to Green's function is

$$G(\vec{r},\vec{r}\,') = -\frac{e^{i\vec{k}.(\vec{r} - \vec{r}\,')}}{4\pi|\vec{r} - \vec{r}\,'|}, \tag{14.14}$$

where 4π arises from the angular integral since $\nabla^2(1/r) = -4\pi\delta(r)$. The solution thus is

$$\psi(\vec{r}) = \psi_0(\vec{r}) - \int d^3\vec{r}\,'\frac{e^{i\vec{k}.(\vec{r} - \vec{r}\,')}}{4\pi|\vec{r} - \vec{r}\,'|}S(\vec{r}\,')$$

$$= \psi_0(\vec{r}) - \frac{m}{2\pi\hbar^2}\int d^3\vec{r}\,'\frac{e^{i\vec{k}.(\vec{r} - \vec{r}\,')}}{|\vec{r} - \vec{r}\,'|}U(\vec{r}\,')\psi(\vec{r}\,'). \tag{14.15}$$

Now assuming the potential is localized, we can write down the $\psi(\vec{r})$ far from the potential, by assuming $|\vec{r} - \vec{r}\,'| \approx r$, $\vec{k} \cdot \vec{r} \approx kr$, and get

P14.4–P14.6

$$\psi(\vec{r}) \approx \psi_0(\vec{r}) - \frac{m}{2\pi\hbar^2} \frac{e^{ikr}}{r} \int d^3\vec{r}\,'e^{-i\vec{k}\,\cdot\,\vec{r}\,'} U(\vec{r}\,')\psi(\vec{r}\,'). \tag{14.16}$$

We can approximate $\psi(\vec{r}\,')$ inside the integral with a similar plane wave with a different wavevector $\exp[i\vec{k}\,'\cdot\vec{r}\,']$. Now, by comparing Eq. (14.16) with Eq. (14.8), we can easily identify that f is given by the Fourier transform of the scattering potential

$$f(\theta,\phi) \approx -\frac{m}{2\pi\hbar^2} \underbrace{\int e^{-i(\vec{k}-\vec{k}\,')\cdot\vec{r}} U(\vec{r})d^3r}_{M_{\mathrm{if}}}. \tag{14.17}$$

Using $d\sigma/d\Omega = |f(\theta,\phi)|^2$, where $(\vec{k}-\vec{k}')\cdot\vec{r} = 2k\sin(\theta/2)$ as shown earlier for elastic scattering, we can then integrate over the solid angle $d\Omega = 2\pi\sin\theta d\theta$ and then use Eq. (14.6) to show that $1/\tau = (2\pi/\hbar)|M_{if}|^2 D$, i.e., Fermi's Golden Rule $\boxed{\text{P14.7}}$. But Fermi's Golden Rule is more general, as it does not limit us to thermal velocity, or to parabolic bands invoked in connecting with the scattering cross-section. It may need to be revisited for correlated noise however $\boxed{\text{P14.8}}$.

14.5 Revisiting Scattering Times

Based on our knowledge of scattering cross-sections, let us try to understand the scattering times we derived earlier. The ability for electrons to bypass the scattering center is quantified by their thermal velocity obtained by equating the kinetic energy with a thermal energy $k_B T/2$ for each mode (the equipartition theorem that arises from the Boltzmann distribution and resulting Gaussian integrals). For kinetic energies much greater than potential energy, the electron escapes scattering while for much lower, they get trapped, so that the cross-section radius r_c is roughly obtained by transition energy where kinetic and potential energies are equal. This gives

$$\frac{3k_B T}{2} = \frac{q^2}{4\pi\epsilon_0 r_C} = \frac{mv_{\mathrm{th}}^2}{2}$$

$$\Longrightarrow \sigma_c = \pi r_C^2 \propto T^{-2}$$

$$\frac{1}{\tau_{\mathrm{imp}}} = \underbrace{v_{\mathrm{th}}}_{\sim T^{1/2}} \underbrace{\sigma_c}_{\sim T^{-2}} N_I \propto T^{-3/2}. \tag{14.18}$$

For phonon scattering, with increasing temperature the total scattering over all centers available increases, $\sigma N_{ph} \propto T$. We then get

$$\frac{1}{\tau_{ph}} = \underbrace{v_{th}}_{\sim T^{1/2}} \underbrace{\sigma N_{ph}}_{\sim T} \propto T^{3/2} \tag{14.19}$$

Thus, the net scattering rate has a temperature dependence of the form

$$1/\mu \propto 1/\tau \approx \underbrace{AT^{3/2}}_{\text{impurity}} + \underbrace{BT^{-3/2}}_{\text{phonon}}. \tag{14.20}$$

With the understanding in place, scientists figured out how to increase the mobility in III–V heterostructures through modulation doping. The ionized impurities were embedded in the barriers of a GaAs/AlGaAs heterojunction, so that the electrons tunnel into the triangular potential wall formed by band-bending at the junction. Since these electrons are physically separated from the parent ions, they lose the ionized impurity scattering term at low temperature, leading to a dramatic increase in mobility in these material systems. The measured temperature dependence in these modulation doped heterostructures shows $\mu \approx T^{-3/2}$, meaning that only phonon scattering persists in these systems. For transistors, there is a complicated field-temperature-doping dependence of the mobility. An empirical way to include doping dependence is the Caughey–Thomas model $\boxed{\text{P16.3}}$

$$\mu = \mu_{min} + \frac{\mu_0}{1 + (N_I/N_0)^\alpha} \tag{14.21}$$

where the parameters $\mu_{0,min}$, N_0 and α are doping and temperature-dependent quantities fitted to experimental data. This equation captures the doping dependence, while simpler expressions like the one for impurity and phonon above capture the temperature dependence. Along with this, there is a field-dependence, which states that at high fields $\sim 10^5$ V/cm, the carriers are drawn to the surface and suffer more surface scattering that cause their velocities to saturate. Figure 14.2 summarizes the empirical rules.

$$\mu = \frac{\mu_0}{\left[1 + \left(\dfrac{\mathcal{E}}{\mathcal{E}_0}\right)^\beta\right]^{1/\beta}}, \quad (\beta \approx 1 \text{ for electrons, 2 for holes}). \tag{14.22}$$

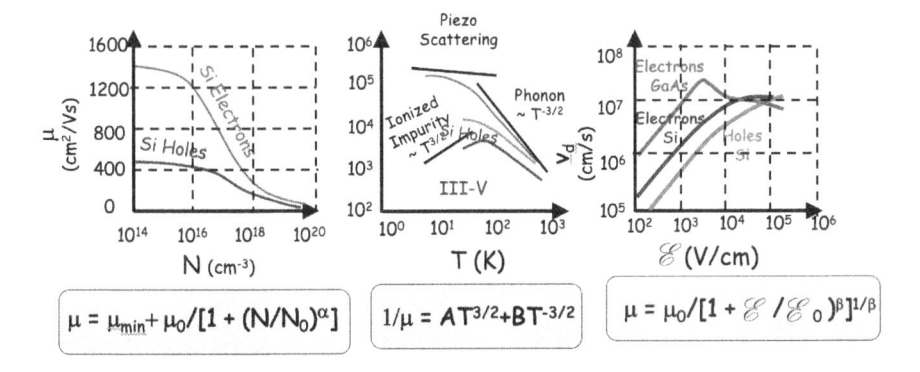

Fig. 14.2 Doping, temperature and field-dependent semiconductor mobilities.

Homework

P14.1 **Hard sphere cross-section.** Show that for a hard sphere, $b = R\sin\alpha$, where $\theta = \pi - 2\alpha$. Use this to get the cross-sectional area.

P14.2 **Coulomb scattering.** Let's find the Fourier transform of the Debye potential, $U(\vec{r}) = q^2 e^{-\kappa r}/4\pi\epsilon_0 r$. We want to calculate

$$\int e^{i(\vec{k} - \vec{k}')\cdot\vec{r}} U(\vec{r}) d^3\vec{r} \tag{14.23}$$

(a) Write $(\vec{k} - \vec{k}')\cdot\vec{r} = \Delta kr\cos\theta$, and $d^3\vec{r} = r^2 dr d\phi d(\cos\theta)$, and show that the result is $q^2/\epsilon_0(\Delta k^2 + \kappa^2)$.

(b) For an unscreened Coulomb potential ($\kappa = 0$), show that we get Rutherford scattering

$$\frac{d\sigma}{d\Omega} = \left(\frac{q^2}{8\pi\epsilon_0 m v_0^2}\right)^2 \csc^4\theta/2. \tag{14.24}$$

Integrating over the solid angle $d\Omega$, show that the cross-section is infinity, because the Coulomb potential has a long range.

(c) We can solve this problem classically, using Kepler's laws for the radial motion of a particle in a Coulomb potential (similar to the effective 1D problem we solved for a hydrogen atom). Solving the Kepler equation gives us the impact parameter

$$b = \frac{q^2}{4\pi\epsilon_0 m v_0^2}\cot\theta/2. \tag{14.25}$$

Show that we get the same answer as above.

(d) Let us now look at the overscreened potential. For this, show that

$$\frac{d\sigma}{d\Omega} = \left(\frac{q^2}{8\pi\epsilon_0 m v_0^2}\right)^2 \frac{1}{\left(\sin^2\theta/2 + \hbar^2\kappa^2/4m^2 v_0^2\right)^2}. \tag{14.26}$$

Do the angular integral, and show that

$$\sigma = \pi r_{\text{eff}}^2, \quad \text{where} \quad \frac{q^2}{8\pi\epsilon_0 r_{\text{eff}}} = \underbrace{\frac{\hbar^2\kappa^2}{2m}}_{\text{screening potential}} \gg \frac{m v_0^2}{2}. \tag{14.27}$$

How is this related to the atomic size? Show that the screening length $\lambda_{sc} = 1/\kappa$ is the geometric mean between r_{eff} and the Bohr radius a_0. If the screening cuts off precisely at the Bohr radius, $\lambda_{sc} = a_0$, then $\sigma = \pi a_0^2$.

P14.3 **Plane wave vs spherical wave.** The actual solution to the wave equation depends on the overall symmetry and boundary conditions.

(a) Consider the wave equation $\nabla^2 \psi + k^2 \psi = 0$ in Cartesian coordinates. Show that $\psi = e^{i\vec{k} \cdot \vec{r}}$ is a solution. Show that the equation for the wavefront (constant phase) represents the equation for a plane. What is the perpendicular to the plane i.e., direction along which the phase change happens and the wavefront propagates?

(b) For a spherical wave, use polar coordinates to solve $\nabla^2 \psi + k^2 \psi = 0$. Show that it is in effect a 1D wave equation for $r \times \psi(r)$. Show now that the wavefront with constant phase has the equation for a sphere.

P14.4 **Cross-section and phase shift.** Recall the radial solution to a 3D free electron (P6.7), $R_l(r) = 2kj_l(kr)$, a spherical Bessel. We can expand the incident plane in terms of these orbitals

$$e^{ikz} = e^{ikr \cos \theta} = \sum_l C_l j_l(kr) P_l(\cos \theta). \tag{14.28}$$

(a) Find C_l by Taylor expanding and matching identical powers of $\cos \theta$.
(b) Far from $r = 0$, use $j_l(kr) \approx \sin(kr - l\pi/2)/kr$. Expand using Euler's formula, and add a scattering phase-shift η_l pre-factor to the *outgoing* wave. Work out $f(\theta)$, and then integrating, show that

$$\sigma = \int d\Omega |f(\theta)|^2 = \sum_l \sigma_l, \qquad \sigma_l = \frac{\pi}{k^2}(2l+1)|1 - \eta_l|^2. \tag{14.29}$$

We can interpret σ_l as the area of a ring of particles $\pi[r_{l+1}^2 - r_l^2]$ between angular momentum l and $l+1$, where $\hbar k r_l = \hbar l$ and $\hbar k r_{l+1} = \hbar(l+1)$.

(c) Writing η_l as a phase shift, $\eta_l = e^{2i\delta_l}$, show that

$$\sigma_l = \frac{4\pi}{k^2}(2l+1)\sin^2 \delta_l. \tag{14.30}$$

P14.5 **Spherical potential.** For a spherical potential of radius R at low energy $R \gg \lambda$, focus on $l = 0$. From **P14.4**, the phase shift gives

$$\frac{R(r)}{r} = \begin{cases} \dfrac{i}{(2\pi)^{3/2}2kr}\left(e^{-ikr} - \eta_0 e^{ikr}\right) = \dfrac{e^{i\delta_0}}{(2\pi)^{3/2}kr}\sin\left(kr + \delta_0\right), & r > R \\[3mm] \dfrac{c_+ e^{ik'r} + c_- e^{-ik'r}}{kr}, & r < R, \end{cases}$$

$$(14.31)$$

where $k = \sqrt{2mE/\hbar^2}$ and $k' = \sqrt{2m(E + U_0)/\hbar^2}$. At $r = 0$ this function must be finite, which means $c_- = -c_+$ and the function looks like $a\sin k'r$.

(a) Using the continuity of wavefunction and derivative at $r = R$, solve for δ_0. Show that for low energies compared to the potential U_0, $\sigma \approx 4\pi R^2$. Note that this is the entire surface of the sphere, as long wavelength electrons 'feel' their way around its entire surface.

(b) For a high energy scattering $kR \gg 1$, we can use Eq. (14.30) summed over l upto a maximum angular momentum $l_{\max} \approx kR$, and then replace $\sin^2 \delta$ by its average, $1/2$. Show that this gives $\sigma = 2\pi R^2$, still twice the classical result because of a 'shadow'. (To recover the classical result of πR^2, we will need a non-abrupt potential so that the wavelength of the electron is shorter than the variation length of the potential — recall how we recovered the classical result for particle on a step.)

P14.6 **Hard sphere.** For a spherically symmetric potential, show

$$f(\theta) = -\frac{2m}{\hbar^2}\int_0^\infty r^2 V(r)j_0(\Delta kr)dr \qquad (14.32)$$

where $\Delta k = 2k\sin\left(\theta/2\right)$, θ being the angle between $\vec{k}$ and $\vec{k}'$, assuming elastic scattering, and j_l is the spherical Bessel function of order l. Note that $j_0(x) = \sin x/x$. By assuming a hard sphere of potential U_0 for $r < a$ and zero outside, show that $f(\theta) = -(2mU_0 a^2/\hbar^2\Delta k)j_1(\Delta ka)$.

P14.7 **Trap lifetime and Golden Rule.** Show that the two equations

$$\frac{1}{\tau} = \sigma N_T v_{\text{th}} = \frac{2\pi}{\hbar}|M_{if}|^2 D, \qquad (14.33)$$

are identical for a classical particle. Start with Eq. (14.17) for f, and do the angular integral to get σ. Also relate the 3D density of states with thermal velocity by assuming $v_{\text{th}} = \sqrt{2E/m}$.

P14.8 **Colored noise and correlation.** The drift–diffusion equation can be formally derived from the overdamped Newton equation with uncorrelated white (thermal) noise (NEMV, Chapter 9). Deterministic forces that cause drift are, however, a special case of 'noise' that is fully correlated. Thus, a general correlated noise should, in principle, capture both.

Let us start with the overdamped, noisy Newton equation in a correlated noisy system, dropping the inertial term amd including a damping coefficient γ

$$m\cancel{\frac{d^2x}{dt^2}}^{\,0} + \gamma\frac{dx}{dt} = F_T(t), \tag{14.34}$$

where we assume a Gaussian noise with a non-zero correlation time τ

$$\langle F_T(t)\rangle = 0, \quad \langle F_T(t)F_T(t')\rangle = \gamma^2\frac{\mathcal{D}}{\tau}C\left(\frac{|t-t'|}{\tau}\right), \tag{14.35}$$

where frequently the correlation function $C(x) = \exp[-x]$ and $\mathcal{D}$ is the diffusion coefficient. For $\tau = 0$, the correlation $\langle F_T(t)F_T(t')\rangle = \gamma^2\mathcal{D}\delta(|t-t'|)$ and we recover thermal white noise.

By integrating the equation by parts show that

$$\langle x^2(t)\rangle = \begin{cases} 2\mathcal{D}C(0)t^2/\tau, & (t \ll \tau,\ \text{drift}), \\ 2\mathcal{D}\left[t - \mu_1\tau\right], & (t \gg \tau,\ \text{diffusion}), \end{cases} \tag{14.36}$$

where μ_1 is the first moment, $\mu_1 = \int_0^\infty Y C(Y)dY$. This will require you to change variables from t, t' to $t \pm t'$ and think through the limits carefully, taking care of the mod function in the argument of C (draw out the coordinates to identify the limits).

In other words, for deterministic systems with long correlation times, we are dominated by drift where $\langle x^2\rangle \propto t^2$, while for very noisy systems with vanishing correlation times, we are dominated by diffusion where $\langle x^2\rangle \propto t$.

Chapter 15

Physics of Recombination–Generation

While drift–diffusion describes the forces driving electrons away from equilibrium, restorative forces try to bring the system back to equilibrium. We need to account for both drive and restoration in order to get steady-state response of a semiconductor (e.g., to get a DC current under an applied bias instead of transient noise or a run-away). In quasi-ballistic channels, this restoration is done by the contacts that quickly siphon charges away and cool them down (as we see with the Landauer equation later in this book). In larger channels with large intrinsic resistances that make the contacts less relevant, restoration is imposed by internal scattering processes that try to bring the charges back to equilibrium. Since equilibrium in a non-degenerate semiconductor is described by the equation $np = n_i^2$, RG recombination rate must depend on how far from equilibrium we deviate, We expect the rate of approach to equilibrium to be driven by how far the system ventured away from equilibrium, i.e.,

$$R = R_0(np - n_i^2). \tag{15.1}$$

This implies two things: (a) We need to replace Eq. (12.9) with a modified version that ensures $np \neq n_i^2$

$$n = n_i e^{\beta(E_{Fn} - E_i)}, \quad p = n_i e^{\beta(E_i - E_{Fp})}, \tag{15.2}$$

with two different quasi-Fermi levels $E_{Fn,p}$ (or two different thermal factors $\beta_{1,2}$ in a thermally driven current for instance). (b) For bipolar devices where the physics is primarily driven by changes in minority carrier concentration — say of electrons in p-dype Si, $p \approx p_0$ (the equilibrium concentration, in the so-called *low level injection* approximation), the factor

$np - n_i^2$ becomes in effect $\Delta n_p = n - n_0$, and the rate is given by $\Delta n_p/\tau_n$, known as the *relaxation time approximation*, with the minority carrier lifetime τ_n (Eq. (14.6)) set by the time it takes for the minority carriers to recombine with opposite charges. For indirect semiconductors like silicon, this requires electrons to thermally diffuse around till they find mid-level traps that allow them to dump their momentum, before relaxing and recombining with holes and emitting photons in the process. In ballistic conductors, the restoration is given by terms like $\Delta n, p/\tau_{n,p} \sim \gamma_{n,p}(N - f_{1,2})$. In this chapter, we explore the physical mechanisms for restoration, and work out the maths behind these restorative forces.

15.1   Basic Recombination Processes

Figure 15.1 shows a cartoon of various scattering processes — both in energy-space and in real space, that end in an electron and a hole recombining. For direct bandgap materials like III–V semiconductors, this involves direct recombination between an electron in the conduction band and a hole in the valence band, emitting a photon that takes away the energy lost equal to the bandgap. The recombination rate R_0 depends on the matrix element in Fermi's Golden Rule, in other words, the effective masses and field-dependent symmetries of the Bloch wavefunctions in the conduction and valence band, including in their confined states in III–V quantum wells, for instance. In lower dimensional semiconductors such as organics, nanowires or quantum dots, there is a higher probability for the electron and hole to bind together like an artificial hydrogen atom — solving Schrödinger

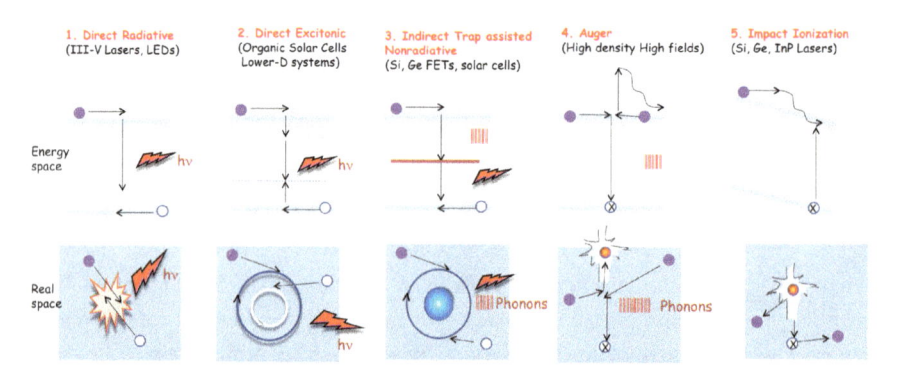

Fig. 15.1   Recombination–generation processes shown schematically in energy and real space.

equation $\boxed{\text{P6.7}}$, we find the binding energy in 2D is higher by about a factor of four, because confinement in the third dimension makes it harder for the two charges to avoid each other.

15.2 Indirect Processes: Shockley–Read Hall

Direct recombination becomes harder in indirect bandgap materials like silicon, where the electron at the bottom of the conduction band near the X-point has a sizeable crystal momentum Δk compared to the hole at the top of the valence band near the Γ point, momentum that it needs to offload somehow (picture it jumping off a running bus), in addition to the energy difference ΔE roughly at the visible frequency. Keep in mind that the velocities of photons are orders of magnitude higher than the velocity $\Delta E/\hbar \Delta k$ needed to conserve both energy and momentum (demanded by temporal and spatial translation invariance). In other words, photo-excited transitions are in effect, *vertical* in a semiconductor E–k diagram because the corresponding photonic E–k is much steeper, and thus quite unable to connect the indirect band edges in k-space. To absorb the momentum Δk, an electron needs to diffuse around till it finds a trap that can take away that momentum (a bump that slows down the bus) for a subsequent vertical energy relaxation. For a semiconductor with trap density N_T, this will involve a delicate dance between number of positive, electron capturing traps p_T, vs negatively charged, hole capturing traps n_T (with $p_T + n_T = N_T$) capable of trapping electrons n in the conduction band and holes p in the valence band.

Let us work out the steady-state value for the trap filling n_T, p_T associated with a given p, n combination — the latter externally imposed by doping and drift–diffusion. The evacuation of a trap state is driven by competing processes. If only the conduction band were involved, the trap emptying fraction p_T/N_T would be set by electrons escaping from trap to conduction band n_1, as opposed to electrons escaping from the Fermi surface to conduction band n. The expression for n_1 is obtained from Eq. (12.9), assuming the trap pins the Fermi energy to itself at equilibrium

$$n_1 = n_i e^{\beta(E_T - E_i)}. \tag{15.3}$$

Strong interactions at the trap complicate this with a degeneracy factor g which shifts E_T a bit, but we will ignore that here. If the valence band were involved, the trap filling fraction $n_T/N_T = 1 - p_T/N_T$ would be electrons moving from valence band to trap, i.e., holes moving from trap to valence band p_1, vs from Fermi energy to valence band p. These two separate cases co-exist, their weights set by electron and hole capture coefficients $c_n = 1/\sigma_n v_{th}$ and $c_p = 1/\sigma_p v_{th}$. In other words,

$$\frac{p_T}{N_T} = \begin{cases} \dfrac{n_1}{n + n_1} & \text{if only trap to CB, } c_n \gg c_p \\[3mm] 1 - \dfrac{p_1}{p + p_1} = \dfrac{p}{p + p_1} & \text{if only VB to trap, } c_n \ll c_p, \end{cases} \tag{15.4}$$

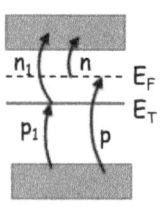

so the fraction of empty or filled traps at steady-state are

$$\frac{p_T}{N_T} = \frac{c_n n_1 + c_p p}{c_n(n + n_1) + c_p(p + p_1)},$$

$$\frac{n_T}{N_T} = 1 - \frac{p_T}{N_T} = \frac{c_n n + c_p p_1}{c_n(n + n_1) + c_p(p + p_1)}.$$

The recombination rate is now obtained by looking at the probability of an empty trap $\sim p_T$ capturing conduction band electrons $\sim n$, minus a filled trap $\sim n_T$ expelling its electron count $\sim n_1$,

$$r_N = c_n\left(p_T n - n_T n_1\right) = \frac{np - \overbrace{n_1 p_1}^{n_i^2}}{\tau_p(n + n_1) + \tau_n(p + p_1)}, \tag{15.5}$$

where $\tau_{n,p} = 1/c_{n,p}N_T = 1/\sigma_n v_{th} N_T$, same as Eq. (14.6). It is easy to check that this rate is the same as the rate r_P for a filled trap $\sim n_T$ to capture a valence band hole $\sim p$ minus an empty trap p_T to emit its hole $\sim p_1$, i.e., $r_P = c_p(n_T p - p_T p_1)$. The equality of the two rates $r_N = r_P$ is what gives us the steady-state result $\boxed{P15.1}$.

15.3 Low-Level Injection & Relaxation Time Approximation

For most minority carrier devices, it is fair to assume one dopant type dominates (say $n \gg p$), and furthermore the minority carrier density is much smaller than the dominant majority carrier density. If we write n, p

in terms of their deviation from equilibrium n_0, p_0, we get

$$n = n_0 + \Delta n, p = p_0 + \Delta p, \quad np \approx \underbrace{n_0 p_0}_{n_i^2} + n_0 \Delta p + p_0 \Delta n + \underbrace{O(\Delta^2)}_{\text{drop}}. \quad (15.6)$$

For low level injection in a p-type material, $p_0 \gg n_0 \gg \Delta p, \Delta n$, so we replace the numerator in Eq. (15.5) by $p_0 \Delta n$ and denominator with $\tau_n p_0$, so

$$r_N \approx \frac{\Delta n_p}{\tau_n} \text{ in } p\text{-type material,} \quad r_P \approx \frac{\Delta p_n}{\tau_n} \text{ in } n\text{-type,} \quad (15.7)$$

which basically says the rate-limiting process is the number of extra minority carriers you inject into a region, Δn_p into a p region (Δp_n in n-type), and the lifetime for those carriers to diffuse around and eventually recombine with majority carriers indirectly using traps that can absorb their sizeable momentum in an indirect semiconductor.

Traps are created by defects in the solid — for instance, broken bonds or vacancies, or foreign atoms. They differ from donor and acceptors primarily because they are deep level, near mid-gap, since they need to be optimally placed in energy to service both the conduction and the valence bond by connecting them. It is straightforward to show that mid-gap traps are in fact the most efficient for recombination–generation (RG) processes $\boxed{\text{P15.2}}$. From here on, we will assume the trap is roughly mid-gap, implying thereby

$$n_1 \approx p_1 \approx n_i, \quad E_T \approx E_i \quad (15.8)$$

15.4 Auger Processes and Impact Ionization

Finally, in materials with high carrier density/large electric fields, the slamming of one high energy electron into a slower one can halt the former by transferring energy to the latter. As a result, the original electron loses energy and falls into a hole in the valence band, sending the latter soaring in energy till it eventually cools down through non-radiative (thermal) events. The recombination terms for such Auger processes would look like $\sim n(np - n_i^2)$ or $\sim p(np - n_i^2)$, with an extra charge to account for the three particle process $\boxed{\text{P15.3 and P15.4}}$. The reverse process is called impact ionization, where an energized electron drops to the band-bottom by pulling an electron straight out of the valence band, leaving behind a hole there and generating a carrier multiplication from one to two electrons in the conduction band. Such a multiplication process is at the heart of high sensitivity detectors such as avalanche photo-diodes, as we will see later in the chapter on noise.

Homework

P15.1

Shockley–Reed–Hall. Show that $r_N = r_P$ in the Shockley–Reed–Hall process.

P15.2

Deep traps. From the general steady-state R representation, confirm that bandgap centers with E_T near E_i make the best RG centers (assume an n-type semiconductor, low level injection and $\tau_n = \tau_p = \tau$. Consider how R would vary with E_T under the given conditions and conclude with a rough sketch of R vs E_T across the bandgap).

P15.3

Auger RG. Let us work out the Auger RG rate. We have two recombination processes — an 'eeh' process involving two electrons and one hole, and an 'ehh' process involving two holes and one electron. The corresponding generation rates are proportional to the number of electrons and to the number of holes. We can thus write the following equation at steady-state:

$$dn/dt = dp/dt = -c_n n^2 p - c_p p^2 n + e_n n + e_p p. \tag{15.9}$$

(a) Use detailed balance within each 'eeh' or 'ehh' process and get expressions for e_n and e_p.

(b) Substitute back to get the expression for the Auger RG rate at steady-state, and show that the non-equilibrium term $np - n_i^2$ factors out naturally.

(c) What are the units of the Auger coefficients c_n and c_p?

P15.4

Auger coefficient. Let us estimate the probability of an Auger transition. In this picture, two electrons scatter and one goes up in energy while the other goes down in energy. Write down the momentum and energy conservation rules, and solve for E_2' in terms of k_1, k_1' and k_2' (eliminate k_2 from momentum conservation), and the energy gap E_G. Now, find the lowest energy $E_{2,\min}'$ by setting its partial derivatives with respect to k_1

and k_1' to zero. Show then by eliminating k_2' that $k_2 = k_1$. Show also that $E_{2,\min}' = E_G(2\mu^{-1} + 1)/(\mu^{-1} + 1)$, where $\mu = m_c/m_v$.

Let us understand the equations. For small m_c, argue that $k_1' = 0$, and find k_1, k_2 in terms of k_2' and argue that $E_{2,\min}' = 2\Delta E$. If now $m_v = 0$, then $k_1 = k_2 = 0$. Show that this gives $k_1' = -k_2'$, and $E_{2,\min}' = \Delta E$.

Chapter 16

Engineering Barriers: The PN Junction Diode

16.1 Depletion Width and Built-in Potential

Basically an electronic switch involves the creation of a barrier to current flow, and then its abrupt removal by applying an external perturbation — such as a current or a field. A perfect example is a PN junction (Fig. 16.1), where a barrier is created between an N-type silicon channel with an excess of mobile electrons (e.g., doped with phosphorus), and a P-type silicon channel with a surplus of holes (doped with boron). The lack of states for electrons on the P side and holes on the N side poses a barrier in the form of a *built-in potential* V_{bi}, spread out over a *depletion region* of width W_d.

To reduce the barrier, we need to 'forward bias' it, where the positive terminal of a battery connects to the p-doped side, and the negative terminal to the n-side. This will push the majority charges closer to the interface away from the battery terminals, and reduce the depletion region, creating a large, exponentially growing current. For the opposite reverse bias, the depletion region and the barrier increase and the current drops. In other words, we get a solid state rectifier. A switch can then be created with a third terminal to modulate the conductive properties of this rectifier.

When we put a P-type semiconductor together with N-type (Fig. 16.2), the charge gradients drive a diffusion current of electrons toward P and holes toward N. However, the ionized parent dopant atoms pull back the charges through electrostatic forces, so that the charges can only move away by a finite distance, the depletion width. Within this region, the mobile carriers have diffused away, leaving the ionized dopants exposed and

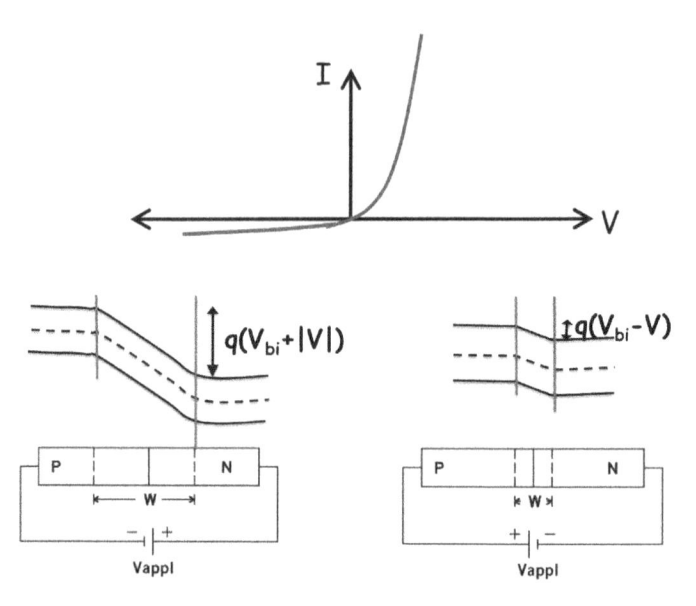

Fig. 16.1 A PN junction diode blocks current for reverse bias when (bottom left) the applied bias increases the barrier and depletion width between the P and N sections. For forward bias (bottom right) the voltage reduces the barrier.

uncompensated, so there is a net charge left behind. Outside, the mobile carriers screen the dopants and we have a quasi-neutral region. The barrier height is called the built-in potential V_{bi}, and is given by the amount of deviation between E_i and E_F on either side. Using Eqs. (12.11) and (12.12) for this deviation,

$$V_{bi} = \left(E_i - E_F\right)_P + \left(E_F - E_i\right)_N = \frac{k_BT}{q}\ln\left(\frac{N_A}{n_i}\right) + \frac{k_BT}{q}\ln\left(\frac{N_D}{n_i}\right)$$

$$\Rightarrow \boxed{V_{bi} = \frac{k_BT}{q}\ln\left(\frac{N_A N_D}{n_i^2}\right)}. \tag{16.1}$$

In other words, the barrier depends on how 'different' each doped segment is from each other, quantified by the amount of doping over the intrinsic carrier concentration. We can write this as $(k_BT/q)\ln(p_{p0}/p_{n0}) = (k_BT/q)\ln(n_{n0}/n_{p0})$, given by the ratio of electron (or hole) concentration between majority and minority regions, the former being $N_{D,A}$ and the latter $n_i^2/N_{A,D}$.

Given the built-in potential, we can next calculate the distance over which the potential varies, in other words, the depletion width W_d. Ignoring the

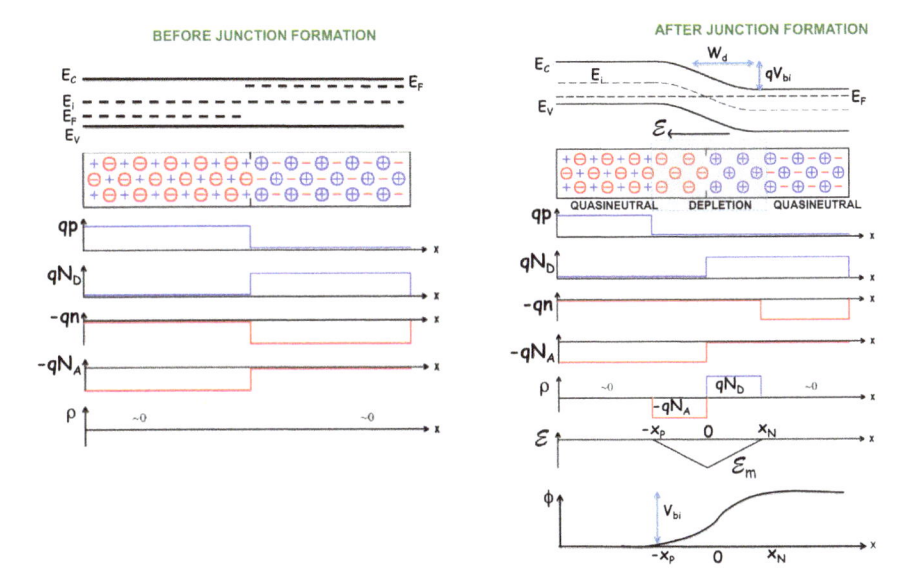

Fig. 16.2 Band-diagrams and charge distributions (red positive, blue negative) across the junction before and after junction formation, leading to creation of a potential barrier ϕ across a depletion width where mobile charges are flushed away (depleted) by the built-in electric field $\mathcal{E}$.

slight rounding off of the edges of the charged regions, we can plot the net charge distributions ρ which equal $-qN_A$ in the P region and qN_D in the N region (Fig. 16.2). Using Gauss's Law, $d\mathcal{E}/dx = \rho/\epsilon_{si}$, we can find the electric field. For constant charges, the integrated field has a linear variation, like a tent. Charge neutrality allows us to calculate the maximum electric field at the center of the depletion region, $\mathcal{E}_m = -qN_Ax_P/\epsilon_{si} = -qN_Dx_N/\epsilon_{si}$.

We now need the depletion width $W_d = x_P + x_N$. We can extract it using the simple principle of ratios that if $a/b = c/d$ that each must equal $(a+b)/(c+d)$. Using this on the expression for $\mathcal{E}_m$ above, we get

$$\mathcal{E}_m = -\frac{q}{\epsilon_{si}}\left(\overbrace{\frac{x_P + x_N}{N_A^{-1} + N_D^{-1}}}^{W_d}\right). \tag{16.2}$$

To get the built-in potential V_{bi}, we need one more integral, since $\mathcal{E} = -\partial\phi/\partial x$, with band energy $U = -q\phi$. This boils down to the area of the tent-like triangle in the $\mathcal{E} - x$ plot (Fig. 16.2), with height $|\mathcal{E}_m|$ and base

W_d, so that $V_{\text{bi}} = W_d|\mathcal{E}_m|/2$, from which we get

$$
W_d = \sqrt{\frac{2\epsilon_{\text{si}}(N_A^{-1} + N_D^{-1})V_{\text{bi}}}{q}} \quad \text{(depletion width)}. \tag{16.3}
$$

Upon applying a bias V, the potential barrier changes to $V_{\text{bi}} \mp V$ ($-$ for forward, $+$ for reverse), and so does the depletion width

$$
W_d(\pm V) = \sqrt{\frac{2\epsilon_{\text{si}}(N_A^{-1} + N_D^{-1})(V_{\text{bi}} \mp V)}{q}}. \tag{16.4}
$$

Finally, we can go back to the $\mathcal{E}_m$ ratio equation $x_P/N_A^{-1} = x_N/N_D^{-1} = W_d/(N_A^{-1} + N_D^{-1})$ to get the individual breakdowns of the depletion width between the two regions. We can see that

$$
x_{N,P} = \frac{N_{A,D}}{N_A + N_D} W_d \tag{16.5}
$$

proportional to the doping on the other side. Most of the depletion region lies on the *lower doped* side. In retrospect, this is obvious because of charge neutrality. Reducing the doping means we need to go out a longer distance to get the same charge as the opposite side, $qN_A x_P = qN_D x_N$.

16.2 Debye Length

Finally, we can calculate the Debye length L_D, which is the smallest distance over which a potential can vary at room temperature. We can get it by combining Poisson's equation with Boltzmann's equation coupling electrostatics and thermodynamics. For a small deviation of potential $\delta\phi$ from equilibrium, and a corresponding deviation in charge $\delta\rho$ from the equilibrium value ρ_0, we get

$$
\frac{\partial^2 \delta\phi}{\partial x^2} = -\frac{\delta\rho}{\epsilon_{\text{si}}} \quad \text{(Poisson's equation)}
$$

$$
\delta\rho = q \int dE D(E - \underbrace{U}_{-q\delta\phi})f(E - E_F) \quad \text{(substitute } E \to E + U)
$$

$$
= q \int dE D(E)f(E - E_F - q\delta\phi) \approx q^2\delta\phi \int dE D(E)(-\partial f/\partial E). \tag{16.6}
$$

For a non-degenerate semiconductor, we can replace $f(E - E_F) \to e^{-(E-E_F)/k_BT}$, whereupon $-\partial f/\partial E = f/k_BT$. Using the integral $\int dE D(E) f(E - E_F) = \rho_0$, we get

$$\delta\rho \approx q^2 \rho_0 \delta\phi/k_BT$$

$$\Rightarrow \left(\frac{\partial^2}{\partial x^2} + \frac{1}{L_D^2}\right)\delta\phi = 0 \quad \text{(from 1st line of Eq. (16.6))}$$

$$\Rightarrow \delta\phi = \delta\phi_0 e^{-x/L_D} \quad \text{(Debye–Hueckel Theory))},$$

$$\boxed{L_D = \sqrt{\frac{\epsilon_{si}(k_BT/q)}{q\rho_0}} \quad \text{(non-degenerate semiconductors)}}. \quad (16.7)$$

Some numbers: For a doping $N_A = 10^{16}\,\text{cm}^{-3}$, the Debye length

$$L_D = \sqrt{\frac{12.9 \times 8.854 \times 10^{-12}\text{F/m} \times 0.025\,\text{V}}{1.6 \times 10^{-19}C \times 10^{16}\,\text{cm}^{-3}}} \approx 42\,\text{nm}. \quad (16.8)$$

For a 1-sided PN junction with the same doping,

$$V_{\text{bi}} = 0.025\,\text{V} \times \ln\frac{10^{16}\,\text{cm}^{-3}}{10^{10}\,\text{cm}^{-3}} = 0.345\,\text{V},$$

$$W_d = \sqrt{\frac{2 \times 12.9 \times 8.854 \times 10^{-12}\text{F/m} \times 0.345\,\text{V}}{1.6 \times 10^{-19}C \times 10^{16}\,\text{cm}^{-3}}} \approx 222\,\text{nm}. \quad (16.9)$$

Compared with the expression for depletion width W_d, we see that the Debye length is analogous, with $V_{\text{bi}} \to k_BT/2q$ and carrier density $N_{A,D} \to \rho_0$. In contrast for strongly degenerate semiconductors, semi-metals or metals where the Fermi energy overlaps with the density of states, we must use the full Fermi–Dirac distribution, whereupon $-\partial f/\partial E \approx \delta(E - E_F)$, and $\delta\rho \approx q\delta\phi D(E_F)$. Compared to the non-degenerate case we see that we need to make the substitution $\rho_0/k_BT \to D(E_F)$, and

$$\boxed{L_D \approx \sqrt{\epsilon_{si}/q^2 D(E_F)} \quad \text{(degenerate semiconductors/metals)}}. \quad (16.10)$$

16.3 Charge Control Model and Shockley Equation

When we forward bias a PN junction, there is a deluge of charges that are pushed across the shrinking depletion width. These carriers cross-over and

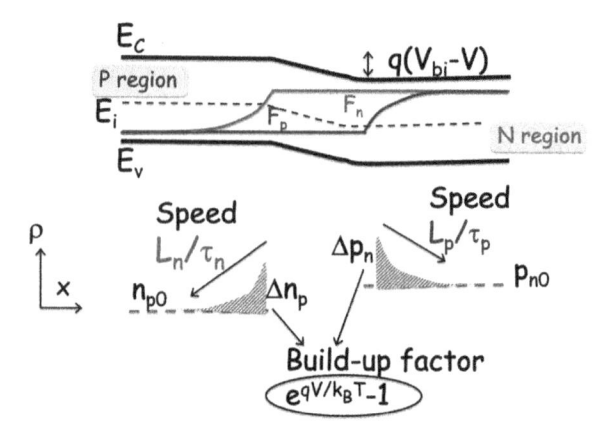

Fig. 16.3 Forward bias reduces depletion width, creating excess minority carrier buildup Δp_n and Δn_p across its ends, diffusing away at speeds $L_{n,p}/\tau_{n,p}$. The bias separated quasi-Fermi levels $F_{n,p}$ are flat in the depletion region and convergent deep in the quasi-neutral regions.

create an excess of minority carriers on the other side that now diffuse slowly through recombination–generation. A quick way to calculate the current is to estimate the excess minority carrier density on each side and multiply by their diffusion speed. Since the bands are shifted by the applied bias V across the entire width of the depletion region, the minority carriers at its edges are inflated by e^{qV/k_BT} over their equilibrium values (this follows from Eq. (15.2) and Fig. 16.3). We assume here low-level injection, meaning the electrostatics primarily affects the minority carrier concentrations. We then get *the law of the junction* — a cool name for a sphaghetti Western!

$$\Delta p_n|_{edge} = p_{n0}(e^{qV/k_BT} - 1), \quad \Delta n_p|_{edge} = n_{p0}(e^{qV/k_BT} - 1). \quad (16.11)$$

The solution to the steady-state minority carrier diffusion equation (Eq. (13.21)) in each quasi-neutral region, example in the n-region

$$\frac{\partial \Delta p_n}{\partial t}^{\,0} = \mathcal{D}_p \frac{\partial^2 \Delta p_n}{\partial x^2} - \frac{\Delta p_n}{\tau_p}$$

$$\Rightarrow \Delta p_n(x) = \Delta p_n|_{edge} e^{-x/L_p}, \quad L_p = \sqrt{\mathcal{D}_p \tau_p}, \quad (16.12)$$

implying that these excess minority holes decay exponentially into the quasi-neutral N region, so that the diffusion currents $-\mathcal{D}_p \partial \Delta p_n/\partial x$ lead to speeds $\mathcal{D}_p/L_p = L_p/\tau_p$. In other words, the minority carrier currents on the two sides are given by

$$I_P|_{n\text{-side}}(x) = qA\,\overbrace{p_{n0}(e^{qV/k_BT} - 1)}^{\text{excess charge }\Delta p_n(x)}\,\overbrace{\frac{L_p}{\tau_p}}^{\text{speed}},$$

$$I_N|_{p\text{-side}}(x') = qAn_{p0}(e^{qV/k_BT} - 1)e^{-x'/L_n}\frac{L_n}{\tau_n}, \qquad (16.13)$$

with x and x' pointing in opposite directions from the edges of the n and p side depletion regions deep into their bulk quasi-neutral regions.

But how do we calculate the total current from here? After all, we only got the minority carrier currents at separate ends of the PN junction. Since Kirchhoff's law stipulates the total current is position-independent, we need both the majority and the minority carrier currents at any one point. It turns out, we do know the majority carriers at the edges of the depletion region $(x, x' = 0)$. Assuming there is no recombination–generation inside the depletion region, the minority carrier on the P side continues over to then become the majority carrier on the n-side. In other words, at the edge of the depletion region on the n-side in addition to the minority carrier $I_P(x = 0)$, we now also have the majority carrier $I_N(x' = 0)$. The result is the Shockley equation for the PN junction current $\boxed{\text{P16.1–P16.4}}$

$$\boxed{I = qA\left(p_{n0}\frac{L_p}{\tau_p} + n_{p0}\frac{L_n}{\tau_n}\right)(e^{qV/k_BT} - 1) = I_{F0}(e^{qV/k_BT} - 1)}.$$

$$(16.14)$$

For reverse bias, we get a constant current $I(V \to -\infty) = -I_{F0}$, because unlike forward bias where the minority charge near the depletion region keeps increasing with voltage, for negative bias the minority carrier density drops to zero, meaning the charge deficit $|\Delta p_n|$ gets pinned to p_{n0} and is voltage-independent. The reverse bias current components are simply given by the equilibrium minority carrier densities and their diffusion velocities from the bulk equilibrium to the fully depleted regions.

16.4 Non-Idealities

Comparing experiments with theory, we see some clear discrepancies, some of which can actually be utilized. For reverse bias, we see a current that does not quite saturate, and in fact diverges at a large negative bias.

For forward bias, the slope of the exponential changes from voltage to voltage. There are several origins of this deviation, bulleted as follows.

- **RG in the depletion region:** To start with, there is RG in the depletion region that was ignored earlier when we flat-lined $F_{n,p}$ in Fig. 16.3. For reverse bias, the mobile charge density drops to zero, but the width of the depletion region increases with reverse bias roughly as its square root, so the current acquires a slope $\propto \sqrt{V}$ instead of saturating.

$$I_{\text{RG}} = qA \int dx \left. \frac{\partial n}{\partial t} \right|_{\text{RG}} = qA \int_{-x_p}^{x_n} dx \frac{np^{\,0} - n_i^2}{\tau_p(n^{\,0} + n_1) + \tau_n(p^{\,0} + p_1)}$$

$$\approx -qA \frac{n_i^2}{\tau_p n_1 + \tau_n p_1} W_d. \tag{16.15}$$

For optimal RG with deep level traps, $n_1 \approx p_1 \approx n_i$, in which case

$$I_{\text{RG}} \approx -qA \frac{n_i W_d}{\tau_p + \tau_n}, \quad W_d \propto \sqrt{V_{bi} + V} \approx \sqrt{V} \tag{16.16}$$

- **Zener tunneling:** Next, we have Zener tunneling whereby the conduction band of the N region slips past the valence band of the P region. This gives us a field-dependent tunneling probabilty across the junction. Zener tunneling happens when $q(V + V_{bi}) > E_G$. For symmetric doping $N_D = N_A = 10^{17} \text{ cm}^{-3}$, we get $V_{bi} \approx 0.8\,\text{V}$, whereupon we only need a reverse bias of $\sim 0.3\,\text{V}$ to get Zener tunneling initiated.
- **Impact ionization:** Finally, we have the onset of impact ionization, whereby under high reverse bias fields, the electrons get accelerated and slam into other atoms, delivering energies that exceed the binding energy of those atoms, liberating further electrons in a cascade process, somewhat like a photomultiplier (but less deterministic). The maximum electric field $|\mathcal{E}_m| \sim V/[W_d/2]$ grows until a critical field $\mathcal{E}_0$, so that the impact ionization starts at a breakdown voltage $V_{\text{BR}} \sim \epsilon_{\text{si}}\mathcal{E}_0^2/2qN_D$. The breakdown field for Si is $\mathcal{E}_0 \sim 3 \times 10^5 \text{ V/cm}$, which give $V_{\text{BR}} \sim 3\,\text{V}$ for a doping $N_D \sim 10^{17} \text{ cm}^{-3}$. We account for impact ionization with a phenomenological carrier multiplication factor M.
- **RG in the depletion region:** For forward bias there are again RG processes in the depletion layer, which contribute even though the depletion width is small, because the minority carrier concentrations np are exponentially higher. Assuming comparable minority carrier lifetimes

$\tau_n = \tau_p \approx \tau_0$, we then get

$$I_{\mathrm{RG}} = qA \int dx \frac{n_i^2 \left(e^{qV/k_B T} - 1\right)}{\tau_0(n + p + 2n_i)}. \tag{16.17}$$

Under non-equilibrium conditions, we have $np = n_i^2 e^{qV/k_B T}$, and we assume the enhancement works on both charge sectors, so that $n \approx p \approx n_i e^{qV/2k_B T}$, whereupon

$$I_{\mathrm{RG}} \approx \frac{qAn_i W_d}{2\tau_0} \left(\frac{e^{qV/k_B T} - 1}{e^{qV/2k_B T} + 1}\right). \tag{16.18}$$

In other words, the current increases with voltage with an exponent $\sim 1/2k_B T$ instead of $1/k_B T$.

- **High-level injection:** At mid-ranged positive voltages the standard Shockley theory works. At higher bias, we initially have *high-level injection*, meaning that the majority and minority carrier concentrations become comparable, each satisfying the law of the junction. This means only half the applied voltage sits across their current path for each carrier, i.e., at the boundary, for example, $\Delta p_n(x = 0) \approx \Delta p_{n0}(e^{qV/2k_B T} - 1)$ and this shows up in the current too as reduced slope.

- **Series resistance:** Finally at high bias, the series resistance starts to matter. The nonlinear I–V means the resistance of the channel suddenly drops and the voltage division starts to have a larger drop across the contacts. We can capture the corresponding stretching out of the I–V by replacing the x-axis, $V \to V - IR$ and replotting the I–V accordingly.

  All these effects can be lumped into an *ideality factor* η, which gives us

$$I_F = I_{F0}(e^{qV/\eta k_B T} - 1) \tag{16.19}$$

with $\eta \sim 1 - 2$.

16.5 RC Model and Small Signal AC Response

For reverse bias, an ideal PN junction diode has a constant saturation current, meaning that the differential impedance $Z = (\partial I/\partial V)^{-1}$ is infinity. The system acts like an open circuit, a voltage-dependent junction capacitor $C_J = \epsilon_{si} A/W_d$ $\boxed{\text{P13.,3}}$. Since we are dealing with ultrafast majority charges that don't cross-over the enlarged depletion region but shuttle between battery and quasi-neutral sections, the response to a small-signal AC perturbation around a reverse bias voltage looks quasi-static.

For forward bias however, the AC response depends on the input frequency compared to the slower minority carrier lifetime of the diffusing charges. Going back to the drift–diffusion equation but using an AC response rather than a steady-state DC response, we now get

$$\underbrace{\frac{\partial \Delta p_n}{\partial t}}_{i\omega \Delta p_n} = \mathcal{D}_P \frac{\partial^2 \Delta p_n}{\partial x^2} - \frac{\Delta p_n}{\tau_p} \implies 0 = \mathcal{D}_P \frac{\partial^2 \Delta p_n}{\partial x^2} - \frac{\Delta p_n}{\tau_p}(1 + i\omega \tau_p),$$

(16.20)

which mimics the DC solution if we replace $\tau_p \to \tau_p/(1 + i\omega \tau_p)$. This means the effective velocity $\sqrt{\mathcal{D}_p/\tau_p}$ and thus the overall current and admittance pick up an extra factor of $\sqrt{1 + i\omega \tau_p}$.

$$Y = G_0 \sqrt{1 + i\omega \tau_p} = G + i\omega C,$$

(16.21)

where $G_0 = 1/R_0$ is the conductance $\partial I/\partial V$ at the forward bias DC operation point. We can now take the real and imaginary parts using phasors, in order to get $G = \mathrm{Re}(G_0 \sqrt{1 + i\omega \tau_p})$ and $\omega C = \mathrm{Im}(G_0 \sqrt{1 + i\omega \tau_p})$. For low frequency $\omega \tau_p \ll 1$, we can use $\sqrt{1 + i\omega \tau_p} \approx 1 + i\omega \tau_p/2$, whereupon $G = G_0$ and $C = C_0 = G_0 \tau_p/2$, i.e., $\tau_p = 2R_0 C$, set by the RC time constant. For large frequency $\omega \tau_p \gg 1$, we can drop the unity and use $\sqrt{i} = \sqrt{e^{i\pi/2}} = (1 + i)/\sqrt{2}$. We then get

$$G(\omega \tau_p \gg 1) \approx G_0 \sqrt{\omega \tau_p/2},$$

$$C(\omega \tau_p \gg 1) \approx C_0 \sqrt{2/\omega \tau_p}.$$

(16.22)

More generally, $G, \omega C = G_0 \left[\sqrt{1 + \omega^2 \tau_p^2} \pm 1\right]^{1/2}/\sqrt{2}$. In other words, we have a leaky capacitance that keeps decreasing with frequency, while the conductance increases due to leakage currents that cannot keep up with the high frequency AC signal.

16.6 Schottky Diodes: Metal–Semiconductor Contacts

For a metal–semiconductor interface, there is a fixed barrier height on the metallic side, set by the difference $\phi_{\mathrm{ms}} = \phi_m - \phi_s$ in metal–semiconductor workfunction (separation between vacuum and Fermi energies far from the junction). The barrier on the semiconductor side is given by the built-in potential for a one-sided PN junction, i.e., $V_{bi} = (k_B T/q) \ln(N_D/n_i)$. The net barrier to flow is thus given by $\phi_{\mathrm{BN}} = \phi_{\mathrm{ms}} - V_{bi}$. When a voltage is applied, the action is limited to the semiconductor side because the screening Debye length L_D and the corresponding depletion width are miniscule on the metallic side with high density of mobile carriers (think of it as a one-sided PN junction). The current from metal to semiconductor side is thus pinned, and the voltage-dependence is entirely from the variation in current from the semiconductor to metal side over the changing barrier $\phi_{BN} - V$. The density of states for the electrons is given by the lumped density of states N_C times the exponential change of barrier with voltage, while the current is given by its product with the average thermal velocity in one direction. The average thermal speed in 1D is given by $v_{th} = \langle |v| \rangle = \int_{-\infty}^{\infty} |v| dv P(v) = \sqrt{2k_B T/\pi m}$ for a normalized 1D Maxwell distribution $P(v) = e^{-mv^2/2k_B T}/\sqrt{2\pi k_B T/m}$. Clearly the speed in 1 direction is the integral above over one half of the velocity distribution $\int_0^{\infty} |v| dv P(v) = v_{th}/2$. Thus,

$$J = qN_C e^{-q\phi_{BN}/k_B T}\left(e^{qV/k_B T} - 1\right) \times \frac{v_{th}}{2}. \qquad (16.23)$$

Using $N_C = 2/\lambda^3 = 2(2\pi m k_B T/h^2)^{3/2}$, we get Richardson's equation

$$\boxed{J_{MS} = \underbrace{\frac{4\pi m q k_B^2}{h^3}}_{A^{**} = 120 A/cm^2 K^2} T^2 e^{-q\phi_{BN}/k_B T}(e^{qV/k_B T} - 1).} \qquad (16.24)$$

16.7 Comparison between MS and PN Junction Diodes

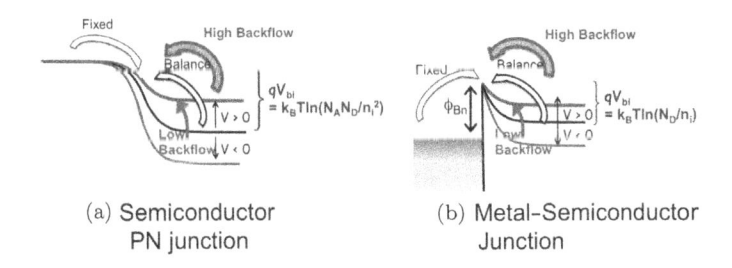

(a) Semiconductor PN junction

(b) Metal-Semiconductor Junction

The results look superficially like PN junctions, except being unipolar majority carrier rectifiers, they depend on free electron mass m jumping above the barrier rather than at the bottom of the band where we would instead have used the effective semiconductor mass m^* (this is why the Richardson constant A^{**} is a universal constant). Also, they do not involve RG or high level injection processes that would have necessitated an ideality factor η — for an MS junction $\eta = 1$. Finally, the effective thermal velocity is a lot faster than the minority carrier diffusion rate determined by the trap density. Thus, the PN junction current is much smaller than that from a metal–semiconductor Schottky diode.

Some numbers: A quick back-of-the-envelope estimate gives us some useful numbers. For a one-sided Silicon ($n_i \sim 10^{10}$ cm^{-3}) PN junction with doping density $N_D \sim 10^{17}$ cm^{-3}, the minority carrier density $n_i^2/N_D \sim 10^3$ cm^{-3}. The carrier lifetime $\tau_p = 1/\sigma v_{\text{th}} N_T$, with trap cross-section set by an average atomic radius $\sim 1 \text{Å}$, thermal velocity $\sim \sqrt{k_B T/m} \sim 10^5$ m/s and trap density $\sim 10^{12}$ cm^{-3} which gives a lifetime $\sim 300 \, \mu$s. Carrier mobility for single crystal silicon is ~ 450 cm^2/Vs for holes, which gives a diffusion constant $D_p = \mu k_B T/q \sim 10$ cm^2/s, so that the diffusion speed $\sqrt{D_p/\tau_p}$ is roughly 200 cm/s. This gives us a reverse bias saturation current density $qnv \sim 3 \times 10^{-14}$ A/cm^2.

In comparison, for an MS junction the speed is the thermal velocity $\sim 10^5$ m/s. The room temperature Debye length is ~ 4–5 nm, so that the lumped electron density of states is $\sim 10^{18}$ cm^{-3}. For Au/Si interface, the barrier height $\phi_{\text{BN}} = 0.8$ eV, which through an added Boltzmann factor $\exp(-q\phi_{\text{BN}}/k_B T)$ reduces the equilibrium carrier density to about $\sim 5 \times 10^4$ cm^{-3}. In other words, the speed is 10,000X higher while the equilibrium carrier density is 50X higher, and thus the reverse bias saturation current, i.e., the Shockley pre-factor, is about **a million times higher** for the Au/Si Schottky diode compared to a moderately doped Si PN junction diode.

If we dope the semiconductor very high, then its depletion width can shrink to the point that electrons can tunnel from the metal contact. The tunneling current can be estimated using the Simmons model (NEMV, Section 14.2), which integrates the WKB-based Landauer current (later chapters) over transverse modes to give a parallel tunneling current over and above the Richardson component

$$\boxed{J_{\text{tunn}} = J_0[\langle\phi\rangle e^{-\mathcal{A}\langle\phi\rangle^{1/2}} - (\langle\phi\rangle + qV)e^{-\mathcal{A}(\langle\phi\rangle + qV)^{1/2}}], \quad J = J_{\text{tunn}} + J_{\text{MS}}},$$

$$(16.25)$$

where $J_0 = q/2\pi h(\Delta s)^2$, Δs is the tunnel barrier thickness (here, the depletion width in the semiconductor), $\mathcal{A} = 2\Delta s\sqrt{2m}/\hbar$, and $\langle\phi\rangle$ is the spatially averaged barrier height. At this point, the current looks Ohmic (linear) rather than activated. In most semiconductor devices, we add highly doped regions right below the metals to make them act as Ohmic contacts $\boxed{\text{P16.5}}$.

Homework

P16.1 **Solar cell: Short-Circuit Current vs Open-Circuit Voltage.** We go back to the DC drift–diffusion equation with a generation term

$$\frac{\partial \Delta p_n}{\partial t} = D_p \frac{\partial^2 \Delta p_n}{\partial x^2} - \frac{\Delta p_n}{\tau_p} + G_L. \qquad (16.26)$$

(a) As we saw earlier P13.6, this changes the boundary condition in the quasi-neutral region, $\Delta p_n(x \to \infty) = G_L \tau_p$ instead of zero, while the enhancement at the depletion region boundary $x = 0$ is still the same, $\propto \exp(qV/k_B T) - 1$, meaning there is now an oppositely directed current corresponding to the reduced charge gradient in x. Find the charge density $\Delta p_n(x)$ and the diffusion current $J_p(x)$, and show that it looks a vertically shifted $I - V$ with a short circuit current $I_{SC} < 0$ at $V = 0$. Show that this current depends on the electron–hole pairs generated within a certain volume (what volume? Remember to add photo-generation inside the depletion width W_d as well, which we are currently not tracking).

(b) We can also calculate the open circuit voltage. Making it open circuit means the electron–hole photo-generated pairs will drift to their respective terminals *down gradient* along the built-in fields, opposite to what they do under bias (hence a negative current). Since these charges cannot escape into the battery, they build up at each end and increase the local electron and hole quasi-Fermi levels at the ends until a reverse diffusion current starts to arise that ultimately cancels the short circuit current. The voltage built-up at this point where the total steady-state current is zero is the open-circuit voltage V_{OC}. Show that

$$V_{OC} = \frac{k_B T}{q} \ln\left(1 + \frac{|I_{SC}|}{I_R}\right). \qquad (16.27)$$

The shaded area between I_{SC} and V_{OC} is the power harnessed by photo-generation, and can be obtained by integrating $\int_0^{V_{OC}} I(V)dV$. The *fill factor* tells us how large this is compared to the maximum power we could get $|I_{SC}|V_{OC}$ if the shaded area was perfectly rectangular and the I–V was infinitely sharp (ideality factor $\eta \to 0$). By simple integration of the exponential, show that

$$FF = \frac{\int_0^{V_{OC}} I(V)dV}{|I|_{SC}V_{OC}} = 1 - \frac{k_BT}{qV_{OC}}\frac{|I_{SC}|}{\left(|I_{SC}| + I_R\right)}. \tag{16.28}$$

P16.2   **Solar cell Shockley–Queisser limit.** Let us estimate the efficiency — $\eta = P_m/P_{in}$, where P_m is the maximum power output from the shifted I–V (largest area rectangle $I \times V$ one can fit into the shaded region). The sun acts as a blackbody source, with its photons following a Bose–Einstein distribution $N_{BE}(E, T_S)$ at temperature $T_S \sim 6000\,\mathrm{K}$.

(a) **Bandgap limits.** The efficiency can be estimated by assuming that out of all incident photons at temperature T_S, all photons at energies less than the bandgap E_G are removed (energy filtering), while all higher energy electron–hole pairs are instantly relaxed by non-radiative recombination to the bandedge E_G (thermalization), generating only a smaller open circuit voltage $V_{OC}^0 = E_G/q$. In other words,

$$\eta(x_g) = \frac{E_G \int_{E_G}^\infty dE D_{\mathrm{ph}}(E)N_{BE}(E, T_S)}{\int_0^\infty dE E D_{\mathrm{ph}}(E)N_{BE}(E, T_S)}. \tag{16.29}$$

Simplify in terms of dimensionless units $x = E/k_BT_S$ and $x_g = E_G/k_BT_S$, using the photon density of states $D_{\mathrm{ph}}(E) = E^2/\pi^2\hbar^3c^3 \propto E^2$ [P11.8 last problem]. Optimizing η with x_g, show that the maximum efficiency at this stage is $\eta = 0.44$ at $x_g = 2$, for $E_G = 1\,\mathrm{eV}$.

(b) **Earth temperature.** So far, we assume the earth is at zero temperature. At finite earth temperature $T_C = 300\,\mathrm{K}$, the open-circuit voltage reduces as part of the photons get re-radiated when the electrons and holes recombine radiatively. Let us go back to the ratio of the short-circuit to saturation current which determines the open-circuit voltage. The additional shrinkage of the current ratio is given by

$$\frac{|I_{SC}|}{I_R} = \frac{E_G \int_{E_G}^\infty dE D_{\mathrm{ph}}(E)N_{BE}(E, T_S)}{E_G \int_{E_G}^\infty dE D_{\mathrm{ph}}(E)N_{BE}(E, T_C)}$$

Reexpress in terms of $x_{g,c} = E_G/k_B T_{S,C}$ and show that for the silicon band-gap E_G, this ratio $\sim 1.4 \times 10^{20}$, sitting inside the logarithm (Eq. (16.27)).

(c) **Angular narrowing.** Finally, there is geometry loss, due to the capture of only a small angular slice of the radiated photons from the sun. To get the loss factor f_S, we account for the fact that a small fraction of the incident photons actually reaches a planar solar cell. This is given by the ratio of the area of the sphere $4\pi L^2$ over which the solar radiation is distributed, to the area of the cross-section πD^2 for the earth. Assuming L is 149 million kilometers and D is 1.39 million, show that $f_S = D^2/4L^2 \sim 2.18 \times 10^{-5}$.

(d) The new open-circuit voltage is then given by $V_{\text{OC}} = (k_B T_c/q)$ $\ln\left(1 + 1.4 \times 10^{20} \times 2.18 \times 10^{-5}\right)$ while the older one was $V_{\text{OC}}^0 = E_G/q$, so that the extra shrinkage in V_{OC} and thus the efficiency is given by the ratio, which for silicon is about 0.8. In other words, the efficiency for a single silicon pn junction (whose efficiency was 0.41, not the optimized 0.44) with unconcentrated light is $0.41 \times 0.8 \approx 0.33$. The following figure shows the efficiency plot, blue from losses in part (a), red including losses in parts (b) and (c). This is the celebrated Shockley–Queisser limit for a single junction solar cell. We can bypass these limits with multiple tandem junctions of varying bandgaps to capture various energy slices of the solar spectrum, concentrated light to avoid narrowing, electron and phonon band-engineering to slow down and capture hot carriers before they thermalize, and upconversion so one photon can generate multiple electron–hole pairs. Each choice though has its own challenges.

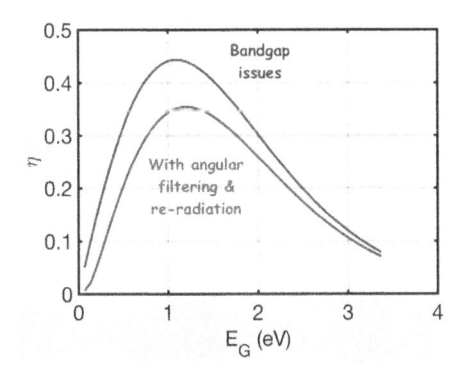

P16.3 **Diode sensor.** Some commercial temperature sensors use the voltage drop across a forward biased diode driven by a constant current source to measure the temperature. Consider an ideal p^+n step junction Si diode with cross-sectional area $A = 10^{-4}\,\text{cm}^2$, $N_D = 10^{16}/\text{cm}^3$ and $\tau_p = 1\,\mu\text{s}$, independent of temperature. Think of all terms that go into the Shockley equation, and their temperature-dependence (which has the highest temperature dependence?). For mobility, use the following empirical equation for mobility in units of cm^2/Vs, doping in units of cm^{-3} and temperature T in Kelvins normalized to $t = T/300$

$$\mu_p(N_D, T) = A_0 t^\alpha + B_0 t^\beta / \left[1 + (N_D t^{-\gamma}/N_0)^\Delta\right], \quad \Delta = C_0 t^\delta, \qquad (16.30)$$

$A_0 = 54.3$, $\alpha = -0.57$, $\beta = -2.23$, $B_0 = 407$, $N_0 = 2.35 \times 10^{17}$, $C_0 = 0.88$, $\gamma = 2.4$, $\delta = -0.146$.

(a) Calculate the voltage as a function of temperature and plot it for a temperature range of $0 < T < 50^\circ\text{C}$. Use a current of 10^{-4} Amps.
(b) Also plot the behavior of the temperature sensor below $100\,\text{K}$. Do you expect any of the physics to change fundamentally?

P16.4 **Small signal.** For the following circuit, find the operating room-T DC voltage V_0 at which the small signal currents in the two parallel vertical branches are equal. The saturation current $I_s = 3 \times 10^{-15}\text{A}$, ideality factor $\eta = 2$, and resistances $R_1 = 1\,\text{k}\Omega$ and $R_2 = 100\,\Omega$. Also calculate the DC currents I_A and I_B and thus the input voltage V_S (ignore the small voltage vs at the operating voltage V_0 that you just calculated).

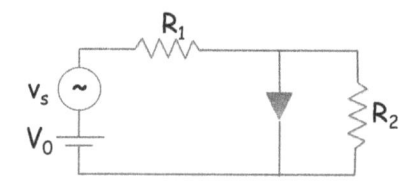

P16.5 **Ohmic contacts.** Calculate the Richardson and Simmons current densities (Eqs. (16.24) and (16.25)) for Au/Si at various dopings above, and show that for $N_D \approx 10^{20}/\text{cm}^3$, tunneling takes over and we get an Ohmic contact. See figure.

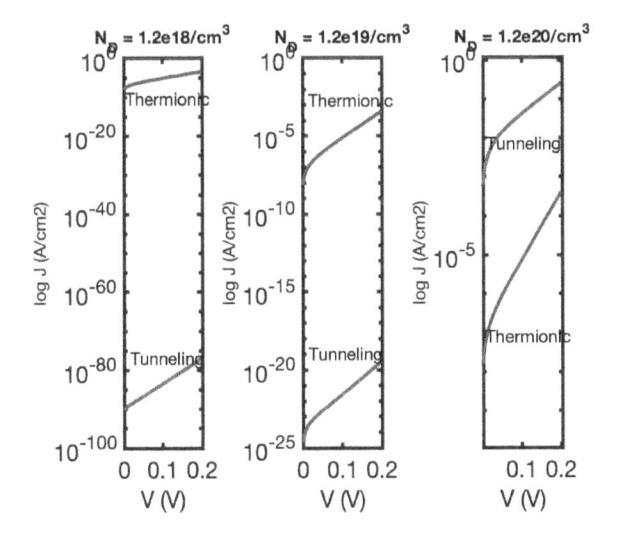

Chapter 17

A Simple Switch: Bipolar Junction Transistor

17.1 Basic Operation and Gain

A bipolar junction transistor (BJT) consists of two back-to-back PN junction diodes, say a PNP structure, with a thin base region (N in this case) that electrically couples the two diodes. The first (emitter-base) junction is forward biased to create a large hole current. It is also asymmetrically doped so that the hole current far exceeds the electron current (in fact, that ratio is the BJT gain). A small amount of holes recombine with electrons in the base, but the rest of the holes diffuse across the thin base to reach the collector, where they are promptly whisked away by the large electric field in the second collector-base reverse-biased PN junction (Fig. 17.1).

The output characteristic of the BJT thus looks like the reverse-bias saturation current for a PN junction. However, the carrier concentration in the collector can be modulated by varying the much smaller electron current in the base, much like a solar cell where injecting charges through photo-induced pair generation increases the saturation current in a reverse-biased PN junction. The BJT thus shows gain if the input were the small base current and the output were the considerable collector current, in other words, in a *common emitter* configuration. The gain is given by the ratio of *collector-to-base currents* P17.1 and P17.2

$$\beta_{\mathrm{DC}} = \frac{I_{Cp}}{I_{Bn}} \gg 1. \tag{17.1}$$

How do we understand the gain? Imagine the doping in the emitter is 1000 times higher than in the base, $N_E/N_B = 1000$. For every 1000 holes entering from emitter to base, 1 electron is back-injected from base to emitter. Assuming the base is very thin, there is very little added recombination and most (say 999) holes reach the collector. If we now inject an extra

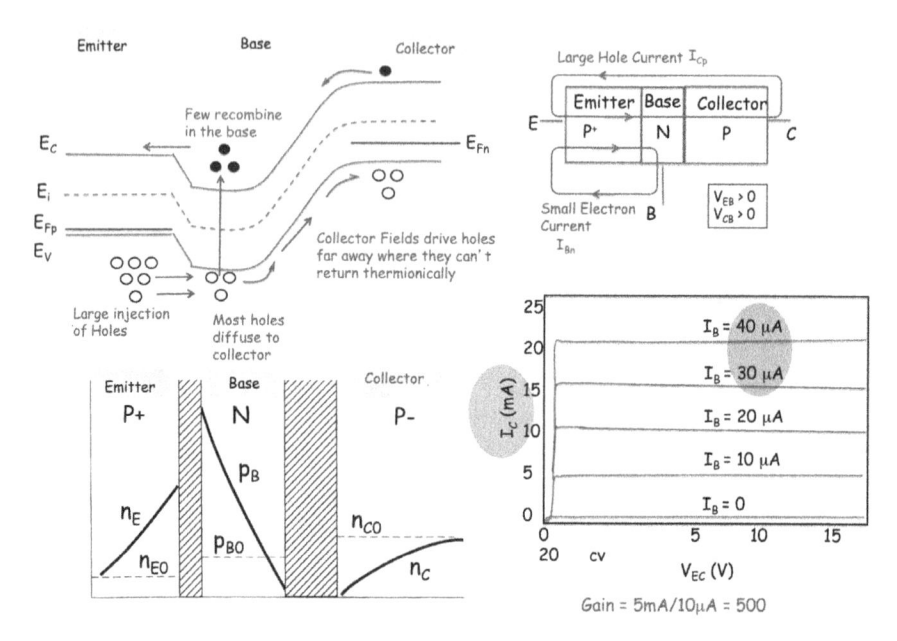

Fig. 17.1   (Top left) Non-equilibrium P^+NP band-diagram and (top right) gain current pathways, (bottom left) charge distributions and (bottom right) transistor characteristics of a P^+NP BJT. A large emitter doping causes a deluge of holes to cross the forward biased P^+N junction and enter a thin base, diffuse through with little recombination with the few base electrons, then get whisked away by the large collector fields in the reversed biased NP junction. In a common emitter configuration where the emitter is grounded, a small recombinent base electron current ($I_{Bn} \sim 10\,\mu A$) at the base input can then control the large collector hole current ($I_{Cp} \sim 5\,mA$) at the collector output, giving a common emitter gain $\beta_{DC} \approx 500$.

electron into the base in common-emitter model, what do we expect to see? Since we are calculating results at a constant output voltage V_{EC}, and emitter voltage is grounded for common emitter, that means the only way to inject more electrons into the base is to reduce the emitter-base barrier by increasing the base quasi-Fermi energy. This will allow more electrons to jump over the barrier by a factor of 2X, giving us the extra electron. But the same barrier reduction also allows 2X increase in holes injected from emitter to collector (an avalanche of sorts since there are many more holes from the higher-doped emitter), which means for each extra electron added in the base, extra 999 holes go to the collector. This is the common-emitter gain, in this case, 999. A small electron current in the emitter-base input circuit controls a large hole current in the emitter-collector circuit (the fabled tail wagging the dog naturally comes to mind).

As we discussed in the previous chapter, we can easily calculate minority carrier currents in the quasi-neutral regions. We can also get the majority carrier currents at the edge of the depletion regions by crossing over to the other side and simply recording the minority carrier current at the boundary, since we assume there is no added recombination–generation in the depletion region itself. Thus, $I_{Ep}|_{\text{EB}} = I_{Bp}|_{\text{EB}}$ at the boundaries of the emitter-base depletion region, and similarly $I_{Bn}|_{\text{CB}} = I_{Cp}|_{\text{CB}}$ at the boundaries of the collector-base depletion region. These currents at the *EB* edge are obtained from the Shockley equation (Eqs. (16.13) and (16.14))

$$I_{Bn} = I_{En} = qAn_{E0}\left(e^{qV_{\text{EB}}/k_BT} - 1\right)\frac{\mathcal{D}_E}{L_E},$$

$$I_{Ep} = I_{Bp} \approx qAp_{B0}\left(e^{qV_{\text{EB}}/k_BT} - e^{qV_{\text{CB}}/k_BT}\right)\frac{\mathcal{D}_B}{W}, \quad (W \ll L_B), \quad (17.2)$$

where for a thin base with a small quasi-neutral region $W \ll L_B$, the minority carrier extraction speed $\mathcal{D}_B/W$ is faster than usual $\mathcal{D}_B/L_B$. A thin base also implies near perfect transmission of holes to the collector, so that $I_{Cp} \approx I_{Ep}$. Assuming forward bias at the *EB* junction, $qV_{\text{EB}}/k_BT \gg 1$ and reverse bias at the *CB* junction $qV_{CB}/k_BT \ll 0$, each current retains just the first exponential, whereupon we get the common-emitter gain

$$\beta_{\text{DC}} = \frac{I_{Cp}}{I_{Bn}} \approx \frac{I_{Ep}}{I_{Bn}} \approx \frac{\overbrace{p_{B0}}^{n_i^2/N_B}}{\underbrace{n_{E0}}_{n_i^2/N_E}}\frac{\mathcal{D}_B/W}{\mathcal{D}_E/L_E} = \frac{N_E\mathcal{D}_BL_E}{N_B\mathcal{D}_EW}. \quad (17.3)$$

In other words, to get a large gain, we need high emitter doping and low base thickness. Note also that we wrote the speed as $\mathcal{D}/L$ instead of L/τ (the difference matters for the narrow width when we replace L with W for the base). This is because shrinking the base would alter the effective τ and L (mean free path and extraction time) but diffusion constant is expected to be unaffected as it is a material property.

It is important to realize however that this is only in the small base width limit, where almost all holes from the emitter reach the collector, making the two junctions strongly coupled. If we increase W, we should get back to the trivial limit of two back-to-back decoupled PN junctions. The base transport factor for holes between emitter to collector will drop exponentially, whereupon the gain goes to zero. Indeed, if we worked through the expression for I_{Ep} away from the thin base approximation, we would need to make the transformation $Ax \rightarrow A \sinh x + \cosh x - 1$

$$\beta_{\mathrm{DC}} = \cfrac{1}{\underbrace{\cfrac{N_B \mathcal{D}_E L_B}{N_E \mathcal{D}_B L_E} \times \cfrac{W}{L_B}}_{\text{narrow base}}} \rightarrow \cfrac{1}{\cfrac{N_B \mathcal{D}_E L_B}{N_E \mathcal{D}_B L_E} \sinh\left(\cfrac{W}{L_B}\right) + \cosh\left(\cfrac{W}{L_B}\right) - 1}$$

(17.4)

so that for $W \ll L_B$ we recover the thin base common-emitter gain, while for $W \gg L_B$ the gain drops exponentially to zero.

17.2   Ebers Moll Model and Small Signal

The BJT can be thought of as two back-to-back diodes plus two current sources, the latter signifying the charges that cross-over without recombining to the other side (the so-called common base gain α_{DC}, i.e., ratio of *collector-to-emitter currents*, always less than unity). The reverse-biased collector keeps siphoning holes toward the battery, which would normally give a very small reverse-bias current (for a single junction, the battery being the only source of charge). But for a BJT, the collector gets a lot of holes coming in from the 'other' side — the input circuit, i.e., the forward biased emitter, as long as the base is thin. This gives the large hole current that is sensitive to the emitter-base voltage and thus the base current.

Kirchhoff's Law suggests $I_E = I_C + I_B$, so that

$$\beta_{\mathrm{DC}} = \frac{I_C}{I_B} = \frac{1}{\alpha_{\mathrm{DC}}^{-1} - 1}, \quad \alpha_{\mathrm{DC}} = \frac{I_C}{I_E} = \frac{1}{\beta_{\mathrm{DC}}^{-1} + 1}. \quad (17.5)$$

For large $W \gg L_B$, α_{DC} also drops to zero as no charges make it to the collector, and the diodes get decoupled. We can use it to conveniently express two-port parameters by eliminating one current and one voltage using Kirchhoff's laws P17.3 .

Figure 17.2 summarizes the circuit representation of two diodes and two current sources parametrized by the common-base gain α_{DC}. It allows us to describe two port parameters (e.g., for common emitter configuration, base-emitter and collector-emitter I–Vs) by eliminating two out of the six voltages and currents using the two Kirchhoff laws.

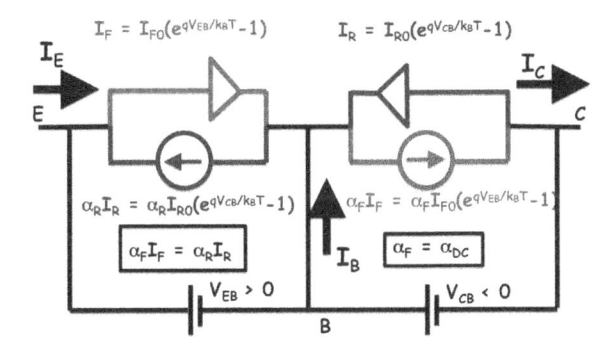

Fig. 17.2 Ebers Moll model: A BJT can be described as two back-to-back diodes plus two current sources arising from charge cross-over. This circuit helps express two-port parameters at input and output out of six currents $I_{E,C,B}$ plus voltages V_{EB}, V_{EC}, V_{CB} using current and voltage Kirchhoff laws.

17.3 Early Effect

One problem with a BJT is the doping in the base. To get a high emitter injection efficiency, we need a low base doping N_B to minimize minority carrier back-flow into the emitter. However, the reverse-biased base-collector end then has a large depletion region entering the base, reducing the quasi-neutral width of the base W and making it dependent on the output voltage $V_{BC} = V_{EC} - V_{EB}$. In the extreme case, the base can become fully depleted (a case called *punchthrough*) whereupon the gain shoots up dramatically with the complete elimination of any recombination in the base. Since the gain β_{DC} increases as W shrinks (Eq. (17.3)), we get an output voltage-dependent gain, i.e., the currents don't saturate. Instead they converge back to a negative voltage and this is called the 'Early effect' $\boxed{\text{P17.4}}$. Non-saturating currents ultimately hurt resistor–transistor logic output by compromising any gain at the circuit level. It is the functional equivalent of 'short-channel effects' that we will see for field effect transistors.

To get the quasi-neutral part of the base, we need to subtract from the metallurgical width of the base the portions of the depletion width on each side, $W = W_B - x_{EB} - x_{BC}$. We can ignore x_{EB} since it is a thin barrier at forward bias. The reverse-biased base-collector part however is big,

$$x_{BC} = \underbrace{\sqrt{\frac{2\epsilon_{Si}(N_B^{-1} + N_C^{-1})(V_{BC}^{bi} + V_{BC})}{q}}}_{W_D} \times \frac{N_C}{N_B + N_C} \approx A\sqrt{V_{BC}} \quad (17.6)$$

since typically $V_{BC} \gg V_{BC}^{bi}$. In other words, the common-emitter gain $\beta_{DC} \propto 1/W \sim 1/(W_B - A\sqrt{V_{BC}})$ increases with output voltage V_{BC}, giving quasi-Ohmic I–Vs.

17.4 Heterojunction Bipolar Transistor

How can we maintain a high gain and yet keep it independent of output voltage? We will need to increase the base doping to shrink the depletion region at the base-collector region. At the same time, we must ensure that the increased number of electrons in the higher doped base do not back inject into the emitter under forward-bias, else the gain would plummet. A heterojunction bipolar transistor (HBT) solves this problem by using two different band-gapped materials at the EB interface, so that the positive electron band-offset prevents the electrons from entering the emitter while the negative hole band-offset increases hole injection into the base. Different bandgaps correspond to different intrinsic carrier concentrations n_{iE}, n_{iB}, whereupon the thin base gain, assuming comparable lumped densities of states for the two materials, is

$$\beta_{DC} = \frac{\overbrace{p_{B0}}^{n_{iB}^2/N_B}}{\underbrace{n_{E0}}_{n_{iE}^2/N_E}} \frac{\mathcal{D}_B/W}{\mathcal{D}_E/L_E} = \frac{N_E \mathcal{D}_B L_E}{N_B \mathcal{D}_E W} \times \underbrace{e(E_G^E - E_G^B)/k_B T}_{n_{iB}^2/n_{iE}^2}. \tag{17.7}$$

In other words, we can increase the base doping N_B to eliminate punchthrough and yet use the bandgap offset to maintain a large gain. The higher base dopings also help to reduce their resistivity and make the HBTs very fast.

The main problem with BJTs is that they rely on current gain, which means there is a constant amount of power dissipation from the currents, especially in resistor-transistor logic (RTL) [P17.5]. In contrast, a field-effect transistor (FET, next chapter) has zero DC current input at its gate. In a CMOS set-up this provides voltage gain without steady-state currents, which makes it very attractive from an energy dissipation point of view and allows large Fanout. The BJT transconductance is $g_m = \partial I_C / \partial V_{BE} = I_C/(k_B T/q)$, compared to an FET with $g_m = \partial I_D / \partial V_G = I_D/[(V_G - V_T)/2]$. The lower denominator in BJTs tends to make them faster, but today's MOSFETs also have $I_D \gg I_C$ due to near ballistic operation, making them much more mainstream.

Homework

P17.1 Current-mirror.

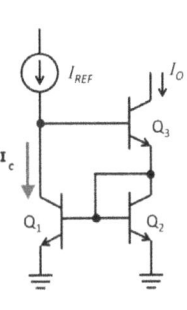

(a) Consider the *Wilson current mirror* circuit here. The two BJTs at the bottom are identical, so their base currents $I_{B1,B2}$ are the same. What are the two identical base currents $I_{B1} = I_{B2}$ for BJTs Q_1 and Q_2 in terms of I_C and β?

(b) Using Kirchhoff's law, what is the current in the collector of Q_3 in terms of I_C and β?

(c) Write I_0 and I_{REF} in terms of I_C and β and find their ratio.

(d) Show that for $\beta = 100$, the two currents $I_{0,REF}$ are very close to each other. In other words, the ratio I_{REF}/I_0 is off from unity by a factor $\approx 1/\beta^2$.

(e) Show that a regular current mirror, one without Q_3, matches I_{REF}/I_0 to a poorer extent, by working out the ratio.

P17.2 Darlington pair.

The picture shows a Darlington pair, which consists of 2 npn BJTs in the active mode. We assume that each has a base emitter voltage drop $V_{BE} = 0.7\,\text{V}$. Find the four currents shown.

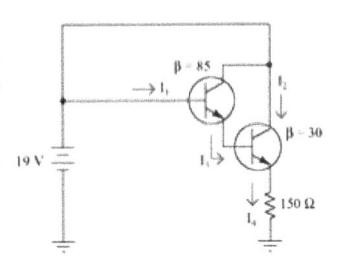

P17.3 Ebers Moll.

Starting with the Ebers Moll model for active mode operation ($V_{EB} > 0$, $V_{CB} < 0$, $e^{qV_{CB}/k_BT} \ll 1$), derive

(a) the common base input characteristic $I_E(V_{EB}, V_{CB})$,
(b) the common base output characteristic $I_C(V_{CB}, I_E)$,
(c) the common emitter input characteristic $I_B(V_{EB}, V_{EC})$,
(d) the common emitter output characteristic $I_C(V_{EC}, I_B)$.

Show all algebraic steps.

P17.4 Early voltage.

Let us get a rough expression for the early voltage, arising primarily out of the dependence of the gain $\partial I_C/\partial I_B$. In our calculations, we will hold V_{BE} constant, so let us assume our common output characteristics are being plotted as I_C vs V_{CB}. The early effect will show up as a slope on this curve, which when extrapolated to negative V_{CB} should converge to a point.

We start with the common emitter output characteristic $I_C = gI_B + I_{C0}$.

(1) We had derived a simple expression for the common emitter gain g in class, assuming $W \ll L_B$. Let W_B be the physical width of the base, while x_{CB} is the depletion width at the reverse-biased collector-base junction. Write down an expression for W in terms of x_{CB} and W_B to leading order in $x_{CB} \ll W_B$.

(2) We now need to examine the V_{CB} dependence of the depletion width and the resulting base width modulation. Let V_{CB}^{bi} be the built-in voltage at the collector-base junction. Since the early slope turns on right above the saturation point near $k_B T/q$, we will assume the collector-based applied bias $V_{CB} \ll V_{CB}^{bi}$ to extract the slope.

Write down x_{CB} in terms of V_{CB}^{bi}, x_0 (the equilibrium depletion width for V_{CB}^{bi}) and V_{CB}, to leading order in V_{CB}.

(3) We will ignore the depletion at the emitter-base x_{EB}. Why is this justified?

(4) Write down the gain g in terms of g_0 (the gain for $x_{CB} = 0$), x_0, W_B, V_{CB} and V_{CB}^{bi}.

(5) Ignoring the I_{C0} term, find the early voltage $V_{CB} = V_A$ where $I_C = 0$. Prove that all the current curves converge to this same point.

P17.5 **Resistor-transistor-logic.** To get a logic gate out of a BJT, such as an inverter, we can use resistor–transistor logic. This is similar to CMOS in Chapter 1, except the BJT ON/OFF resistance is compared against a fixed resistance, as opposed to complementary resistances that make CMOS always off at rest.

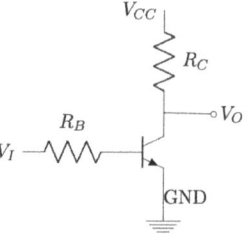

A BJT in common emitter mode has an input voltage and output voltage shown as follows. The cut-off voltage is 0.7 V, while the saturation voltage is 0.2 V. The common emitter gain is 100. The input resistance RB is 50 KΩ while the output resistance R_C is 1 KΩ. $V_{CC} = 5.2$ V.

Plot the expected V_o-V_{in} graph as quantitatively as possible based on the above information, including slopes, transition voltages, input output values, etc.

Chapter 18

Voltage Gating: The Field Effect Transistor

18.1 Basic Operation and Threshold for Inversion

For an N-field effect transistor (FET), we start with a P-type semiconductor channel between N^+ contacts. A positive gate would attract electrons into the channel by lowering the potential, shorting the channel and turning on its conductivity. Applying a drain bias would now allow the shorted channel to drive a current. The current increases with drain bias because the charges in the inverted channel flowing back from the drain to the source are decreased by reducing the relative voltage between gate and the drain end of the channel (the latter increasing with drain bias). Beyond a certain drain bias, this backflow shrinks to zero (a condition called 'pinch-off'), whereupon the only current from the source to drain saturates (Fig. 18.1).

The Fermi energy of the P-type material is at an energy $\psi_B = (k_B T/q) \ln (N_D/n_i)$ above the intrinsic energy E_i, signaling the extent of the doping. To invert the channel with a gate voltage and turn it n-type, we need enough band-bending so that at the channel surface, we have a strongly n-type channel, compensating for the p-type bulk material. In fact, we need E_F to be as far below E_i as it started above, to give a strong inversion. This means the local surface potential ψ_S needs to match $2\psi_B$. The corresponding depletion width also grows as the bending increases, till it reaches a value of $W_{\text{dm}} = \sqrt{2\epsilon_{si}(2\psi_B)/qN_A}$. The actual gate voltage applied to achieve inversion is the transistor threshold voltage V_T, and can

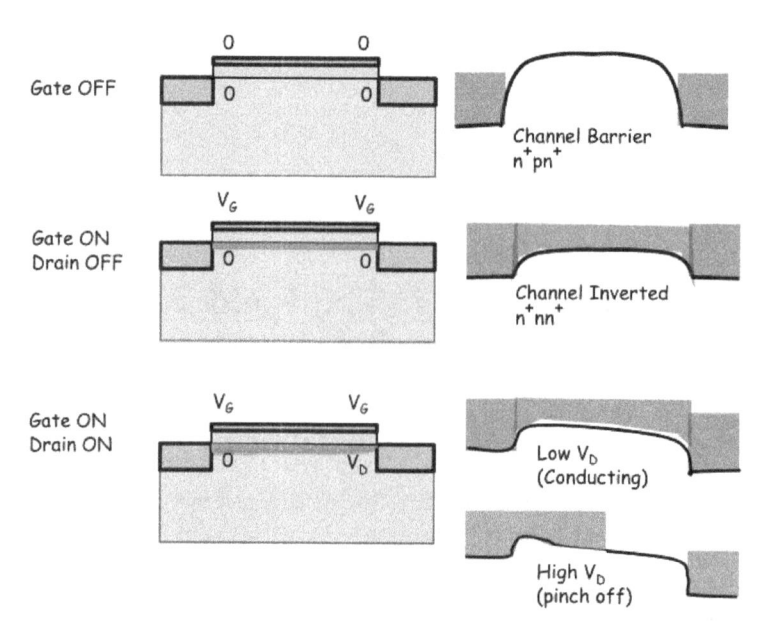

Fig. 18.1 Off state n-FET has a n^+pn^+ source-channel-drain region with a large barrier that is lowered with a positive gate bias that inverts the channel with dominant n-type carriers. Under drain bias, a current starts flowing, until the drain end gets uninverted (zero back-flow) and the current saturates with bias.

be related to $2\psi_B$ using a simple voltage division across the oxide capacitor

$$V_T = \underbrace{\overbrace{\frac{Q_s}{C_{ox}}}^{\text{C/cm}^2}}_{\psi_{ox}} + \underbrace{2\psi_B}_{\psi_S}, \quad Q_s = qN_AW_{dm}, \quad C_{ox} = \frac{\epsilon_{ox}}{t_{ox}}. \qquad (18.1)$$

As expected, the two voltages are related by the electrostatic boundary condition at the interface on continuity of electric displacement field at zero interfacial charge

$$\frac{\epsilon_{ox}\psi_{ox}}{t_{ox}} = \frac{\epsilon_{Si}\psi_S}{W_{dm}/2}, \qquad (18.2)$$

where the field across the inversion layer corresponds to a thickness of $W_{dm}/2$ because, as we will see shortly, the potential varies quadratically with depth $\boxed{\text{P18.1}}$.

Some numbers: For a p-type Si with doping $10^{17}\,\text{cm}^{-3}$ and an SiO_2 oxide of thickness $10\,\text{nm}$, assuming flatband initial conditions, the built-in potential, depletion width and threshold voltage

$$\psi_B = \underbrace{\frac{k_B T}{q}}_{0.025\,\text{V}} \ln \left(\frac{\overbrace{N_A}^{10^{17}\,\text{cm}^{-3}}}{\underbrace{n_i}_{10^{10}\,\text{cm}^{-3}}} \right) \approx 0.403\,\text{V}$$

$$W_{dm} = \sqrt{\frac{2\epsilon_{si}(2\psi_B)}{qN_A}} = \sqrt{\frac{2 \times 12.9 \times 8.854 \times 10^{-12}\,\text{F/m} \times 2 \times 0.403\,\text{V}}{1.6 \times 10^{-19}C \times 10^{17}\,\text{cm}^{-3}}}$$

$$\approx 107\,\text{nm}$$

$$V_T = \underbrace{2\psi_B}_{0.608\,\text{V}} + \frac{\overbrace{1.6 \times 10^{-19}C \times 10^{17}\,\text{cm}^{-3} \times 107\,\text{nm} \times 10\,\text{nm}}^{qN_A W_{dm} t_{ox}}}{\underbrace{3.9 \times 8.854 \times 10^{-12}\,\text{F/m}}_{\epsilon_{ox}}}$$

$$\approx 1.102\,\text{V} \tag{18.3}$$

After inversion, there will be a discontinuity in the displacement field as both the inversion charge and the corresponding V_{ox} grow roughly proportionally, much like a parallel plate capacitor. Any electrostatics inside the original depletion layer is now screened out by the inversion charges, and the layer stops growing beyond W_{dm} (it actually grows a bit — logarithmically with V_G).

18.2 Electrostatics Near Threshold

Let us spend a bit of time discussing the approach to inversion and its aftermath. The best way to do so is coupling Poisson–Boltzmann equations. The fixed and mobile charges in the semiconductor and on the electrodes determine the electron bands through Poisson's equation, while the distorted bands in turn determine the charges through thermodynamics, electrons sinking like stones to the band bottoms and holes floating like bubbles the other way. In the p-doped region for instance, we can look at the balance between majority holes (p_p), minority electrons (n_p) as well as the background donors (N_D, few in this case) and acceptors (N_A, the dominant source of charge). Poisson's equation for the potential, or equivalent Gauss's Law for the fields, gives us

$$\frac{\partial \mathcal{E}}{\partial x} = \frac{q}{\epsilon_{si}} \overbrace{\left(p_p + N_D - n_p - N_A \right)}^{\rho}. \tag{18.4}$$

By bending the bands down by ψ we increase the electron density to $n_{p0} \exp\left(q\psi/k_B T\right)$ and reduce the hole density to $p_{p0} \exp\left(-q\psi/k_B T\right)$ exponentially, according to Eq. (12.5). The rest of the terms in ρ conspire together to make sure that at $\psi = 0$, there is no net charge, i.e., $N_D + p_{p0} = N_A + n_{p0}$. Furthermore, we can write $\partial \mathcal{E}/\partial x = (\partial \mathcal{E}/\partial \psi) \times (\partial \psi/\partial x)$, and then replace the second factor with $-\mathcal{E}$. This gives us the field

$$\underbrace{\frac{\mathcal{E} \partial \mathcal{E}}{d\psi}}_{= \partial(\mathcal{E}^2)/2\partial\psi} = \frac{q p_{p0}}{\epsilon_{si}} \underbrace{\left[\left(1 - e^{-q\psi/k_B T} \right) + \frac{n_{p0}}{p_{p0}} \left(e^{q\psi/k_B T} - 1 \right) \right]}_{\Phi(\psi, n_{p0}/p_{p0})}, \tag{18.5}$$

integrating which between $x = 0$ ($\psi = \psi_s$, $\mathcal{E} = \mathcal{E}_S$) and ∞ ($\psi = \mathcal{E} = 0$), we get the surface electric field. This field $\mathcal{E}_S$ in turn determines the areal surface charge density Q_S (Coulombs/sq cm) from Gauss's law, applied to a parallel plate capacitor

$$Q_S = \epsilon_{si} \mathcal{E}_S = \pm \epsilon_{si} \sqrt{\frac{2 q p_{p0}}{\epsilon_{si}} \int_0^{\psi_s} \Phi(\psi, n_{p0}/p_{p0}) d\psi}$$

$$= \pm \sqrt{2} \epsilon_{si} \underbrace{\frac{(k_B T/q)}{L_D}}_{V/m} \left[\left(\underbrace{e^{-q\psi_s/k_B T} - 1 + \frac{q\psi_s}{k_B T}}_{\text{accumulation}} \right. \right. \underbrace{}_{\text{depletion}} \tag{18.6}$$

$$\left. \left. + \frac{n_{p0}}{p_{p0}} \underbrace{\left(e^{q\psi_s/k_B T} - 1 - \frac{q\psi_s}{k_B T} \right)}_{\text{inversion}} \right]^{1/2}.$$

We can see that for negative gate bias, we are in the accumulation regime where $|Q_S|$ increases exponentially as $\sim \exp\left(-q\psi_s/2k_B T\right)$. For depletion, the charge grows slower, $\sim \sqrt{\psi_s}$. Finally, since $\psi_B = (k_B T/q) \ln\left(N_A/n_i\right)$, we can write the ratio n_{p0}/p_{p0} as $n_i^2/N_A^2 = e^{-2q\psi_B/k_B T}$. For $\psi_S > 2\psi_B$, the inversion term dominates other terms and grows as $\sim e^{q(\psi_S - 2\psi_B)/2k_B T}$.

A lot of physics is hidden in the above equations $\boxed{\text{P18.1}}$. For instance, concentrating on just depletion, we can show that $\psi(x)$ varies quadratically until the band-bending vanishes at $x = W_d$. The surface electric field $\mathcal{E}_S = |-\partial\psi/\partial x|_{x=0} = \psi_S/(W_d/2)$, as we assumed earlier. Note that this is still approximate, since we dropped a few terms from F above. A similar analysis can give us insights into the inversion layer, including its depth $\boxed{\text{P18.2}}$. To understand the dynamics around inversion, we can look at the voltage division at the channel ψ_S and the oxide ψ_{ox} as a function of V_G (Fig. 18.2). We can see that below threshold, much of the initial voltage drop happens in the channel (blue) with minimal across the oxide (red). After inversion, the free charges at the surface screen out the dynamics inside the channel, pinning its potential close to the inversion threshold of $2\psi_B$, growing only logarithmically thereafter. At this point onwards, the bulk of $V_G - V_T$ drops across the oxide, following

$$\boxed{Q_{inv} \approx C_{ox}\left(V_G - V_T\right)}. \tag{18.7}$$

This equation follows by integrating Q_S over x from 0 to infinity. While the depth integrated areal density of inversion charge Q_{inv} increases exponentially with $\exp[q(\psi_S - 2\psi_B)/2k_B T]$, ψ_S increases over $2\psi_B$ only logarithmically with $V_G - V_T$, and the exponential of a log gives us a linear variation.

We also plotted here the effective width of the total charge (Fig. 18.2), by taking the ratio $W_{eff} = \psi_S/\mathcal{E}_S$. We see that as V_G grows, this width increases until $\sim W_{dm}/2$ (the half because of quadratic variation of $\psi(x)$), and then decreases rapidly to t_{inv}. The semiconductor capacitance ϵ_{si}/W_{eff} shows a corresponding non-monotonic behavior, decreasing until we reach inversion and then growing back as the 'action' moves from deep inside the depletion layer back toward the edge of the oxide at the inversion layer.

In an MOS capacitor (sans source-drain contacts), this non-monotonic capacitance–voltage $(C-V)$ curve (Fig. 18.2, $C \propto$ inverse electrostatic thickness) may not appear in a high frequency scan, as the inversion charges can only come from the slowly moving minority carriers from the semiconductor bulk. In a MOSFET, however, they jump straight out of the highly doped contact majority carriers, and the non-monotonic $C-V$ is easy to measure.

$$V_G = \psi_s + (\varepsilon_{si}t_{ox}/\varepsilon_{ox})\sqrt{(2k_B TN_A/\varepsilon_{si})}[\beta\psi_s + e^{\beta(\psi_s-2\psi_B)}]^{1/2}$$

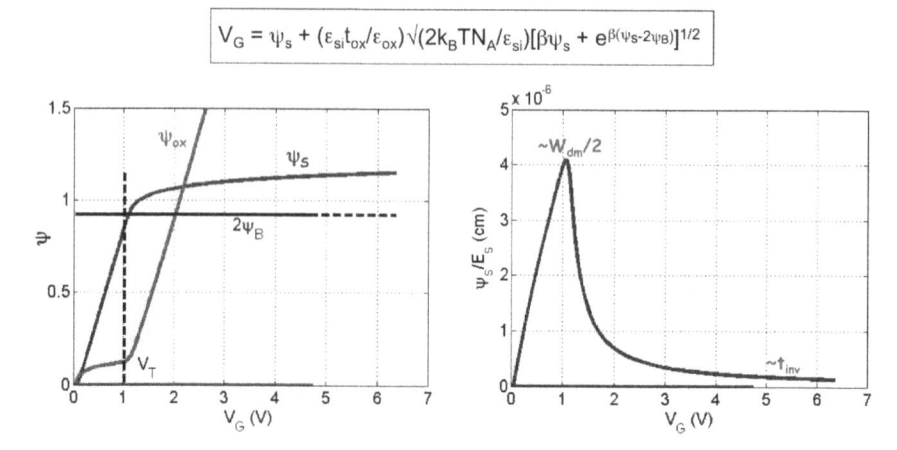

Fig. 18.2 FET surface potential ψ_S grows initially with gate bias until inversion at V_T, following which the drop moves primarily to the oxide while the former saturates. The electrostatic thickness $\psi_S/\mathcal{E}_S$ in the semiconductor increases to a maximum depletion width at inversion, after which action moves to a rapidly thinning inversion layer.

$$N_A = 1.67 \times 10^{15}$$

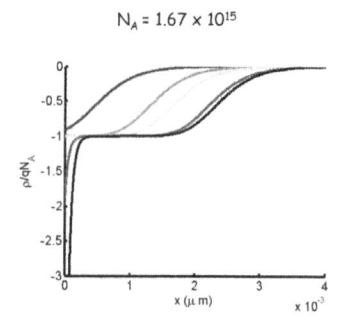

Fig. 18.3 Charge density across the FET shows increasing depletion with gate bias, until an inversion layer opens at the surface like a sinkhole.

We can plot the entire charge distribution, including depletion and inversion charges, by numerical integration.

$$\rho = qn_i[e^{U_B}(e^{-U} - 1) - e^{-U_B}(e^U - 1)],$$

$$\int_U^{U_S} dU'[F(U', U_B)]^{-1} = \pm\frac{x}{L_D},$$

$$F = e^{U_B}(e^{-U} - 1 + U) + e^{-U_B}(e^U - 1 - U), \qquad (18.8)$$

where $U = q\psi/k_B T$ and $U_B = q\psi_B/k_B T$. Figure 18.3 shows the charge density profile (vary U, solve for x and ρ and plot) for various gate voltages, i.e., surface potentials U_S. We can see how the depleted charge gets pushed out further and further until an abrupt inversion channel opens at the surface like a sinkhole, and the depletion width stops growing.

18.3 Square Law Theory

The charge density above threshold varies linearly with gate overdrive, while the velocity depends on the electric field which in turn varies linearly with voltage. Thus, the saturation current is a quadratic function of gate voltage — the square law theory for a traditional MOSFET $\boxed{\text{P18.3}}$.

Let us first discuss the physics of current saturation. We can calculate the current density by finding the charge density $\rho = q n_{\text{inv}} = Q_{\text{inv}}/t_{\text{inv}}$, multiply by drift velocity $\mu_{\text{eff}}\mathcal{E}$ and multiply by the cross-sectional area $A = W t_{\text{inv}}$, which gives us

$$I_D = W Q_{\text{inv}} \mu_{\text{eff}} \mathcal{E} = W C_{\text{ox}} \left(V_G - V_T - V(y) \right) \mu_{\text{eff}} \frac{\partial V}{\partial y}. \tag{18.9}$$

To enforce Kirchhoff's law of current conservation at steady-state, we integrate both sides over y from 0 to L and use the fact that I_D is independent of y and can be pulled out of the integral. The right-hand side simplifies to an exact differential in V which can be integrated to give us a quadratic term. This then gives us the current I_D which looks like an inverse parabola when plotted vs V_D, with a peak at $V_{DSat} = V_G - V_T$. However, the reduction in current along the second half of the inverted parabola is unphysical, and arises because we are assuming the inversion channel persists beyond the maximum point when in reality it does not. The way around it is to then assume that once we reach the maximum current, the latter continues unchanged thereafter as a saturation current. In other words,

$$I_D = \frac{C_{\text{ox}} W \mu_{\text{eff}}}{L} \left[(V_G - V_T) V_D - \frac{V_D^2}{2} \right], \quad V_D < V_{DSat} = V_G - V_T$$

$$= \frac{C_{\text{ox}} W \mu_{\text{eff}}}{2L} (V_G - V_T)^2, \quad V_D > V_{DSat} = V_G - V_T. \tag{18.10}$$

Some numbers: For a $1\,\mu\text{m} \times 1\,\mu\text{m}$ Si FET with oxide thickness $10\,\text{nm}$ and $V_T = 1\,\text{V}$, hole mobility is roughly $450\,\text{cm}^2/\text{Vs}$. The saturation current at $V_G = 2\,\text{V}$

$$I_{\text{sat}} = \frac{\dfrac{3.9 \times 8.854 \times 10^{-12}\,\text{F/m}}{10\,\text{nm}} \times 1\,\mu\text{m} \times 450\,\text{cm}^2/\text{Vs}}{2 \times 1\,\mu\text{m}}$$
$$\times (2\,\text{V} - 1\,\text{V})^2 \approx 156\,\mu\text{A} \qquad (18.11)$$

18.4 Small Signal Characteristics and Inverter Gain

Keeping the CMOS structure in mind, the signal input is at a gate while the output is at the drain. In DC, there is infinite current gain because the gate leaks no current and has infinite resistance. However, as we crank up the AC frequency, the impedance $Z = (\partial I / \partial V)^{-1} = (i\omega \partial Q / \partial V)^{-1} = 1/i\omega C$ of the gate capacitor decreases, and the current gain, needed to drive one stage by a previous one without any degradation, also decreases accordingly. The maximum frequency where the current gain is still greater than unity is called the *cut-off frequency*. The following figure shows a circuit describing

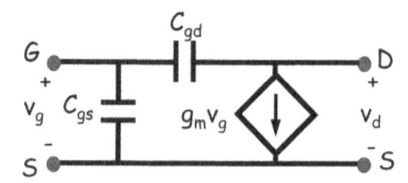

the MOSFET capacitive gate input and resistive drain output. The conductance $g_D = \partial I_D / \partial V_D$ and the transconductance $g_m = \partial I_D / \partial V_G$ can be readily computed by linearizing the transistor characteristics Eq. (18.10). For the linear regime, we get an Ohmic conductance $g_D = \sigma A / L$, where $A = W t_{\text{inv}}$, and the conductivity $\sigma = q n_{\text{inv}} v_d$ with $q n_{\text{inv}} = Q_{\text{inv}} / t_{\text{inv}}$. In the saturation regime, $g_D = 0$, and we get infinite output resistance. From the figure, we find the gain

$$\text{Gain} = \left| \frac{I_{\text{out}}}{I_{\text{in}}} \right| = \left| \frac{\overbrace{\dfrac{W \mu_{\text{eff}} C_{\text{ox}} V_D}{L}}^{g_m}}{i\omega \underbrace{\left(C_{gs} + C_{gd} \right)}_{= C_{\text{ox}} W L} v_g} v_g \right| = \frac{\mu_{\text{eff}} V_D}{2\pi f L^2}. \qquad (18.12)$$

Setting gain to unity sets the cut-off frequency for unity current gain

$$f_T = \frac{\mu_{\text{eff}} V_D}{2\pi L^2}, \quad \omega_{\max} = 2\pi f_{\max} = \frac{\overbrace{\mu_{\text{eff}} V_D/L}^{\mu_{\text{eff}} \mathcal{E} = v_d}}{L}, \tag{18.13}$$

which as we see is set by the *transit time* L/v_d of the electron.

There is also a cut-off frequency $f_{\max}$ for unity *power gain*, where we include not just the voltage ratio but the current (and thus impedance ratio). Which one we care about depends on the application. For current amplifiers, we need a high f_T, while for oscillators where one stage needs to drive the next, we typically need a high $f_{\max}$.

Some numbers: For a 50 nm $\times 1\,\mu$m p-MOSFET with 1 nm oxide at 1 V, the velocity is $v_d = 450\,\text{cm}^2/Vs \times \dfrac{1\,\text{V}}{50\,\text{nm}} \approx 900,000\,\text{m/s}$ (ignoring velocity saturation from electron–phonon coupling), so that the transit time $L/v_d \approx 5.5 \times 10^{-14}$ s and the cut-off frequency is 100 THz. This is way more than the clock frequency at a few GHz. The dissipated power $C_{\text{ox}} W L V^2 f = \dfrac{3.9 \times 8.854 \times 10^{-12}\,\text{F/m} \times 50\,\text{nm} \times 1\,\mu\text{m}}{1\,\text{nm}} \times 1\,\text{V}^2 \times 100\,\text{THz} = 0.17\,\text{W}$ at a power density of $0.17/WL = 345\,\text{MW/cm}^2$, way over any thermal removal capacity. If instead we choose to run it at a clock frequency $f = 1\,\text{GHz}$ and only activate 10% gates at a time, the power density is $34.5\,\text{KW/cm}^2$, which is still too high (a rocket nozzle)!

18.5 Short Channel Effects and MOSFET Scaling

As the channel length degrades, the pinch-off point advances closer to the top of the barrier that controls the charge injection from the source end (Fig. 18.1). At one point, the drain bias makes incursions into that region and the barrier height, set roughly by the potentials at the various terminals, ceases to be completely dominated by the gate alone and starts listening to the drain, shifting down with it. This will immediately increase the charge injection from the source, and add a slope to the current, like the early effect in BJTs (Section 17.3). We can estimate the voltage at the top of the barrier as $V_b \approx \alpha_S V_S + \alpha_D V_D + \alpha_G V_G$, where the transfer factors $\alpha_{S,D,G} = C_{S,D,G}/C_E$ are set by the source, drain and gate capacitances sitting in parallel, and $C_E = C_S + C_D + C_G$. In Eq. (18.9), we can now use

$Q_{\text{inv}} = C_{\text{ox}}(V_b - V_T)$, and the electric field $\mathcal{E} = -\partial V_b/\partial x$,

$$I_D = \frac{C_{\text{ox}} W \mu_{\text{eff}}}{2L}(V_G - V_T)^2(1 + \lambda V_D). \tag{18.14}$$

The factor λ is called the body-coefficient, describing the competition between multiple contacts over control of charge in the MOSFET channel. We can get it automatically if we modified Eq. (18.10) by replacing $C_{\text{ox}}(V_G - V_T - V(y)) \to C_{\text{ox}}(V_G - V_T) - (C_{\text{ox}} + C_{\text{dep}})V(y)$, where the depletion capacitance $C_{\text{dep}} = \epsilon_{\text{si}}/W_{\text{dm}}$. This extra term comes when we replace the surface potential $\psi_S = 2\psi_B \to \psi_S + V(y)$ in the equation for the gate induced threshold voltage Eq. (18.1), sitting both in the equation for the inversion voltage, and the oxide voltage drop involving W_{dm}. Linearizing the corresponding I–V, we get the equation above with the coefficient $\lambda = C_{\text{dep}}/(C_{\text{ox}} + C_{\text{dep}})$. The body effect arises when the drain bias counters inversion locally around pinch-off, unscreening the depletion region around the drain that starts to grow again with voltage above W_{dm}. When C_{dep} is a substantial fraction of C_{ox}, such as in short channels, then we get the phenomena of drain induced barrier lowering (DIBL), where the top-of-the barrier is no longer held fixed by a dominant gate capacitance. DIBL compromises saturation in I–V_D, because a fixed barrier would keep the inversion charge injected above the barrier from the source fixed, while progressively reducing the charge backinjected from the drain to zero. As we will see shortly, the CMOS gain is also compromised as a result. This also impacts the I–V_G by making the threshold voltage V_T dependent on V_D. A good transistor should have a DIBL in the range of $\sim 1\,\text{mV/V}$.

Experimental data on charge control in a MOSFET has led to an empirical relation on the minimum length L_{min} for long-channel behavior $\boxed{\text{P18.4 and P18.5}}$ in terms of r_j the junction depth, t_{ox} is the oxide thickness, and $W_{S,D}$ the source and drain depletion widths. We can get a sense for these factors by looking at charge control in the geometry above (Fig. 18.4). With a little bit of algebra, we can calculate the shift in threshold voltage

$$\Delta V_T = -\underbrace{\frac{qN_A W_{\text{dm}}}{C_{\text{ox}}}}_{\psi_{\text{ox}} = V_T - 2\psi_B} \times \frac{r_j}{L} \times \left(\sqrt{1 + \frac{2W_{\text{dm}}}{r_j}} - 1\right). \tag{18.15}$$

This means we reduce short-channel effects with smaller doping, higher gate capacitance (thinner oxide) and smaller junction depth. The latter,

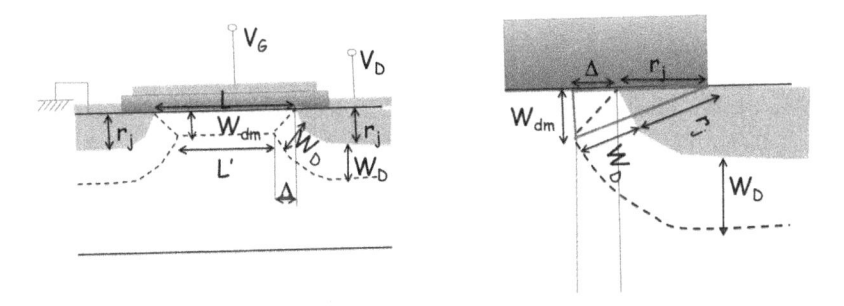

Fig. 18.4 (Left) Short channel MOSFET showing charge controlled by gate vs source/drain. (Right) Blow-up diagram of edge of control zone.

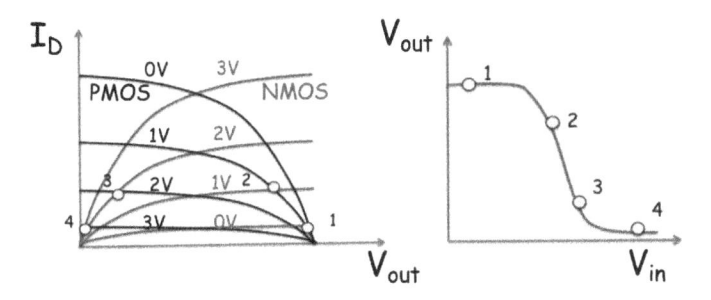

Fig. 18.5 Overlaid output characteristics of NMOS and PMOS. Their intersecting voltage values V_{out} for complementary V_{in} give us the gain curve.

however, increases contact resistance, leading to the concept of raised source/drain, sticking outside the planar channel.

Finally, let us take a quick look at why a saturating current with high output resistance ($g_D \approx 0$) is needed for good CMOS gain $\boxed{P18.6}$. In the CMOS structure, a high gate bias V_{in} turns on the NMOS and turns off the PMOS. Since the two split the power supply voltage V_{DD} (Fig. 1.2), the NMOS characteristic is obtained by setting $V_G = V_{in}$ and $V_D = V_{out}$ in Eq. (18.10), while the PMOS has $V_G = V_{DD} - V_{in}$, $V_D = V_{DD} - V_{out}$. This means we need to flip the PMOS voltage axis and shift it up by V_{DD}, given the overlapping characteristics shown in Fig. 18.5. As the two need to share the same current by Kirchhoff's law, the intersection points give us the resulting output voltage V_{out}. Plotting V_{in} vs V_{out} gives us the gain curve. It is straightforward to see that when a short channel effect gives a body coefficient with a non-saturating I–V, the gain curve becomes less steep. The slope is given by $|\partial V_{out}/\partial V_{in}| = |\partial I_D/\partial V_G|/|\partial I_D/\partial V_D| = |g_m/g_D|$, so that a non-zero output conductance $g_D \propto \lambda$ compromises the CMOS gain.

18.6 Modern MOSFETs: FinFETs, High-k, Strained Si

Good electrostatics thus needs the gate to control the channel charge over its competing electrodes. (i) High-k oxide technology, introduced in the mid-2000s, allows us to maintain a high gate capacitance ϵ_{ox}/t_{ox} by having a high dielectric constant ϵ_{ox} without thinning t_{ox} to below the tunneling width $\boxed{\text{P16.5}}$. (ii) Non-planar FinFETs since $\sim$2010 allow multiple gates around the structure $\boxed{\text{P18.7}}$, giving once again a higher capacitance with better electrostatic control. (iii) Finally, biaxially strained silicon channels, introduced in the early 2000s by alloying silicon with germanium (larger lattice constant), split the silicon hole bands, pushing up the light hole band by about $\sim$67 meV for 10% Ge content, leading to higher mobility just above threshold. We can see this from the tight binding sp^3s^* Hamiltonian (Eq. (8.20)), where the entries can be related to head-on (σ) and edge-on (π) bonds using the decomposition in $\boxed{\text{P8.11}}$, namely,

$$V_{xx} = l^2(V_{\sigma\sigma} - V_{\pi\pi}) + V_{\pi\pi}, \quad V_{xy} = lm(V_{\sigma\sigma} - V_{\pi\pi}) \qquad (18.16)$$

and similarly V_{yy} involving m^2, $V_{yz} \propto mn$, etc., where the diamond cubic direction cosines $(l, m, n) = (1 + \epsilon, 1 + \epsilon, 1 - 2\epsilon)/\sqrt{3}$ in the presence of small biaxial strain ϵ in the x–y plane and uniaxial compression along z (the trace, representing volume, stays constant). For silicon, $V_{\sigma\sigma} = 4.47$ eV and $V_{\pi\pi} = -1.12$ eV. In presence of a compression along the z-axis, like flattening a tent, the tetragonally placed p_z orbitals rotate and become more π-like (edge-on), reducing overlap $|V_{zz}|$, the correponding bonding-antibonding split and pushing up the light hole band. Simultaneously, the heavy hole V_{xx}, V_{yy} terms grow more σ-like (head on and stronger), increasing overlap and pushing down the heavy hole band $\boxed{\text{P18.8}}$.

Some futuristic MOSFET concepts try to reduce the threshold voltage V_T and thus the dynamic power dissipation $\propto V_T^2$ below their fundamental thermal limit. In subthreshold, $V_G < V_T$, any charge in the channel arises out of thermal Boltzmann excitation over the barrier, $I_D \propto Q_{inv} \propto e^{q\psi_S/k_BT}$. This gives a subthreshold slope in an I–V_G semi-log plot as $S^{-1} = \partial \log_{10} I_D/\partial V_G$, with the Boltzmann mandated minimum voltage swing $S = k_BT \ln 10/\alpha_G q \approx 57$ meV/decade $\boxed{\text{P18.9}}$ for perfect gate control $\alpha_G = \partial\psi_S/\partial V_G = 1$. Reducing S below this limit to get a steeper gate transfer curve is a 'holy grail' in device technology, and requires innovative ideas such as transmission engineering (see NEMV, Chapter 29 on tunnel transistors, mechanical relays, Mott and Klein-FETs) and internal voltage amplification ($\alpha_G > 1$) in negative capacitance devices. Most of these ideas, however, come with trade-offs, and are far from mature technology.

With the slow-down of hardware scaling due to thermal costs, and speed-up of software from machine learning, deep neural nets and AI, we are now in an era of *tech inversion*, replacing hardware driven software of yore (Boolean computing in Von Neumann architecture with separate memory and logic cores, built on CMOS devices) with software driven hardware — accelerators like graphics processing units (GPUs) and application specific integrated circuits (ASICs), edge-of-the-node sensing and in-memory computing. This opens up the field for new devices and materials whose underlying physics directly maps onto a desired computational paradigm. In fact, there is a concerted push for merging front-end energy efficient analog pre-computes on emerging technology with backend digital silicon CMOS for higher accuracy. Time will tell how these trade-offs work out.

Homework

P18.1 **Depletion width.** A lot of physics is hidden in the above equations. For instance, concentrating on just the depletion term, but integrating between x and ∞, show that

$$\mathcal{E}(x) = -\partial\psi/\partial x \approx \sqrt{2k_B T\psi(x)/q}/L_D. \tag{18.17}$$

Multiplying both sides by $\psi^{-1/2}$ and integrating from x to 0 (where $\psi = \psi_S$), show that we have a quadratic variation of ψ with x. Write down the expression for $\psi(x)$. At what distance does $\psi(x)$ vanish?

P18.2 **Inversion layer thickness.** Retaining the inversion part of the integral of F and solving for the first-order differential equation in ψ, show that the charge in the inversion layer varies with depth as a Lorenzian, with the inversion layer thickness set by the Debye length and the amount of overdrive beyond inversion

$$\rho_{\text{inv}}(x) \sim 1/(x + x_0)^2, \quad x_0 \approx L_D\sqrt{2}e^{-q(\psi_S - 2\psi_B)/2k_B T}. \tag{18.18}$$

P18.3 **A traditional MOSFET.** From the following table, calculate (a) the ON-OFF current ratio; (b) The ON current and operating voltage gives an equivalent ON resistance R_{ON}, along with parasitic source drain resistance. Calculate the ratio of source/drain resistance R_{SD} to the total resistance (R_{SD} plus R_{ON}). (c) Similarly, in addition to the ideal gate capacitance, there is a parasitic capacitance that adds in parallel to it. Find the percentage of the total gate capacitance that is attributable to parasitic capacitance. (d) Calculate the estimated inversion layer density per cm^2 under ON-current conditions. Remember that both drain and gate biases are set by the power supply voltage. Ignore any potential variations along the channel. (Keep in mind that you need an oxide capacitance per unit area, which is different from the gate capacitances per unit length provided, as that also includes other capacitances in series such as from the depletion layer. But the equivalent oxide thickness should help you get the capacitance you need). (e) There is an additional voltage drop inside the contacts with net resistance $R_{\text{SD}}/2$. What is the inversion charge density if you include it? (f) From the charge density calculated, estimate the velocity at the beginning of the channel under ON-current conditions. (g) Assume the MOSFET voltage scales by a factor $1/k$ and its dimensions scale by $1/\lambda$. Assuming normal MOSFET operation and scaling, what factors do the energy and power dissipated scale by?

Physical gate length	14 nm
Equivalent oxide thickness	0.5 nm
Maximum gate leakage current density	1.43e3 A/cm^2
Power supply voltage	0.9 V
Saturation threshold voltage	112 mV
ON-current	1.762 mA/μm
OFF-current	1.37 μA/μm
Parasitic source-drain resistance	180 Ω-μm
Intrinsic gate capacitance	6.33e-16 F/μm
Extended planar bulk capacitance	7.93e-16 F/μm
Intrinsic device delay	0.4 ps
Channel doping	8.4e18 cm^{-3}
Drain extension junction depth	7 nm
Drain extension junction sheet resistance	1160 Ω/sq
Lateral abruptness of drain extenstion	1.4 nm/decade
Contact junction depth	15.4 nm
Sidewall spacer thickness	15.4 nm
Metal-semiconductor contact resistance	5.6e-8 Ω-cm^2
Contact silicide sheet resistance	17.3 Ω/sq
Shallow trench isolation depth	316 nm

P18.4 **Short channel threshold roll-off.** Refer to Fig. 18.4. The oxide part of the threshold voltage in the MOSFET can be written as total charge controlled by it divided by total capacitance, in other words,

$$\psi_S = V_T - 2\psi_B = qN_AW_{dm}ZL/C_{ox}ZL, \tag{18.19}$$

where Z is the channel width and L the channel length. In long channel devices, Z and L cancel out. For the short channel, the gate only controls charge in the smaller trapezoidal region with sides L and L'.

(a) Show that as a result, $\Delta V_T = -\dfrac{Q_{ox}}{C_{ox}}\left(1 - \dfrac{L+L'}{2L}\right)$, $Q_{ox} = qN_AW_{dm}$

for a p-channel MOSFET.

(b) Use Pythagoras's theorem across the red triangle, assuming $W_D \approx W_{dm}$, to calculate $\Delta = (L - L')/2$, derive Eq. (18.15).

P18.5 **Rounded-junctions.** Consider the cylindrical portion of the p^+n step junction shown. The green hatched area is the depletion region, which lies almost entirely in the weaker doped n side, as expected. In the figure, r_j is the junction depth (radius) and r_k the radius to the n-edge of the depletion region. Assume dopings N_D and N_A, with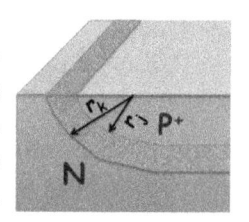
$N_A \gg N_D$. You will first work out the 1D electrostatics of the pn junction, except Gauss's law needs to be solved along the cylindrical coordinate r. Noting that $\vec{\nabla} \cdot \vec{\mathcal{E}} = d[r\mathcal{E}(r)]/rdr$,

(a) obtain a general expression for $\mathcal{E}(r)$ across the entire p^+n junction, from $r < r_j$ to $r > r_k$. Plot it along r.
(b) Obtain a general expression for $V(r)$ and work out the built-in voltage V_{bi} in terms of the dopings, r_j and r_k.
(c) Let us assume conditions such that $r_k = 2r_j$. Find the ratio of the maximum electric field in the cylindrical region (extracting it from the first part of the problem) to the maximum field in the planar region (e.g., proceeding vertically down along the p^+n interface), in other words, $(\mathcal{E}_{\mathrm{max|cyl}}/\mathcal{E}_{\mathrm{max|planar}})$, corresponding to the same built-in voltage V_{bi} in each region. [Note: The depletion widths will be different — the picture gives a pretty good depiction of what to expect.]

P18.6 **Current saturation and voltage gain.** For a CMOS circuit (Fig. 1.2) with supply voltage V_{DD}, the nMOS input $V_{\mathrm{in}} = V_G$ and output $V_{\mathrm{out}} = V_D$. For the pMOS in series, what are its input–output voltages?

We equate the currents $I_{n\mathrm{MOS}} = I_{p\mathrm{MOS}}$ as the transistors are in series. Linearize the I–Vs as $I \approx g_m V_G + V_D/Z$, in terms of the respective transconductances g_m and output resistances Z. Find the inverter gain and show that a strong gain needs good gate control with high transconductance g_m and a saturating current with large output resistance Z.

Assume a transistor with short channel effects, with body coefficient in terms of capacitances (Eq. (18.14)). Find the condition for current saturation and explain why short channels tend to show poorer gain.

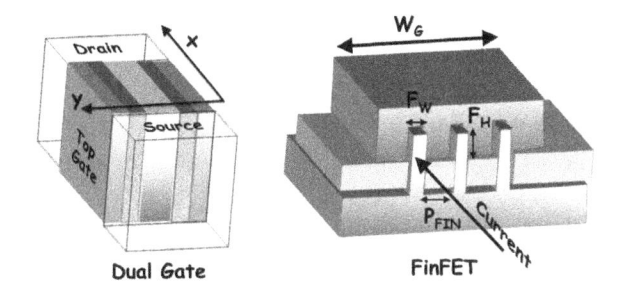

Dual Gate   FinFET

P18.7 **FinFET electrostatics.** Let us look at a dual gate device with gates on both sides of an Si channel of width t_{si} along the y-direction, oxides of width t_{ox} and dielectric constant ϵ_{ox} and transport along x.

(a) Write down the 2D Poisson equation in the x–y plane, for charge density N_{ch} and channel dielectric constant ϵ_{si}.

(b) We assume the confined channel potential behaves parabolically with y, $V(x,y) = a_0(x) + a_1(x)y + a_2(x)y^2$. Assume the boundary condition at the top and bottom are $V(x, y = 0) = V(x, y = t_{\text{si}}) = V_{\text{si}}(x)$. Argue that the transverse electric fields at the two interfaces are $\mathcal{E}_y(x, 0) = -\mathcal{E}_y(x, t_{\text{si}}) = -\left(C_{\text{ox}}/C_{\text{si}}\right)\left[V_G - V_{\text{si}}(x)\right]/t_{\text{si}}$.

(c) Solve for $a_{0,1,2}(x)$ and show that the dual gate Poisson looks like the Debye equation with screening length $\lambda_{\text{sc}} = \sqrt{(\epsilon_{\text{si}}/2\epsilon_{\text{ox}})t_{\text{ox}}t_{\text{si}}}$. Gate control is considered superior when the controlled channel length $L_G \sim (5-10)\lambda_{\text{sc}}$.

(d) Show that the length of the channel controlled by the gate $L_G = W_G(2F_H + F_W)/P_{\text{FIN}}$, where W_G is the planar gate length, P_{FIN} is the fin pitch, $F_{H,W}$ are the height and width, respectively, of the fins.

P18.8 **Strained Silicon.** From the simplified sp^3 Si Hamiltonian (Eq. (8.20)), by varying strain up to $\pm 30\%$, show that V_{zz} controls the light hole band and V_{xx}, V_{yy}, the heavy hole bands. At the Γ-point, show that the valence band eigenvalues $E = E_p - |V_{ii}|$, $i = x, y, z$. Show that in the presence of uniaxial compressive (biaxial tensile) strain $\epsilon = (d_{\text{Ge}} - d_{\text{Si}})/d_{\text{Si}}$ from a SiGe contact, which is the fractional difference in lattice constant (Eq. (18.16)), this strain pushes up the light hole band and pushes down the heavy-hole band.

P18.9 **Subthreshold swing.** The gate threshold voltage sets the minimum dissipated energy CV^2. We will now show that the minimum voltage in turn is set by thermal limits. Above threshold, the inversion charge grows linearly with gate bias as the voltage drops primarily across the oxide, but below threshold the gate action goes into creating an exponentially growing inversion charge, as seen in Fig. 18.2 and Eq. (18.6) though the term $\exp\left[q(\psi_S - 2\psi_B)/k_BT\right] \propto \exp\left[q\alpha_G V_G/k_BT\right]$, where $\alpha_G = C_G/C$ is the gate transfer factor (relevant for MOSFETs with three contacts, not for MOS capacitors). Show that the subthreshold current also varies exponentially with a swing $(d\log_{10} I/dV_G)^{-1} = k_BT\ln 10/\alpha_G q$, which for perfect gate control gives a minimum gate voltage of $V_{\min} \sim 57\,\text{mV/decade}$ change in subthreshold current. Note that we get the same result from Eq. (1.6) by assuming an error rate of $p_{\text{err}} = 10^{-1}$ for each decade.

Part 3

Bottom–Up: Quantum Transport

Chapter 19

Landauer Theory for Quantum Transport

For classical devices, the drift–diffusion equation states $\vec{J} = \sigma \vec{\nabla} \mu / q$. This equation gives us standard Ohm's law with a bulk resistance $R = L/\sigma A$, the conductivity set by density of states per unit volume and diffusion coefficient, $\sigma = q^2 \langle\langle \bar{D}\mathcal{D} \rangle\rangle$, averaged around the Fermi energy. We will now see that a full quantum treatment treats the channel as a waveguide, with resistance and conductivity set by number of modes M, transmission probability $\mathcal{T}$ and scattering length $\lambda_{\rm sc}$. Specifically, $R_Q = h/q^2 M$ and $\sigma = q^2 \langle\langle M\lambda_{\rm sc}/hA \rangle\rangle$.

19.1 Ballistic Flow

Let us start with ballistic current flow through a channel containing M modes or subbands, much fewer than the number M_c in the contacts. We expect to see a 'squeezing' resistance at the contact-channel interface as electrons from M_c modes enter M modes, like traffic from a four lane highway entering a single lane. The probability of entering M modes is $\sim M/(M_c + M)$, while the number of incoming modes is M_c, so the current is proportional to $MM_c/(M + M_c) \propto M$ when $M_c \gg M$. This will yield a quantum of resistance $R_Q \propto 1/M$ at the contact-channel interface.

To estimate R_Q, note that in the absence of scattering in the channel, all right moving electrons arose from the left contact. The right half of its $E - k$ (positive k states) must therefore share the left contact electrochemical potential and be occupied till μ_1. Similarly, the left moving states must arise from the right contact and be occupied up to μ_2. The variation of μ across the entire structure is shown in Fig. 19.1. Note that there is a split between μ of right and left moving electrons in the contacts as well (same current), but given the large number of modes there, plus inelastic

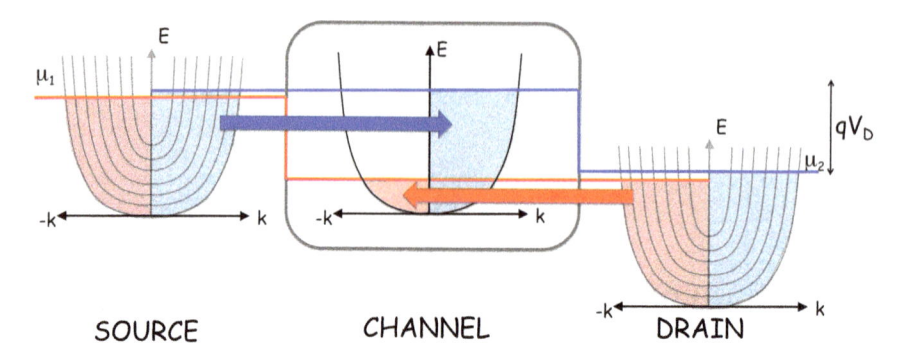

Fig. 19.1 Right moving electrons in a ballistic channel populate the $+k$ (blue) states up to energy μ_1 set by the source Fermi function, while left moving $-k$ electrons (red) originate in the right electrode and fill upto μ_2.

scattering that reorganizes electrons between modes, each contact has a nearly fixed μ and is locally in quasi-equilibrium (recall the merged $F_{n,p}$ deep in the quasi-neutral regions in Fig. 16.3, except the changes here are abrupt).

For ballistic channels therefore, we have a special case of non-equilibrium where the states divide into two lanes that are separately in equilibrium with one contact each. The current is carried by the difference between right and left moving states in the channel. For low bias, we can write

$$I = q \underbrace{\Delta n}_{D(\mu_1 - \mu_2)} / \underbrace{\tau}_{L/v}$$

$$= \frac{q}{h} \underbrace{\frac{hvD}{L}}_{M} \underbrace{(\mu_1 - \mu_2)}_{qV_D} \qquad \Longrightarrow \qquad \boxed{R_Q = \frac{V_D}{I} = \frac{h}{q^2 M}}. \qquad (19.1)$$

The quantity $M = hvD/L$ is the number of modes in the channel. For a 1D subband, recall we had $D = L/\pi\hbar v$, so that $M = 2$ for each single subband, corresponding to two spins. We get the celebrated quantum of resistance $R_Q = h/2q^2 \approx 12.9$ KΩ for a single subband and a corresponding quantized conductance $G_Q = 2q^2/h \approx 77$ μS, where 1 Siemen S is 1 A/V.

19.2 With Momentum Scattering

Let us repeat our calculations including a momentum scatterer in the middle (Fig. 19.2). This scatterer will redirect some of the right moving electrons from the left contact back toward it, thereby increasing the

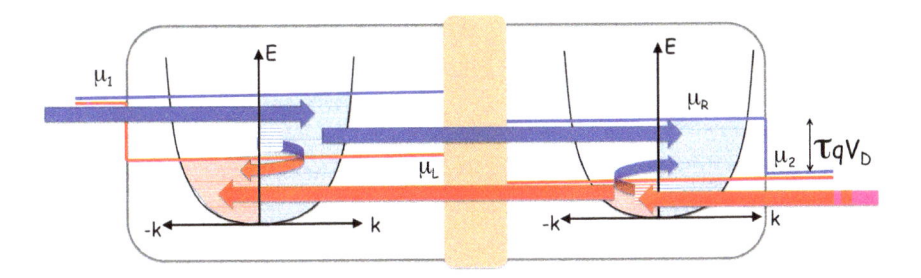

Fig. 19.2 A scatterer redistributes electrons with probability $1 - \mathcal{T}$.

population and thus the electrochemical potential of the left moving electrons at the left end of the device, from μ_2 to μ_L. This also decreases the population of right moving electrons and thus their electrochemical potential at the right end of the device, from μ_1 to μ_R. The reduction in each case is proportional to the reflection coefficient $1 - \mathcal{T}$ and generates an internal channel resistance manifested through the local drop in μ. In fact, we can write down two equations connecting them — at the left end of the device, the left moving electrons upto energy μ_L came from left contact electrons at μ_1 reflecting with probability $1 - \mathcal{T}$, or from right contact electrons at μ_2 transmitting with probability $\mathcal{T}$. A similar equation can be written for right moving electrons at the right end of the device. In other words,

$$\mu_L = \mu_1(1 - \mathcal{T}) + \mu_2\mathcal{T} \implies \mu_1 - \mu_L = \mathcal{T}(\mu_1 - \mu_2),$$
$$\mu_R = \mu_1\mathcal{T} + \mu_2(1 - \mathcal{T}) \implies \mu_R - \mu_2 = \mathcal{T}(\mu_1 - \mu_2). \quad (19.2)$$

The current can be evaluated at either the left or the right end

$$I = \frac{q}{h}M(\mu_1 - \mu_L) = \frac{q}{h}M(\mu_R - \mu_2)$$

$$= \frac{q}{h}M\mathcal{T}\underbrace{(\mu_1 - \mu_2)}_{qV_D} \implies \boxed{R = \frac{V_D}{I} = \frac{h}{q^2 M\mathcal{T}}}. \quad (19.3)$$

At finite temperature and voltage, the window of electron transport expands

$$\boxed{I = \frac{q}{h}\int dE M(E)\mathcal{T}(E)\Big(f_1(E) - f_2(E)\Big)} \quad \text{(Landauer equation)}$$

$$(19.4)$$

with $f_{1,2}(E) = 1/[1 + \exp{(E - \mu_{1,2})/k_B T}]$ smearing the Fermi functions over an energy range $\sim k_B T \approx 25$ meV around $\mu_{1,2}$ at room temperature.

In the literature, the Landauer equation historically involved the transmission summed over all the modes $\boxed{\text{P19.1–P19.4}}$. Alternately, we can write it as the number of modes times the *average* transmission per mode

$$\sum_i T_i(E) = T(E)M(E). \tag{19.5}$$

For notational simplicity, we will instead refer to the product TM as an energy-resolved conductance following Datta

$$I = \frac{1}{q} \int_{-\infty}^{\infty} dE\, G(E)\Big[f_1(E) - f_2(E)\Big]$$

$$\approx V_D \underbrace{\int_{-\infty}^{\infty} dE\, G(E)\left(-\frac{\partial f}{\partial E}\right)}_{G}, \quad \text{(linear response } qV_D \ll k_B T\text{)}, \tag{19.6}$$

where $G(E) = q^2 M(E) T(E)/h$.

19.3   From Conductance to Conductivity

We will shortly see that for dirty samples, $T \approx \lambda_{\text{sc}}/L$, where λ_{sc} is the momentum scattering mean-free path. For low bias at low temperature, we get from the above equation

$$G = \frac{\partial I}{\partial V_D} = \frac{q^2}{h} \int dE\, M(E) \underbrace{\frac{\lambda_{\text{sc}}(E)}{L}}_{T(E)} \left(-\frac{\partial f}{\partial E}\right) = \sigma \frac{A}{L}, \tag{19.7}$$

where the conductivity

$$\boxed{\sigma = \frac{q^2}{h} \int dE\left(-\frac{\partial f}{\partial E}\right) \frac{M(E)}{A} \lambda_{\text{sc}}(E) = q^2 \left\langle\!\!\left\langle \frac{M}{hA}\lambda_{\text{sc}} \right\rangle\!\!\right\rangle}. \tag{19.8}$$

This equation is the equivalent of the earlier one, $\sigma = q^2\langle\!\langle \bar{D}\mathcal{D}\rangle\!\rangle$. We can make the connection once we realize that the number of modes relates to density of states and the mean-free path relates to diffusion coefficient

$$\frac{M}{A} = \frac{h\langle|v_z|\rangle D}{2LA} = \frac{h\langle|v_z|\rangle \bar{D}}{2}, \quad \mathcal{D} = \langle v_z^2 \tau\rangle, \quad \lambda_{\text{sc}} = \frac{\langle v_z^2 \tau\rangle}{2\langle|v_z|\rangle}. \tag{19.9}$$

It is worth emphasizing that these equations ignore quantum tunneling in $\mathcal{T}$, which show up in dirty amorphous semiconductors as Variable Range Hopping, and dirty graphene sheets as quantized conductivity (NEMV, Sections 13.1 and 16.2). When tunneling matters, we must return to basics on calculating transmission. Back to basics is what we do in the next chapter.

Homework

P19.1 Carnot efficiency.

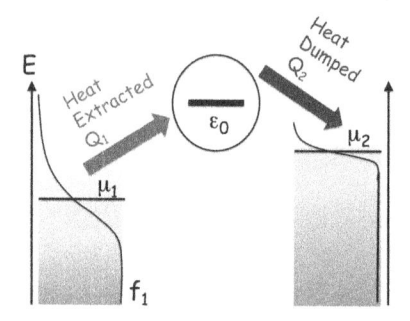

The diagram shows an engine with a high energy level sandwiched between two contacts maintained at two different temperatures $(T_1 > T_2)$. This causes electron flow from hotter to colder contacts, and a voltage build-up in an open circuit configuration. Electrons absorb heat Q_1 from the hot contact to get promoted to ϵ_0, and dump energy Q_2 to the colder contacts while escaping. The rest of the energy builds up voltage that can be harnessed later as useful work.

(a) Electrons come in at μ_1 and leave at μ_2. Write down $Q_{1,2}$.

(b) Condition of zero current at ϵ_0 imposes an equation between Fermi–Dirac functions. Show that this implies $Q_1/T_1 = Q_2/T_2$, i.e. entropy equality (assume no leakage currents or dissipation in the channel).

(c) The work done by the engine $W = qV_{OC} = \mu_2 - \mu_1$. Show that this equals the difference in heat.

(d) Find the thermodynamic efficiency $\eta_{\text{Carnot}} = W/Q_1$. This is the maximum theoretical efficiency of the engine.

P19.2 Linear response coefficients.

(a) Linearizing the Landauer equation for small voltages and temperatures, we get four coefficients Eq. (12.14), G, G_S, G_Q and G_K. Relate the measured response functions — resistance R, Seebeck coefficient S, Peltier coefficient Π and thermal conductance κ — to the four coefficients above.

(b) Write down the four linearized and the four response coefficients in terms of derivatives of Landauer equation. In particular, assuming ballistic flow $\mathcal{T} = 1$, show that

$$S = \frac{\pi^2 k_B^2 T}{3q} \frac{d\ln\left[M(\epsilon)\right]}{d\epsilon}\bigg|_{E_F} \quad \text{(Mott formula)},$$

$$\kappa = \frac{\pi^2 k_B^2 T}{3h} \quad \text{(thermal conductance quantum).} \quad (19.10)$$

Show that for a symmetric mode distribution around the Fermi energy, $S = 0$. For the first part, you may find the Sommerfield expression useful for any well-behaved function $H(E)$ (try your hand at proving it!)

$$\int_{-\infty}^{\infty} H(\epsilon)f(\epsilon)d\epsilon \approx \int_{-\infty}^{E_F} H(\epsilon)d\epsilon + \frac{\pi^2}{6}(k_BT)^2 H'(E_F) + \cdots, \quad (19.11)$$

where $'$ denotes a derivative.

(c) Repeat with κ for phonons, where we use the Bose–Einstein distribution N instead of the Fermi–Dirac equation f.

(d) For a dirty sample where $\mathcal{T} \approx \lambda_{\text{sc}}/L$, rederive these coefficients in terms of modes per unit area M/A and scattering length λ_{sc}.

P19.3 **Thermoelectric ZT.**

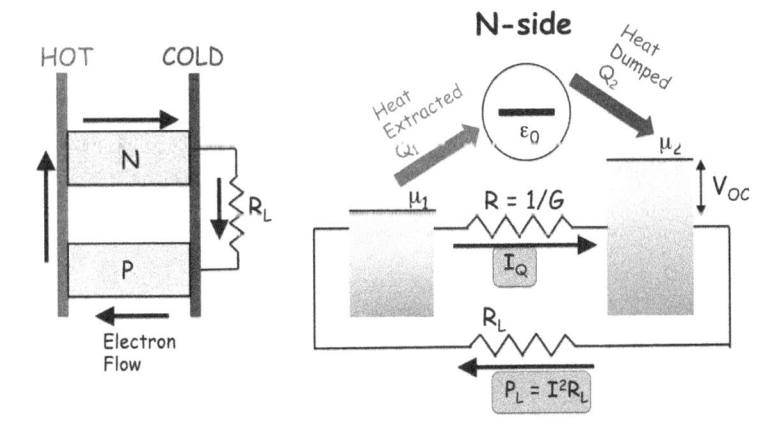

(a) Let us connect this open circuit device of voltage V_{OC} and resistance $R = 1/G$ to drive an external load R_L. The load current is

$I_L = V_{OC}/(R+R_L)$, and dissipated power across the load $P_L = I_L^2 R_L$. Maximizing P_L as a function of R_L, we find the best power transfer to the load will happen when the load resistance equals the channel resistance ($R_L = R = 1/G$). Under that condition with both matched resistors, the maximum power transferred

$$P_{max} = V_{OC}^2 R_L/(R+R_L)^2 = G V_{OC}^2/4 \quad \text{(for } R_L = R = 1/G\text{)}. \quad (19.12)$$

The input power is the heat current, defined as

$$P_{in} = \kappa \Delta T, \quad (19.13)$$

where κ is the thermal conductance (just like charge current is charge per unit time, heat current is heat energy per unit time and energy/time is power).

Show that the thermoelectric efficiency can be written as

$$\eta = \frac{P_{max}}{P_{in}} = \frac{I_L^2 R_L}{\kappa \Delta T} = \frac{1}{4} \underbrace{\left(\frac{S^2 G \bar{T}}{\kappa} \right)}_{Z\bar{T}} \underbrace{\left(\frac{\Delta T}{\bar{T}} \right)}_{\eta_{Carnot}}, \quad (19.14)$$

where $\bar{T} = (T_1 + T_2)/2$ is the average contact temperature.

(b) Let us modify the denominator — i.e., heat absorbed at the hot end, to include Peltier heat ΠI_L and subtract power back-delivered to the load at the hot contact $-I_L^2 R_L/2$

$$\eta = \frac{I_L^2 R_L}{\underbrace{\kappa \Delta T + \Pi I_L}_{I_Q} - I_L^2 R_L/2}, \quad (19.15)$$

with once again $I_L = V_{OC}/(R+R_L)$. Show that for maximum power, $R_L = R$, $\eta = \eta_{Carnot} \times 1/\left[2 + 4/Z\bar{T} - \eta_{Carnot}/2\right]$. For $Z\bar{T} \ll 1$, we recover Eq. (19.14).

(c) Let us look at maximum efficiency $d\eta/d(R/R_L) = 0$, not maximum power ($R/R_L = 1$). Show that now the ratio $R/R_L = \sqrt{1 + Z\bar{T}}$, and the thermodynamic efficiency

$$\frac{P_{max}}{P_{in}} = \eta_{Carnot} \times \frac{\sqrt{1 + Z\bar{T}} - 1}{\sqrt{1 + Z\bar{T}} + T_2/T_1}. \quad (19.16)$$

Note: The term $Z\bar{T} = \Pi I_L/\kappa\Delta T$ is called the *thermoelectric figure of merit*. Thermoelectricity is expected to be a viable source of energy conversion if the number $Z\bar{T} > 3$. This will require making the power factor $S^2 G$ large and κ small. Since G (charge conductance) is usually due to electrons and κ (heat conductance) in insulators is usually due to vibrations (phonons), the trick is to make crystals that are conductive to electrons and insulating to phonons (the so-called *electron-crystal, phonon-glass*).

P19.4 **Landauer for spins.**

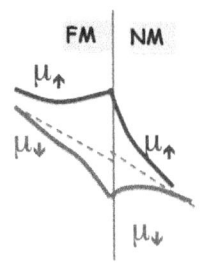

(a) At a **ferromagnet–non-magnet (FM–NM)** interface, we can obtain a steady-state solution to the spin diffusion equation P13.4 . The spin potentials split at the interface due to a spin impedance mismatch and reconnect far from the interface, so we solve the spin diffusion equation P13.4 for solutions that are finite everywhere with asymptotic boundary conditions $\mu_{\text{FM},\uparrow}(-\infty) = \mu_{\text{FM},\downarrow}(-\infty)$, $\mu_{\text{NM},\uparrow}(\infty) = \mu_{\text{NM},\downarrow}(\infty)$. Show that we get electrochemical potentials decaying away from the interfaces albeit with a slope, with notations $\lambda_i = \sqrt{\mathcal{D}_i \tau_i}$, ($i =$ FM,NM)

$$\mu_{\text{FM},S}(x) = \mu_1 - a_1 x + c_{\text{FM},S} e^{x/\lambda_{\text{FM}}}, \quad x < 0,$$
$$\mu_{\text{NM},S}(x) = \mu_2 - a_2 x + c_{\text{NM},S} e^{-x/\lambda_{\text{NM}}}, \quad x > 0. \quad (19.17)$$

Let us now nail down the coefficients $a_{1,2}$, $\mu_{1,2}$, $c_{\text{FM/NM},S}$ in terms of material parameters $\lambda_{\text{FM,NM}}$, $\sigma_{\text{FM},S} = \sigma(1 \pm \beta)$, σ_{NM} and current density J. First we use continuity of charge current $\partial J/\partial x = 0$ at all x to show that $c_{N\downarrow} + c_{N\uparrow} = 0$ and $\sigma_{\text{FM},\uparrow} c_{F\uparrow} = \sigma_{\text{FM},\downarrow} c_{F\downarrow}$. Substitute back into the expression for charge current to show that $J = -2a_2\sigma_{\text{NM}} = -(\sigma_{\text{FM},\uparrow} + \sigma_{\text{FM},\downarrow})a_1$.

Next, we use continuity of potentials at the interface $\mu_{\text{FM},S}(0) = \mu_{\text{NM},S}(0)$ to get c_{FS}, c_{Ns} in terms of $\mu_1 - \mu_2$. Finally, we equate the interfacial spin current densities $J_{\text{FM},S}(0) = J_{\text{NM},S}(0)$, assuming no interfacial spin flip. Define specific conductances $G_{\text{FM}} = \sigma(1-\beta^2)/\lambda_{\text{FM}}$ and $G_{\text{NM}} = \sigma_{\text{NM}}/\lambda_{\text{NM}}$. Show that the spin split potentials at the interface

$$\Delta\mu_S(0) = \mu_{\text{FM},\uparrow}(0) - \mu_{\text{FM},\downarrow}(0) = -\left(\frac{J\beta}{G_{\text{FM}} + G_{\text{NM}}}\right). \quad (19.18)$$

Finally, show that the spin current density $J_\uparrow - J_\downarrow$

$$J_S(x) = \begin{cases} J\beta \left[1 - \dfrac{G_{\text{FM}}}{G_{\text{FM}} + G_{\text{NM}}} e^{x/\lambda_{\text{FM}}}\right], & x < 0, \\[2ex] J\beta \left[\dfrac{G_{\text{NM}}}{G_{\text{FM}} + G_{\text{NM}}} e^{-x/\lambda_{\text{NM}}}\right], & x > 0, \end{cases} \quad (19.19)$$

so that far from the interface, the spin current equals $J\beta$ inside the ferromagnet, and 0 deep in the non-magnet (NM), and is a simple voltage division at the interface where spin is injected upto a distance $\sim \lambda_{\text{NM}}$. In other words,

$$J_S(0) = \overbrace{J_{\text{FM},S}(-\infty)}^{J\beta} \frac{G_{\text{NM}}}{G_{\text{FM}} + G_{\text{NM}}} \quad (19.20)$$

Note also that $G_{\text{FM}} = 2G_\uparrow G_\downarrow/(G_\uparrow + G_\downarrow)$, a series combination of up and down resistances, which renders itself a simple circuit picture.

Here's one way to think of this. Assume the NM has infinite spin scattering length, so that the two spin potentials vary linearly inside it, and stay parallel with equal slopes at $x = 0$. Now, assume $\sigma_{\text{NM}} = \sigma$, so that $\sigma_\downarrow < \sigma_{\text{NM}} < \sigma_\uparrow$. Clearly, constancy of each spin current density at the interface implies that the slope of $\mu_\uparrow$ must decrease to the left, while that of $\mu_\downarrow$ must increase, while eventually converging. This means at the interface, we need $\mu_\uparrow > \mu_\downarrow$. Note that this also means that if the FM were to the right side with the same orientation and NM to the left, then $\mu_\downarrow > \mu_\uparrow$ at the interface.

$$\mu_{\text{NM},\uparrow}(0) - \mu_{\text{NM},\downarrow}(0) = \left(\frac{J\beta}{G_{\text{FM}} + G_{\text{NM}}}\right),$$

$$J_S(0) = \overbrace{J_{\text{FM},S}(\infty)}^{J\beta} \frac{G_{\text{NM}}}{G_{\text{FM}} + G_{\text{NM}}}. \quad (19.21)$$

We will invoke this flipping of interfacial spin polarization next.

(b) At a **ferromagnet–semiconductor–ferromagnet (FM–SC–FM)** interface, i.e., a spin valve, we expect the spins potentials to align as discussed. (Also consider symmetry — figure as seen standing on your head!) Now, inside the semiconductor, the spin diffusion length λ_{SC} is assumed larger than the width L of the semiconducting region, so that the potentials satisfy $\partial^2 \mu_{SC,s}/\partial x^2 \approx 0$, and vary linearly with distance. If the two magnets are parallel, then we expect a large spin current. Because the slopes inside the contacts are $\pm \partial/\partial x$, the polarizations at the two interfaces will be opposite, $[\mu_{FM,\uparrow}(0) - \mu_{FM,\downarrow}(0)] = -[\mu_{FM,\uparrow}(L) - \mu_{FM,\downarrow}(L)]$, and the spin potentials cross in the middle. The majority spin diffusion current is very high, the minority low, corresponding to resistances $2r_{min}$ and $2R_{maj}$ (in fact, it follows the circuit in $\boxed{\text{P22.7}}$ minus the bridge resisor R_{sc}, plus semiconducting resistances $R_{SC\uparrow}$, $R_{SC\downarrow}$ in the two arms). For antiparallel contacts however, the spin polarizations at the interfaces maintain their sign, $\mu_\uparrow$ and $\mu_\downarrow$ are parallel to each other, and the spin current density $J_\uparrow - J_\downarrow = \sigma_{SC}\partial(\mu_\uparrow - \mu_\downarrow)/\partial x$ drops rapidly, with each spin component having the same resistance $R_{maj} + r_{min}$.

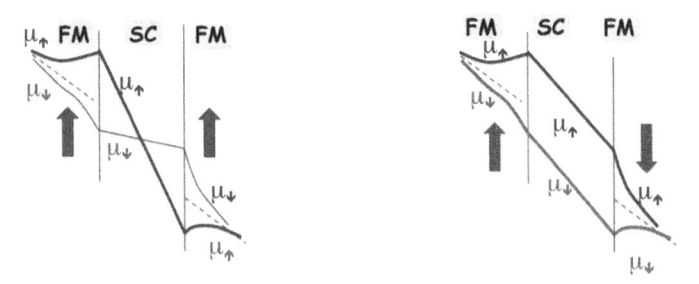

Using similar boundary conditions, show that the constant spin current density in the semiconductor for parallel spin valve with identical contacts

$$J_S^P(0 < x < L) = J\beta \left[\frac{G_{SC}}{G_{FM} + G_{SC}} \right], \tag{19.22}$$

while $J_S^{AP} = 0$, with $G_{SC} = \sigma_{SC}/L$.

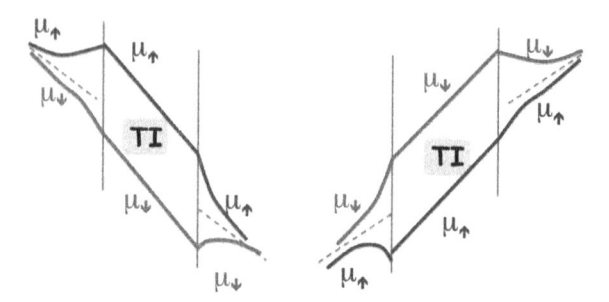

(c) Along the surface of a **topological insulator** or Weyl semi-metal P9.4, and P21.3 , the spin variation looks like a semiconductor (parallel linear drop, albeit with a spin current since $\sigma_{TI,\uparrow} \neq \sigma_{TI,\downarrow}$). However, the spin splitting is not driven by contact polarization — instead, it depends on current direction (witness change in polarization and color above). We can imagine an internal spin polarizing magnetic field that depends on current direction and makes $\sigma_{TI,\uparrow,\downarrow}$ different, while the spin conductivities in the contacts are the same. Work out the spin split interfacial μ and spin current density for the left figure, and show that $\mu_{TI,\uparrow} - \mu_{TI,\downarrow} = J\beta/G_M$, $G_M = \sigma_M/\lambda_M$ is the specific conductance per-unit area of the contact, and $J_S = J_\uparrow - J_\downarrow = J\beta$. The spin–momentum locking arises because β switches signs with current direction, $\beta = |\beta|J/|J|$.

Chapter 20

Quantum Transport with Scattering

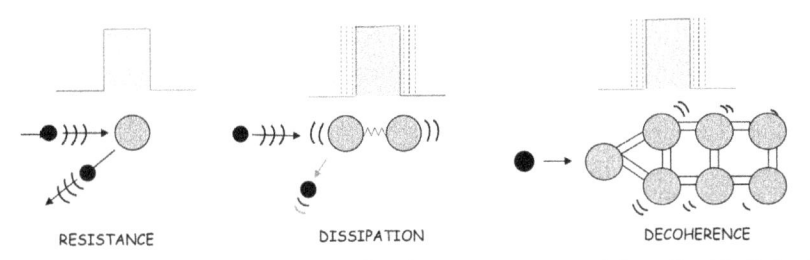

Along the electron's path, it can encounter various disruptions — impurities that can redirect their momenta and add resistance, optical phonons — finite energy short-wavelength vibrations that can dissipate their energy and slow them substantially (or excite them as with photo-absorption), and finally long-wavelength acoustic phonons like heat that can randomize their phase information, causing a lapse in memory and reducing overall coherence (like turning a laser into a light bulb). Let us talk about how to account for these scattering events in the $\mathcal{T}$ term in the Landauer equation.

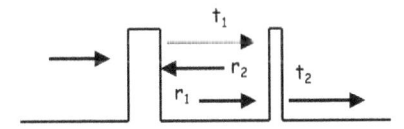

The quantum transmission through a pair of barriers depends on the history of the particle. The electron can go through both barriers with probability amplitude $t_1 \times t_2$, or encounter a pair of reflections inside the well between the barriers with an added probability amplitude $r_2 r_1$, or encounter two sets of reflections with amplitude $(r_2 r_1)^2$, and so on.

The final transmission amplitude involves summing a geometric series

$$t = t_1 t_2 + t_1 r_2 r_1 t_2 + t_1 (r_2 r_1)^2 t_2 + \cdots = \frac{t_1 t_2}{1 - r_1 r_2}. \tag{20.1}$$

The amplitudes are related to their probabilties

$$r_i = \sqrt{\mathcal{R}_i} e^{i\theta_i}, \quad \mathcal{R}_i = |r_i|^2, \quad \mathcal{T}_i = |t_i|^2 = 1 - \mathcal{R}_i, \tag{20.2}$$

so that

$$\mathcal{T} = |t|^2 = \frac{\mathcal{T}_1 \mathcal{T}_2}{1 - 2\sqrt{\mathcal{R}_1 \mathcal{R}_2} \cos(\theta_1 - \theta_2) + \mathcal{R}_1 \mathcal{R}_2}. \tag{20.3}$$

This equation shows how interferences crop up for multiple barriers, and can make the net resistance of two barriers in series non-additive, in violation of Kirchhoff's law. We see these interference effects in Shubnikov–deHaas conductances vs magnetic fields, and Fano interferences for two parallel conducting channels as a function of gate bias. Extending it to a sequence of barriers needs the equivalent of a transfer matrix treatment $\boxed{\text{P4.3}}$.

For below-barrier transport, this equation gives us WKB tunneling, since the net phase $\Delta\theta$, nominally picked up through transit across the well ($\sim 2kL$), becomes imaginary, and the cosine term gets replaced by a hyperbolic cosine and eventually a decaying exponential.

Finally, let us assume there is a phase randomization as for a long sequence of random barriers (spatial incoherence), or random vibrations at each barrier (temporal incoherence). We now have a probability distribution $P(\Delta\theta)$ for the phase picked up in between the barriers.

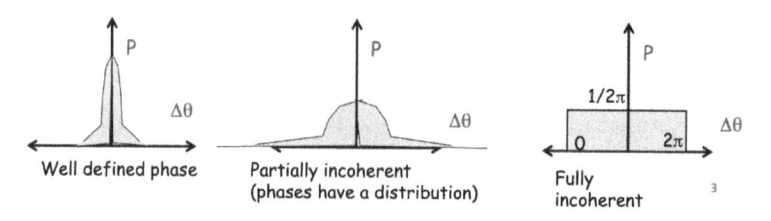

Well defined phase | Partially incoherent (phases have a distribution) | Fully incoherent

In the limit of strong scattering where we get a uniform probability distribution, we use the identity $\int_0^{2\pi} d\theta/(A - B\cos\theta) = 2\pi/\sqrt{A^2 - B^2}$, to get rid of the interference terms and simply add transmission probabilities

$$\mathcal{T} = \frac{1}{2\pi} \int_0^{2\pi} \frac{d\Delta\theta \ \ \mathcal{T}_1\mathcal{T}_2}{1 - 2\sqrt{\mathcal{R}_1\mathcal{R}_2}\cos\Delta\theta + \mathcal{R}_1\mathcal{R}_2} = \frac{\mathcal{T}_1\mathcal{T}_2}{1 - \mathcal{R}_1\mathcal{R}_2}. \tag{20.4}$$

Writing $\mathcal{R}_i = 1 - \mathcal{T}_i$, this equation boils down to *the additivity of inverse transmissions*, i.e., *series resistances*

$$\frac{1}{\mathcal{T}} - 1 = \left(\frac{1}{\mathcal{T}_1} - 1\right) + \left(\frac{1}{\mathcal{T}_2} - 1\right) = \sum_n \frac{\mathcal{R}_n}{\mathcal{T}_n}. \tag{20.5}$$

20.1 Recovering Ohm's Law

The Landauer resistance can be separated out into a quantum contribution from the contacts where numerous contact modes coalesce into a few channel modes, plus an additional series resistance from the channel itself

$$R = \underbrace{\frac{h}{q^2M}}_{R_Q} + \underbrace{\frac{h}{q^2M}\left(\frac{1-\mathcal{T}}{\mathcal{T}}\right)}_{R_{\text{channel}}}. \tag{20.6}$$

As we saw earlier, even this channel resistance can have strong quantum contributions such as coherent interferences Eq. (20.3). However, in the presence of incoherent scattering, this term $(1 - \mathcal{T})/\mathcal{T}$ becomes additive Eq. (20.5), proportional to the channel length, and can be written as L/λ_{sc}, whereupon $\boxed{\text{P20.1}}$

$$\boxed{\mathcal{T} = \frac{\lambda_{\text{sc}}}{\lambda_{\text{sc}} + L}} \tag{20.7}$$

so that the resistance intrinsic to the channel behaves classically, in fact satisfying Ohm's Law

$$R_{\text{channel}} \to R_{\text{classical}} \approx \frac{hL}{q^2M\lambda_{\text{sc}}} = \frac{L}{\sigma A} \quad \text{(Ohm's law)}, \tag{20.8}$$

where $\sigma = q^2(M\lambda_{\text{sc}}/hA)$. This is at zero temperature, where $-\partial f(E)/\partial E = \delta(E - E_F)$ selects a single energy E_F where M and λ_{sc} are evaluated. At finite temperature, we recover Eq. (19.8).

Just as Ehrenfest theorem connects quantum averages with classical expectations, the intrinsic channel resistance in a dirty material smoothly transitions to Ohm's Law.

20.2 Recovering Drift–Diffusion

We can think of drift–diffusion equation $\vec{J} = \sigma \vec{\nabla}\mu/q$ as the continuum equation and Ohm's law $I = V/R$, $R = L/\sigma A$ as its lumped version. In 1D for instance, a forward difference treatment of the spatial derivative at site x_i on a grid of spacing a gives $J_i = \sigma_i[\mu_{i+1} - \mu_i]/qa$. Writing the current $I = JA$, we can interpret this as a voltage drop $\Delta\mu_i/q$ across an Ohmic resistance $R_i = a/\sigma_i A$ Eq. (20.8). The electrochemical potentials in turn couple with the electrostatic potentials ϕ_i through the chemical potentials $\mu_i - q\phi_i$ that determine the local charge build-up Δn_i, which in turn sets ϕ_i through Gauss's Law $\boxed{\text{P20.2}}$.

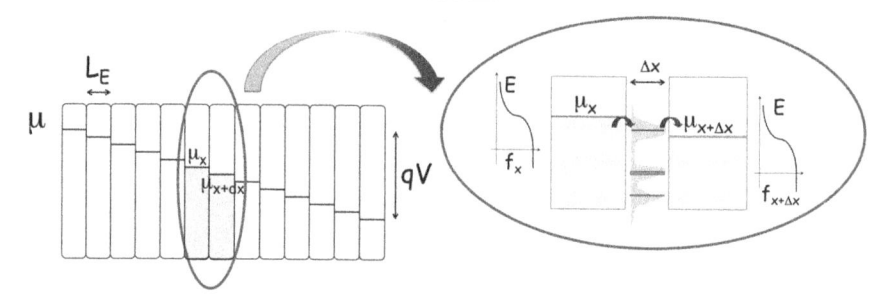

Conversely, we can write Landauer equation across a small section of width Δx across which μ drops, and replace the transmission $\mathcal{T}$ with $\mathcal{T}/(1 - \mathcal{T}) = \lambda_{sc}/\Delta x$ Eqs. (20.6) and (20.7) to eliminate contact resistances and focus on intrinsic resistances. Landauer equation then gives

$$J = \frac{I}{A} = \frac{q}{h}\int dE \frac{M(E)}{A} \underbrace{\frac{\lambda_{sc}}{\Delta x}}_{\mathcal{T}/(1-\mathcal{T})} \times \underbrace{\left(-\frac{\partial f}{\partial E}\right)\frac{\partial\mu}{\partial x}\Delta x}_{\Delta f(E-\mu_x)}$$

$$= \underbrace{\left[q^2\int dE\left(-\frac{\partial f}{\partial E}\right)\frac{M\lambda_{sc}}{hA}\right]}_{\sigma = \langle\langle q^2 M\lambda_{sc}/hA\rangle\rangle}\frac{1}{q}\frac{\partial\mu}{\partial x} \quad \text{(drift–diffusion).} \quad (20.9)$$

20.3 Inelastic Scattering: Mixing Energies

Inelastic scattering allows us to exchange energy between the electron and the channel during transport. This is relevant if the electron lingers long enough to scatter from sources of inelastic scattering such as localized vibrations (*vibrons*) in molecules, extended phonons in solids, or photons absorbed (we already encountered this earlier in Problem $\boxed{\text{P12.5}}$). We can extend the conventional Landauer formalism to accommodate such processes. We can visualize the scattering event as one where an electron comes in from a contact at one energy and escapes into the other contact at a different energy, the energy difference going to the inelastic scatterer. It is important of course that the first contact have electrons at the original energy, with probability f, while the second contact has empty states to escape into at the final energy $\boxed{\text{P12.5}}$. In other words

$$I_{\text{sc}} = \frac{q}{h} \int dE_1 dE_2 \big[\mathcal{T}_{1\to 2} f_1(E_1)[1 - f_2(E_2)] - \mathcal{T}_{2\to 1} f_2(E_2)[1 - f_1(E_1)] \big].$$

(20.10)

At equilibrium, we expect $I_{\text{sc}} = 0$, and $f_1 = f_2 = f_0$. At this point, we invoke the concept of *detailed balance*, which states that the vanishing of current happens energy by energy, so that the integrand is zero as well. This gives us

$$\frac{\mathcal{T}_{2\to 1}}{\mathcal{T}_{1\to 2}} = \frac{f_0(E_1)[1 - f_0(E_2)]}{f_0(E_2)[1 - f_0(E_1)]} = e^{\beta(E_2 - E_1)}, \quad \beta = 1/k_B T, \quad (20.11)$$

where we used the identity for Fermi–Dirac distributions that $[1 - f_0(E)]/f_0(E) = e^{\beta(E - E_F)}$. For $E_2 > E_1$, the transition from 2 to 1 (energy lowering) is thermodynamically much more efficient than the opposite. Electrons prefer to lower their energy when given a chance.

Let us write this energy difference $E_2 - E_1 = \hbar\omega$, the energy of the scatterer (e.g., phonon). Let us also assume the phonons are always kept in equilibrium, coupled to a thermal bath. The phonon distribution function is then $N_\omega = 1/\big[\exp(\hbar\omega/k_B T) - 1\big]$. We can write the ratio above as

$$\frac{\mathcal{T}_{2\to 1}}{\mathcal{T}_{1\to 2}} = e^{\beta\hbar\omega} = \frac{N_\omega + 1}{N_\omega}. \quad (20.12)$$

We can thus write

$$\mathcal{T}_{2\to 1} = A\big(N_\omega + 1\big), \quad \text{(phonon emission)},$$

$$\mathcal{T}_{1\to 2} = A N_\omega, \quad \text{(phonon absorption)}. \quad (20.13)$$

The integrals over energy $\int dE_1 \int dE_2$ transform into one over E_1 and another over the difference $\hbar\omega$, weighted by the phonon density of states $D_{ph}(\omega)$. Since phonons only exist over $\omega > 0$, and $N_{-\omega}/(N_{-\omega} + 1) = (N_\omega + 1)/N_\omega$, we can rewrite it over only the positive frequency phonons as

$$I_{sc} = \frac{q}{h} \int_{-\infty}^{\infty} dE M(E) \int_0^\infty d\omega D_{ph}(\omega) A\big[(N_\omega + 1)f_{1E}\bar{f}_{2,E-\hbar\omega}$$

$$+N_\omega f_{1E}\bar{f}_{2,E+\hbar\omega} - (N_\omega + 1)f_{2E}\bar{f}_{1,E-\hbar\omega} - N_\omega f_{2E}\bar{f}_{1,E+\hbar\omega}\big], \quad (20.14)$$

where $\bar{f} = 1 - f$, and the energy arguments went to the subscripts, for notational compactness.

Equipped with this modified equation, we can look at its application to a variety of problems involving scattering.

20.4 A Crude Approximation: Büttiker Probes

For coherent transport, the electron runs through the channel in one step. The modified Landauer equation allows us to include incoherent scattering, where the electron 'pauses' inside the channel before exiting. Let us capture this with a cruder model. We refer to the following figure. We have a density of states $D(E)$ and two contacts with injection rates $\gamma_{1,2}$ and electrochemical potentials $\mu_{1,2}$ that define their Fermi functions $f_{1,2}(E)$. We will treat the scattering center 's' as a third virtual 'contact' with its own Fermi function $f_s(E)$ and injection rate γ_s. The transmission function between any two contacts can then be written as $T_{ij}(E) = 2\pi\gamma_i\gamma_j D(E)/(\gamma_i + \gamma_j)$. We are ignoring magnetic fields, which would actually break reciprocity, $T_{ij} \neq T_{ji}$.

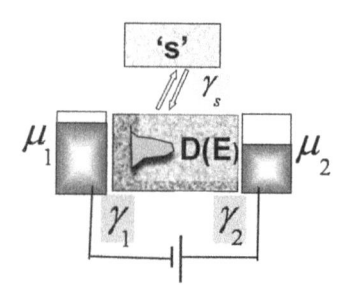

We can then calculate current at the mth terminal (m = '1', '2' or 's'), by summing all outgoing currents through it (this is a generalization of the two-terminal Landauer formula, called the Landauer–Büttiker formula).

$$I_m = (q/h) \int dE \left[\mathcal{T}_{m1}(f_m - f_1) + \mathcal{T}_{m2}(f_m - f_2) + \cdots \right], \quad (20.15)$$

$$= (q/h) \sum_{mn} \int dE \mathcal{T}_{mn}(f_m - f_n). \quad (20.16)$$

Let us assume we know γ_s (determined by some microscopic process), but we still do not know $f_s(E)$. The contact 's' acts like a regular contact, except it does not draw any net current (as there is no real terminal there). Any electron withdrawn at the contact is reinjected — but in the process it loses memory ('coherence'). For instance, if an electron enters contact 's' from '1', it is promptly reinjected back into the channel, but now the injected electron can go either way (from contact 's' back to '1' or onwards to '2'). In other words, just by adding a virtual contact at 's' we have made the electron pause, and lose its memory of where it was going.

If we impose the condition that the contact 's' does not draw any net current, $I_S = 0$, we get an integral relation between f_1, f_2 and f_s. However, to get to $f_s(E)$ from there, we need to make some additional assumptions.

First, let us assume that although the electron loses its memory, it still does not lose energy (in other words, the scattering center is unable to add to or remove any energy from the reinjected electron). This is called *elastic scattering*. Thus, energy levels between electrons do not mix. This means that you should be able to pull the integrands from the above equation out of the integral. The resulting f_s becomes a weighted average of $f_{1,2}$ at the same energy'

$$f_s(E) = \frac{f_1(E)\mathcal{T}_{1s}(E) + f_2(E)\mathcal{T}_{2s}(E)}{\mathcal{T}_{1s}(E) + \mathcal{T}_{2s}(E)}. \quad (20.17)$$

Clearly, for $qV_D > k_B T$, this does not look like an equilibrium Fermi–Dirac distribution.

If we now calculate the terminal current $I_1(E)$ using Eq. (20.16), substituting the expression for $f_s(E)$ from Eq. (20.17), we find that the current now has the same form as a two-terminal current, with an effective transmission

$$\mathcal{T}_{\text{elastic}} = \underbrace{\mathcal{T}_{12}}_{\text{coherent}} + \underbrace{\frac{\mathcal{T}_{1s}\mathcal{T}_{s2}}{\mathcal{T}_{1s} + \mathcal{T}_{s2}}}_{\text{incoherent}}. \quad (20.18)$$

This can be written in a very intuitive circuit form involving a ballistic resistance in parallel with a series sum of two scattering resistances between the real and virtual contacts $\boxed{\text{P20.3}}$. *Note, however, the prediction is that unlike coherent elastic scattering, incoherent elastic scattering always increases transmission by adding a parallel resistance path and opening new transport channels.*

Let us look at the opposite limit where we have strong inelastic scattering, leading to *perfect thermalization* that imposes a Fermi–Dirac form on f_s in local equilibrium with the channel electrons

$$f_S(E) = \frac{1}{e^{(E-\mu_S)/k_B T} + 1} \tag{20.19}$$

We then enforce the integral equation by varying μ_S values until you get $I_S = 0$, and use that μ_S to get $f_S(E)$ and ultimately the current $\boxed{\text{P20.3}}$.

In practice, the degree of thermalization depends on the strength of electron–phonon coupling A, and also the degree of phonon equilibriation determining how close to N_ω the phonon distribution actually exists. This in turn depends on the coupling of phonons to an external bath as well as nonlinearity (anharmonicity) among the phonons (i.e., phonon–phonon scattering), determining the so-called Q-factor describing phonon equilibriation.

20.5 A Series of Büttiker Probes: The D'Amato Pastawski Model

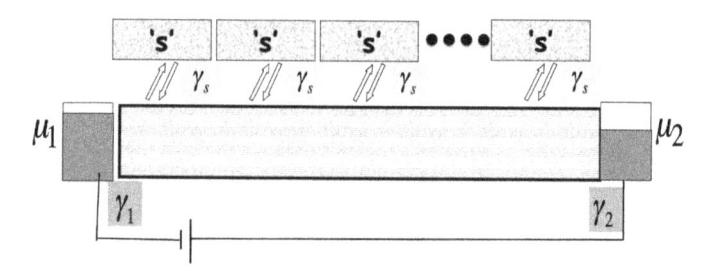

If we have a series of incoherent scatterers along the channel, we can write a current equation similar to earlier. In the example above, currents at all nodes from 3 to N are zero, and only the ends at 1 and 2 get finite current. Contact 1 is assumed to be grounded, and we also assume a small drain bias, so that $f_i(E) = F_T(E)\delta\mu_i$, where $F_T = \partial f_i/\partial\mu_i = -\partial f_i/\partial E \approx \delta(E - \mu_i)$. The equation then becomes

$$
\begin{pmatrix}
I_1 \\
I_2 \\
\boxed{\begin{matrix} I_3 = 0 \\ \cdots \\ \cdots \\ \cdots \\ \cdots \\ I_{N-1} = 0 \\ I_N = 0 \end{matrix}}
\end{pmatrix}
=
$$

$$
\begin{pmatrix}
\sum_i \mathcal{T}_{1i} & -\mathcal{T}_{12} & -\mathcal{T}_{13} & \cdots & \cdots & -\mathcal{T}_{1N} \\
-\mathcal{T}_{21} & \sum_i \mathcal{T}_{2i} & -\mathcal{T}_{23} & \cdots & \cdots & -\mathcal{T}_{2N} \\
-\mathcal{T}_{31} & -\mathcal{T}_{32} & \boxed{\begin{matrix}\sum_i \mathcal{T}_{3i} & \cdots & \cdots & -\mathcal{T}_{3N} \\ -\mathcal{T}_{43} \\ \cdots & & \mathbf{W} \\ \cdots \\ -\mathcal{T}_{N3} & \cdots & \cdots & \sum_i \mathcal{T}_{Ni}\end{matrix}} \\
\vdots & & & & & \\
-\mathcal{T}_{N1}(E) & \cdots & & & &
\end{pmatrix}
\begin{pmatrix}
\delta\mu_1 = 0 \\
\delta\mu_2 \\
\delta\mu_3 \\
\cdots \\
\cdots \\
\cdots \\
\delta\mu_{N-1} \\
\delta\mu_N
\end{pmatrix}
$$

$$(20.20)$$

where all transmissions are calculated at the respective μs, and we referenced all voltages relative to the left contact with $\delta\mu_1 = 0$ and evaluate all transmissions at $E = \mu_2$. We can then invert these equations to get

$$
\delta\mu_i = \sum_{j=3}^{N} (W^{-1})_{i,j} \mathcal{T}_{j2} \delta\mu_2. \tag{20.21}
$$

Substituting this back in the transport equation, we get

$$
\mathcal{T}_{\text{elastic}} = \underbrace{\mathcal{T}_{12}}_{\text{coherent}} + \underbrace{\sum_{ij \neq 1,2} \mathcal{T}_{1i}\left(W^{-1}\right)_{ij} \mathcal{T}_{j2}}_{\text{incoherent}}, \tag{20.22}
$$

as the generalization of the elastic single Büttiker probe result, Eq. (20.18). As it stands, there is no spatial information (we can swap probes or place them at the same point) — for which we need to incorporate the intuitive result that probes far apart should have low transmission between them. In physical systems, this transmission is set by the *local* density of states which involves wavefunction overlaps — in non-ballistic systems and inside band-gapped materials, these overlaps decay exponentially, so that a sequence of these exponentials enforce a linear (Ohmic) drop in transmission with channel length P20.4. Think of the electrons as jumping across a stream. A single shot (coherent) jump across it gets exponentially harder with increasing stream width. However, a sequence of probes acts like stepping stones that offer frequent respites for electrons to cross the stream easily. The corresponding conductance has the form often seen experimentally

$$G = \frac{q^2}{h} M \mathcal{T}_{\text{elastic}} = G_{\text{coh}} e^{-\alpha N} + G_{\text{incoh}}/N. \qquad (20.23)$$

Homework

P20.1

Graphene I–V.

(a) Write down the Landauer equation for the current, expressed in terms of the electron density of states $D(E)$, the band velocity $v(E)$ and the transmission $T(E)$. Assume a sheet of width W and length L, an energy-independent scattering length λ. Ignore any self-consistent potential U. Assume a temperature T.

(b) Simplify the Landauer expression and do the integrals (include graphene DOS [P10.3]) to calculate the I–V analytically at zero temperature, assuming E_F is positive (n-doped graphene), and letting the drain go from zero to a positive bias value. The velocity $v(E) = v_0$. Assume the scattering length λ is independent of energy, which is what you expect for elastic, uncharged impurity scattering. *Note: Be careful about signs when calculating density of states and doing the integrals!! For instance, remember that DOS cannot be negative.*

(c) Use Matlab to calculate and plot the two-terminal I–V characteristics for a sheet with dimensions $L = W = 1$ μm, and scattering length $\lambda = 100$ nm. Assume the velocity $v_0 = 10^6$ m/s. Assume $E_F = 0.2$ eV, and plot it for V running from 0 V to +0.5 V. Also plot it for gate voltages running in steps of 0.5 V from +5 V to −5 V.

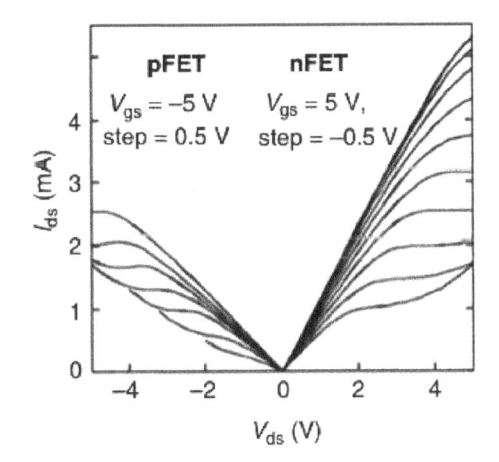

Fig. 20.1 Experimental graphene I–V (P. First, W.A. deHeer, T. Seyller, C. Berger, J.A. Stroscio, J. Moon, *MRS Bulletin* 35, 296, 2010).

P20.2 **Quantum capacitance.** Note that drift–diffusion usually comes coupled with Gauss's Law and equation of continuity. In large bulk channels, the extraction of charges is set by recombination generation. In quasi-ballistic channels the contacts do that job. We can then write down the coupled equations at steady-state as

$$\frac{1}{q}\vec{\nabla}\cdot\left(\sigma\vec{\nabla}\mu\right) = \frac{\Delta n}{\tau},$$

$$\vec{\nabla}\cdot\left(\epsilon\vec{\nabla}\phi\right) = -q\Delta n,$$

$$\Delta n \approx \bar{D}_0(\mu - q\phi), \tag{20.24}$$

where in this case, the extraction rates $1/\tau$ exist only for charges at the channel ends in contact with the source and drain, given by $\tau_{S,D} = \hbar/\gamma_{S,D}$. $\bar{D}_0$ is the local density of states at the Fermi energy. Discretize each equation on a 1D grid, and show that each node simplifies to the network shown, with contact resistances given by the Landauer equation $R_Q = h/q^2 M$. Write down expressions for the classical capacitance C_I, quantum capacitance C_i^Q and classical resistance R_i at each point.

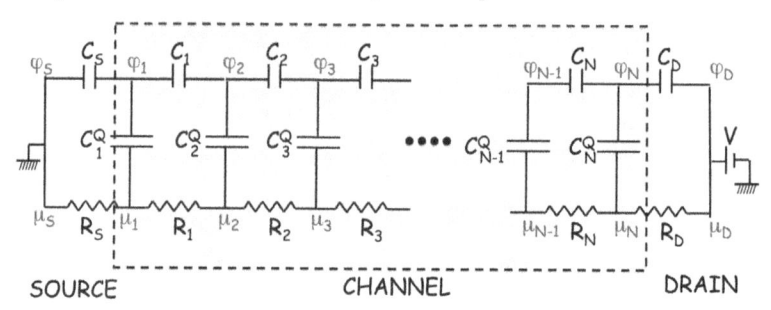

P20.3 **Büttiker probes and thermalization.**

(a) For elastic scattering, show that the transmission can be thought of as a circuit of series and parallel elements, representing the various transmission paths, direct between contacts 1 and 2, as well as indirect from 1 to s and s to 2. Plot $f_s(E)$ for E from -2 to 2 eV and show that it looks different from Fermi–Dirac. Use $\gamma_1 = 0.1$ eV, $\gamma_2 = 0.1$ eV, $\gamma_s = 0.05$ eV, and a Lorenzian density of states $D(E) = (\gamma_1 + \gamma_2)/2\pi[E^2 + (\gamma_1 + \gamma_2)^2/4]$.

(b) For fully thermalized scattering, repeat by numerically calculating I_S for various μ_S values (501 values varying from 0.5 to -1 eV) that you adjust until you get $I_s = 0$, and use that μ_S to plot $f_S(E)$ as well as to compare with the elastic scattering result.

(c) Plot I–V_D with and without scattering for both types of scattering (elastic and thermalized). In each case, assume we start with $E_F = 0.5$ eV, and go to drain bias of $V_D = 1.5$ V to cross the density of states.

P20.4 **Incoherent scattering: A series of probes.** Assume a set of 100 equally spaced Büttiker probes along a 1D channel, with weak couplings ($\gamma_{iS} = 0.01$ eV). Also assume the two contact couplings are $\gamma_{1,2} = 1$ eV, large enough that we can assume linear response. Assume a constant $DOS = 1/$eV, a grounded source and drain bias $V_d = 1.5$ V. To account for the intuitive result that probes sitting far away from each other should have weaker transmission between them (something we don't care about when the channel is treated as a single point), we need to scale $\mathcal{T}_{ij} \propto e^{-|i-j|a/\alpha}$, where a is the grid spacing, and α is the coherent decay length signifying the decreasing overlap of wavefunctions between the two points. In the upcoming NEGF chapters, we will see that this decreasing overlap comes naturally from the product of two corners of a Green function matrix.

Plot the spatial variation of the μ_S for the 100 grid points for both extremes (elastic vs thermalized scattering) and the impact on current. Assume $\alpha = 100a$ (no decay/spatial variations in $\mathcal{T}$, i.e., coherent) and $\alpha = 5a$ (strong decay, i.e., incoherent). Your results for thermalized scattering (Eq. 20.20) should look as follows

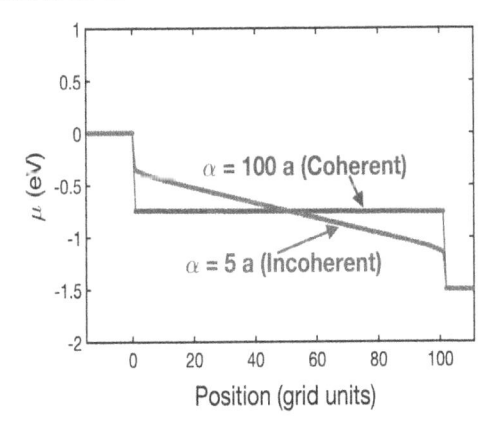

We can assign an overall resistance to the entire channel from the voltage drop $V = IR$, broken into two contact resistances $R_C/2$ each, and a channel resistance $R_{ch} = R_C L/\lambda$, so that the total resistance is $R = R_C(1 + L/\lambda)$.

Chapter 21

Non-Equilibrium Green's Functions for Charge

Earlier, we discussed the Landauer equation for current flow, which gives us the impact of quantization and tunneling on quasi-ballistic transport. In the presence of sufficient momentum scattering, we recover Ohm's law for a lumped structure, and drift–diffusion for an extended system with spatial variations. We will now discuss how collective quantum flow of electrons generalizes classical drift–diffusion to quantum non-equilibrium Green's function (NEGF) equations, and Landauer equation to extended systems with spatial variations. A formal way to do this for any system, including strongly interacting electrons, is using contour ordering. I personally prefer a simple and elegant derivation by Datta (in fact, with some creatively constructed self-energies, we can employ it for many interacting systems as well — NEMV, Chapter 27). We begin by rewriting Schrödinger's equation for an open system. Pay attention to the emergence of two additional boundary terms — an outflow term that depends on the wavefunction inside the device, and an inflow term–a source, that carries thermodynamic information about the outside world and will need to be handled within some assumptions. In fact, we already guessed this equation earlier (Eq. (5.10)) and relation with the DOS (Eq. (11.8)).

21.1 Open Boundary Schrödinger Equation

Let us write down the Schrödinger equation for a device coupled to a single contact. Before the contact is connected to the device, it carries an electron wavefunction ψ_I, satisfying

$$\mathcal{H}_C \psi_I = E \psi_I. \tag{21.1}$$

After coupling, there is a reflection much like particle on a step, and a reflected wave ψ_R returns into the contact, while a transmitted wave ψ enters the device. We can write the Schrödinger equation for the entire device-contact system ($\mathcal{H}_D : N \times N$ in size, $\psi : N \times 1$, $\mathcal{H}_C : N_c \times N_c$, $\tau : N_D \times N_C$)

$$\begin{pmatrix} \mathcal{H}_D & \tau \\ \tau^\dagger & \mathcal{H}_C - i\eta \end{pmatrix} \begin{pmatrix} \psi \\ \psi_I + \psi_R \end{pmatrix} = E \begin{pmatrix} \psi \\ \psi_I + \psi_R \end{pmatrix}. \tag{21.2}$$

The small vanishing imaginary part $i\eta$ is an artifact that we need to introduce to formally turn a closed system into an open one. $\tau = \mathcal{H}_{DC}$ is the device to contact coupling Hamiltonians, typically assumed to be localized.

We wish to write down the equation for the device wavefunction ψ alone, so we need to eliminate ψ_R. By expanding the second line, we get

$$\tau^\dagger \psi + [\mathcal{H}_C - i\eta](\psi_I + \psi_R) = E(\psi_I + \psi_R) \implies \psi_R = \underbrace{(E\mathcal{I} - \mathcal{H}_C + i\eta)^{-1}}_{G_C} \tau^\dagger \psi,$$

$$\tag{21.3}$$

where G_C is the contact Green function (we will connect it with Green's function shortly). Now, substituting in the first equation, we get

$$\mathcal{H}_D \psi + \tau(\psi_I + \psi_R) = E\psi \implies (E\mathcal{I} - \mathcal{H}_D)\psi - \underbrace{\tau G_C \tau^\dagger}_{\Sigma} \psi = \underbrace{\tau \psi_I}_{S}, \tag{21.4}$$

which gives us the open boundary Schrödinger equation (OBSE)

$$\boxed{\left(E\mathcal{I} - \mathcal{H}_D - \underbrace{\Sigma}_{\text{Outflow}} \right)\psi = \underbrace{S}_{\text{Inflow}}}. \tag{21.5}$$

More generally, the time-dependent version of this equation will involve a 'memory' term that integrates over the past history of the wavefunction

$$\underbrace{\left(i\hbar \frac{\partial}{\partial t} - \mathcal{H}_D(t) \right)\psi(t)}_{\text{unitary evolution}} - \underbrace{\int dt' \Sigma(t,t')\psi(t')}_{\text{outflow}} = \underbrace{S(t)}_{\text{inflow}}. \tag{21.6}$$

In essence, opening up the device brings in signatures of the outside world projected onto its energy spectrum through the self-energy Σ. Σ acts like a boundary potential, except it needs to be energy-dependent and complex in order to enforce open boundaries. We saw this for our numerical simulation of an electron on a barrier (Section 5.3, Fig. 5.2). We can see how having an anti-Hermitian part creates a level broadening

$$\boxed{\Gamma = i(\Sigma - \Sigma^\dagger) \quad \text{(broadening)}}. \qquad (21.7)$$

For a single energy level ϵ_0, having a complex self-energy $\Sigma = -i\gamma/2$ gives us a non-unitary evolution of the electron ($\psi^*(t) \neq \psi(-t)$)

$$(i\hbar\partial/\partial t - \epsilon_0 + i\gamma/2)\psi = 0 \Longrightarrow \psi(t) = \psi(0)e^{-i\epsilon_0 t/\hbar}e^{-\gamma t/2\hbar}, \qquad (21.8)$$

which allows the electronic charge $n = |\psi|^2 = |\psi_0|^2 e^{-\gamma t/\hbar}$ to irreversibly leave the device after a lifetime $T = \hbar/\gamma$. Going into Fourier space, we can see the impact of this finite lifetime on the energy spectrum as a broadening

$$n(t) \propto e^{-\gamma t/\hbar} \Longrightarrow n(E) = \frac{1}{2\pi}\int n(t)e^{iEt/\hbar}dt = \frac{\gamma/2\pi}{E^2 + \gamma^2/4}, \qquad (21.9)$$

i.e., Fourier complementarity $\gamma = \hbar/T$ between lifetime and broadening.

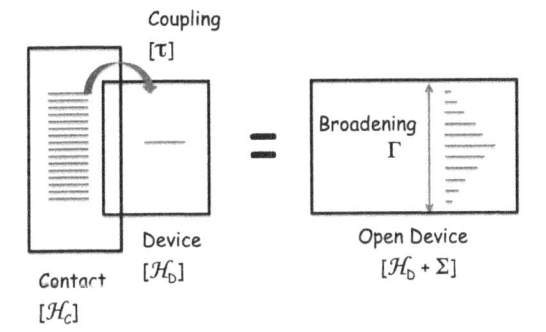

21.2 Solving OBSE with Retarded Green's Function G

Recall that one way to solve this inhomogeneous equation with a source term is to solve for the wavefunction with an impulse response (Section 14.4). The retarded Green's function G solves the equation when the source is a delta function (a spike) so that in presence of any source S decomposable into spikes with coefficients, the solution is simply the sum over Green's functions with the same coefficients (Eq. (14.13)). The formal solution to the open boundary Schrödinger equation can be obtained by inverting Eq. (21.5)

$$\psi = GS, \text{ where } \boxed{G = \left(E\mathcal{I} - \mathcal{H}_D - \Sigma\right)^{-1} \implies \text{Green's function}} \quad (21.10)$$

showing that $\psi = G$ when $S = \mathcal{I}$, which is the delta function on a discrete spatial grid. We also see why we call $G_C = (E\mathcal{I} - \mathcal{H}_C + i\eta)^{-1}$ the contact Green's function.

It is also easy to show that the anti-Hermitian part of Green's function gives the spectral density A, whose diagonal components give us the local density of states and trace gives us the total density of states. In fact, we argued as much in Eq. (11.8)

$$\boxed{D = \text{Trace}(A)/2\pi, \quad A = i(G - G^\dagger) \implies \text{spectral function}}. \quad (21.11)$$

We can easily see this for a device with weak coupling to the contacts, with a vanishingly small imaginary part $\Sigma \approx -i\eta\mathcal{I}$. In the eigenbasis of $\mathcal{H}_D$ with eigenvalues ϵ_n and eigenstates $\{\psi_n\}$, each an $N \times 1$ vector for N spatial grid points, we can write

$$G \underbrace{\psi_n}_{N \times 1} = \frac{\psi_n}{E - \epsilon_n + i\eta}. \quad (21.12)$$

Right multiplying by the $1 \times N$ vector $\psi_n^\dagger$ and using the continuity equation $\sum_n \psi_n \psi_n^\dagger = \mathcal{I}$, the $N \times N$ identity matrix, we get

$$G = \sum_n \frac{\overbrace{\psi_n \psi_n^\dagger}^{N \times N}}{E - \epsilon_n + i\eta}$$

$$\implies \underbrace{i(G - G^\dagger)}_{A} = \sum_n \psi_n \psi_n^\dagger \underbrace{\frac{2\eta}{(E - \epsilon_n)^2 + \eta^2}}_{\approx 2\pi\delta(E - \epsilon_n)}, \quad (21.13)$$

so that the diagonal entries divided by 2π give us spikes at the various eigenenergies weighted by their probability density $|\psi_n(x)|^2$, in other words the local density of states $D(x, E)$, while the trace becomes $\sum_n \delta(E - \epsilon_n) = D(E)$, the total density of states.

For finite broadening, we can write the spectral function in a slightly different but equivalent form involving Γ, the antiHermitian component of

Σ which represents the overall level broadening generated by coupling the discrete device levels with the continuum of states in the contact

$$G^{-1} = EI - \mathcal{H}_D - \Sigma, \quad (G^{\dagger})^{-1} = EI - \mathcal{H}_D - \Sigma^{\dagger}$$
$$\Longrightarrow \Gamma = i(\Sigma - \Sigma^{\dagger}) = -i[G^{-1} - (G^{\dagger})^{-1}]$$
$$\Longrightarrow G\Gamma G^{\dagger} = i(G - G^{\dagger}) = A. \tag{21.14}$$

To recap, given the device Hamiltonian $\mathcal{H}_D$ and the contact terms $\mathcal{H}_C$ and τ, we can estimate the device's Green's function G and thence compute the response ψ to any incident perturbation S. The self-energy $\Sigma = \tau G_C \tau^{\dagger}$ requires knowledge of the contact Hamiltonian $\mathcal{H}_C$ and the device-contact coupling τ. We will see shortly how we can exploit periodicity in the contact to convert the relevant elements of G_C, specifically it's surface element g_C directly coupled with the device, into a recursive equation. But what about the source term S?

21.3 Non-equilibrium Green's Function G^n for Charge

Given S, we know how to calculate ψ, a purely quantum mechanical problem. For current flow however, we are dealing with non-equilibrium quantum statistical mechanics. The source wavefunctions fluctuate as charges enter and leave the device, so that their averages are zero and only bilinear products have non-zero averages relating to the thermal distribution. Thankfully, most response functions involve bilinear products of ψ. The $N \times N$ electron occupancy factor is then given by the Keldysh equation

$$G^n = \langle \psi \psi^{\dagger} \rangle = G \langle SS^{\dagger} \rangle G^{\dagger} = G\Sigma^{in} G^{\dagger}, \quad \text{where } \Sigma^{in} = \langle SS^{\dagger} \rangle,$$
$$n = \int dE Tr[G^n(E)]/2\pi, \tag{21.15}$$

thus separating the charge contribution into a part involving the evolution of the electron under the Hamiltonian through G, and a separate part involving the incoming sources S that create a current noise, quantified by the bilinear average Σ^{in}. For metallurgical contacts, we wrote down $S = \tau \psi_I$, so that the $N \times N$ in-scattering function

$$\Sigma^{in} = \tau \langle \psi_I \psi_I^{\dagger} \rangle \tau^{\dagger}. \tag{21.16}$$

Inside the contact, the electrons are in equilibrium at a local Fermi–Dirac distribution f_0, so that their occupancy is given by the contact spectral function (matrix DOS) times the Fermi function

$$\langle \psi_I \psi_I^\dagger \rangle = A_C f_C. \tag{21.17}$$

We can write $A_C = i(g_C - g_C^\dagger)$, so that using $\Sigma = \tau g_C \tau^\dagger$,

$$\Sigma^{in} = \tau[i(g_C - g_C^\dagger)]\tau^\dagger f_C = i(\Sigma - \Sigma^\dagger)f_C = \Gamma_C f_C. \tag{21.18}$$

If we have two contacts now, we assume they are uncorrelated and maintained at their own separate Fermi functions by an external power source such as a battery (it is this difference in Fermi functions that makes the problem non-equilibrium). We then get

$$\Sigma^{in} = \Gamma_1 f_1 + \Gamma_2 f_2, \quad G^n = G(\Gamma_1 f_1 + \Gamma_2 f_2)G^\dagger. \tag{21.19}$$

Based on Eq. (21.14), it is easy to interpret the diagonal elements of $G\Gamma_{1,2}G^\dagger = A_{1,2}$ as *partial densities of states* (PDOS) from the two contacts, in which case the nonequilibrium electron occupancy G^n diagonal entries are given by the respective PDOS from the contacts filled by their respective Fermi functions, $G^n = A_1 f_1 + A_2 f_2$.

We can also define equations for holes by Hermitian conjugating Eq. (21.2), and replacing $\psi^\dagger \to \psi$ (hole creation equals electron annihilation)

$$\left(\overbrace{\psi}^{1 \times N} \; \psi_I + \psi_R \right) \left(\begin{matrix} \mathcal{H}_D & \tau \\ \tau^\dagger & \mathcal{H}_C + i\eta \end{matrix} \right) = E \left(\psi \; \psi_I + \psi_R \right) \implies \psi = \psi_I \tau^\dagger G^\dagger. \tag{21.20}$$

Defining an $N \times N$ out-scattering function $\Sigma^{out} = \tau \langle \psi_I^\dagger \psi_I \rangle \tau^\dagger = \Gamma_1 \bar{f}_1 + \Gamma_2 \bar{f}_2$ (using fermionic anti-commutation rules for ψ_I), we get the $N \times N$ hole occupancy $G^p = \langle \psi^\dagger \psi \rangle = G\Sigma^{out}G^\dagger = A_1 \bar{f}_1 + A_2 \bar{f}_2$, with $\bar{f} = 1 - f$. Adding electron and hole contributions, we get the contact broadenings for terminal $\alpha = 1, 2$ and total DOS (Eq. (21.14))

$$\Sigma_\alpha^{in} + \Sigma_\alpha^{out} = \Gamma_\alpha(f_\alpha + \bar{f}_\alpha) = \Gamma_\alpha, \quad G^n + G^p = A_1 + A_2 = A. \tag{21.21}$$

21.4 Simple Estimates for Self-Energy: A 1D Conductor

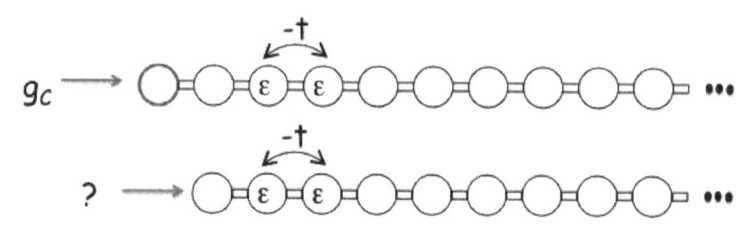

Note that we have already encountered these equations, back in Chapter 5 (Eq. (5.10)). Let us try to get a feel for the self-energies now. The easiest case is for a uniform wire, where each atom looks the same, with onsite energy ϵ_0 and nearest neighbor hopping term $-t$. Following the numbering of the atoms from 1 to N, atom 1 coupled to the left contact which we are now evaluating, we can write its self-energy $\Sigma_C = \tau G_C \tau^\dagger$. Since only the corner entry of τ_C is non-zero in a nearest neighbor coupled system, most of Σ_C is zero, except $\Sigma_C(1,1) = \tau_{C1} g_{11} \tau_{C1}^\dagger = t^2 g_{11}$, involving the contact surface Green's function, it's $(1,1)$th element. Here we note that the matrix multiplication $\tau g \tau^\dagger$ boiled down to multiplication of three scalar elements τ_{C1}, etc., because nearest neighbor coupling means only the g_{11} surface term communicates with the left contact. Now, we can write the Green function of the $(1,1)$th element by stripping that atom away and exposing the next, $(2,2)$th element, and following the definition of the Green function, $g_{11} = [(EI - \mathcal{H}_C)_{11} - \tau_{12} g_{22} \tau_{12}^\dagger]^{-1} = 1/[E - \epsilon_0 - t^2 g_{22}]$. At this point, we recognize that the surface Green's function after stripping away the surface atom stays intact, $g_{22} = g_{11} = g_C$, much like the equivalent resistance in Problem $\boxed{\text{P21.1}}$, which after adding or removing the R_1–R_2 branch stays intact, due to the virtue of its periodic semi-infinite structure. We thus get a recursive quadratic equation for the surface Green's function g_C

$$g_C^{-1} = E - \epsilon_0 - t^2 g_C \quad \text{solve quadratic equation for } g_C$$
$$\implies g_C = \frac{(E - \epsilon_0) - i\sqrt{4t^2 - (E - \epsilon_0)^2}}{2t^2}, \tag{21.22}$$

where we chose the sign of the imaginary part so that the contact surface density of states $D_C = i(g_C - g_C^\dagger)/2\pi$ is positive. For a 1D wire, we can use $E = E_k = \epsilon_0 - 2t \cos ka$, whereupon we get Eq. (5.11)

$$g_C = -e^{ika}/t, \quad \boxed{\Sigma_C(1,1) = t^2 g_C = -t e^{ika}}. \tag{21.23}$$

The broadening relates to transit time Δt through the uncertainty principle

$$\Gamma_C(1,1) = i[\Sigma_C(1,1) - \Sigma_C^\dagger(1,1)] = 2t \sin ka = \underbrace{\frac{\hbar}{a/v}}_{\Delta t}, \tag{21.24}$$

where $v = \partial E_k / \hbar \partial k = 2at \sin ka / \hbar$ is the band velocity. In fact, we can write it in terms of the contact surface density of states $D_C = i(g_C - g_C^\dagger)/2\pi$,

$$\boxed{\Gamma_C(1,1) = i\tau(g_C - g_C^\dagger)\tau^\dagger = 2\pi t^2 D_C \quad \text{Fermi's Golden Rule}}. \tag{21.25}$$

I provide in what follows a simple Matlab code to get you started. You can see the following plots, including the nature of the self-energy matrix. I will derive the equation for coherent transmission in the next chapter (Eq. (22.4)), but here's what it looks like. For a uniform 1D wire, the DOS looks as expected, $\propto 1/\sqrt{E(4t - E)}$ and transmission is unity inside the band. Also note that Σ_R and Γ are related — they are Hilbert transforms of each other

$$\Sigma = H(\Gamma) - i\Gamma/2, \quad H(\chi(\omega)) = \int_{-\infty}^{\infty} d\omega' \chi(\omega')/\pi (\omega - \omega'), \qquad (21.26)$$

a condition needed to ensure that the area under the device DOS curve is preserved upon contact coupling.

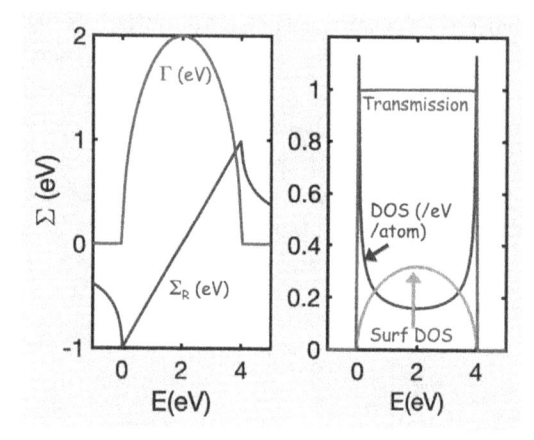

Matlab code:

```
%% Set up [H] of size Nₓ as usual, and initialize sig1, sig2 (not shown)
eta=1e-5;Ne=101;E=linspace(-1,5,Ne);
for ke=1:Ne
    k=acos(1-(E(ke)+i*eta)/(2*t)); s=-t*exp(i*k);
    sig1(1,1)=s; sig2(Nx,Nx)=s;
    gam1=i*(sig1-sig1'); gam2=i*(sig2-sig2');
    G=inv((E(ke)+i*eta)*eye(Nx)-H-sig1-sig2);
    D(ke)=i*trace(G-G')/(2*pi);
    T(ke)=trace(gam1*G*gam2*G');
end
%% Now plot D and T vs E
```

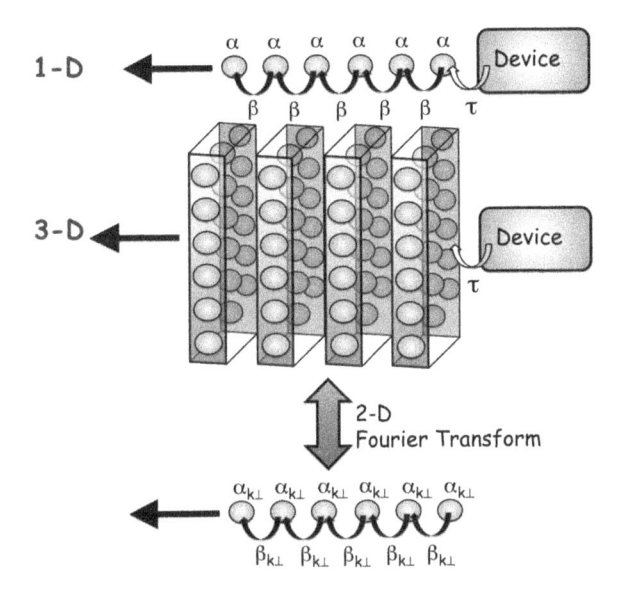

21.5 Self-Energy in a Multimoded Conductor

Equation (21.23) gives us the frequency-dependent complex potential we need to invoke at the ends of a finite conductor to impose open boundary conditions. How do we generalize it to multiple modes P21.2–P21.3? For a semi-infinite solid with a large cross-section over which periodic boundary conditions can be implemented, we can Fourier transform in the transverse direction and treat each $\vec{k}_\perp$ as a 1D chain. As a result, the matrix recursive equation for contact surface Green's function in transverse $\vec{k}_\perp$-space $g_{\vec{k}_\perp}(E)$ becomes the matrix analogue of Eq. (21.22)

$$g_{\vec{k}_\perp}^{-1} = \underbrace{[(E+i\eta)I - H_{\text{on},\vec{k}_\perp}]}_{\alpha_{\vec{k}_\perp}} - \underbrace{H_{\text{off},\vec{k}_\perp}\, g_{\vec{k}_\perp}\, H_{\text{off},\vec{k}_\perp}^{\dagger}}_{\beta_{\vec{k}_\perp}}. \tag{21.27}$$

I present in the following a Matlab code for transmission through a 3D slab of uniform cross-section, including a function that solves the matrix recursion equation. Pay attention to the sequence b or $b^\dagger$ for left vs right contacts. Also note that if the device were finite cross-section, such as a molecule attached to metal contacts, we would first inverse transform $g_{1,2,\vec{k}_\perp}$ to real space, extract $\Sigma_{1,2}$ and then calculate everything in real space. We can include the first few metal layers within our 'molecule', accounting for deviations from bulk such as surface reconstruction, for instance.

Matlab 3-D NEGF code for slab of simple cubic lattice

```
NE=101;E=linspace(-10,10,NE);
T=zeros(1,NE);D=zeros(1,NE);
t=-1;tp=-1;eta=1e-4;
Ha0=0*[-1 0;0 1];Ha1=t*[1 0;0 1];Hb=tp*[1 0;0 1];
Ra0=[0;0]; Ra1=[1 -1 0 0;0 0 1 -1]; Rb=[0;0];


N=51;M=51;
K1=2*pi*[1 0];K2=2*pi*[0 1];
for n=1:N
    for m=1:M
        K=(((n-1)/N)*K1)+(((m-1)/M)*K2);[n m]
        Hak=(Ha0*sum(exp(i*K*Ra0)))+(Ha1*sum(exp(i*K*Ra1)));
        Hbk=Hb*sum(exp(i*K*Rb));

        for ie=1:NE
            a=(E(ie)+i*eta)*eye(2)-Hak;    b=-Hbk;
            g1=grec(a,b');   g2=grec(a,b);
            sig1=b'*g1*b; gam1=i*(sig1-sig1');
            sig2=b*g2*b'; gam2=i*(sig2-sig2');
            G=inv(a-sig1-sig2);
            D(ie)=D(ie)+(i*trace(G-G')/(2*pi*N*M));
            T(ie)=T(ie)+trace(gam1*G*gam2*G');
        end
    end
end
%% Now Plot

function g=grec(alpha,beta)
g=inv(alpha);
it=1;epsil=1;
while(epsil > 1e-6)
    it=it+1;
    gnew=g;
    g=inv(alpha-(beta*gnew*beta'));
    epsil=(sum(sum(abs(g-gnew))))/(sum(sum(abs(g+gnew))));
```

```
    g=(0.5*g)+(0.5*gnew);
    if it > 5000
        epsil=-epsil; epsil
    end
end
```

21.6 A General Self-Energy for the 'Outside' World

The main lesson in this chapter is that we can account for the impact of an outside world through an effective potential. A potential that is complex and energy-dependent, i.e., a self-energy matrix allows us to 'open up the system' and bleed electrons into the outer world. Even when electrons cannot escape, as in non-metallurgical virtual contacts (Büttiker probes), e.g., scattering from a phonon coupled to a thermal bath, a complex self-energy randomizes phase by siphoning it outside, where it gets thermally scrambled by the large number of modes in the environment. We can write down the self-energy approximately for many cases, for instance, within the Born approximation for phonons (next chapter), and the equation of motion technique for strongly correlated electrons (NEMV, Chapter 26–28). Model self-energies for electron–phonon scattering and Coulomb blockade are written as follows. λ is the electron–phonon coupling, U_0 is the single-electron charging energy, n is the electron density, and $\sigma =\uparrow,\downarrow$, $\bar{\sigma}$ is its complement.

$$\Gamma_{el-ph}(E) \propto \langle\lambda\rangle^2[(N_\omega + 1)D(E + \hbar\omega) + N_\omega D(E - \hbar\omega)]$$

$$\Sigma_{CB,\sigma}(E) = U_0 n_{\bar{\sigma}} + U_0^2 n_{\bar{\sigma}}(1 - n_{\bar{\sigma}})/[E - U_0(1 - n_{\bar{\sigma}})]. \qquad (21.28)$$

Homework

P21.1 Recursive network. Let us solve for the equivalent resistance R_∞ of the infinite resistance network shown here. Adding the sequence R_1, R_2 in red to R_∞ preserves the topology and reproduces R_∞, giving us the needed recursive relation for R_∞. Solve it for R_∞.

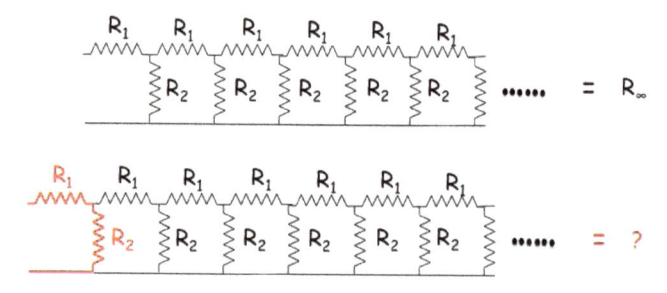

P21.2 Surface states in graphene. Using a nearest neighbor tight binding parametrization for graphene ($t_0 = -3.31$ eV, $t_2 = -0.106$ eV), find the DOS and transmission. Also do bilayer graphene, with an added interlayer coupling between the Bernally stacked members $\gamma_1 = 0.4$ eV. Finally, do a $(9,0)$ armchair edge graphene nanoribbon, where edge strain gives a modification $\Delta t_0 = -0.38$ eV.

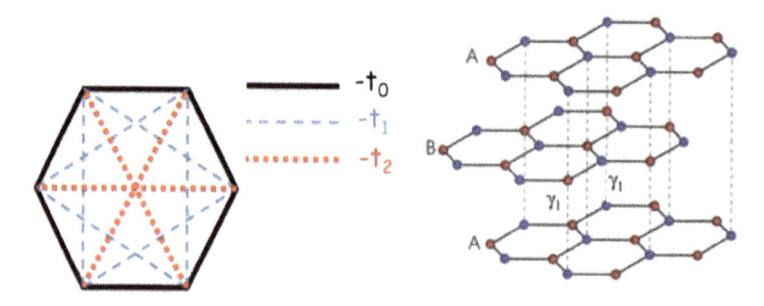

Work out the $E–k$ and DOS for a zigzag edge nanoribbon, and show that it has edge states with zero dispersion connecting the two Dirac points. Repeat for inversion symmetry breaking ($E_A \neq E_B$ for the sublattice) and time reversal symmetry breaking (complex next-nearest neighbor interactions, Eq. (5.8)). What do the edge states look like in the $E–k$? Plot the wavefunctions as well.

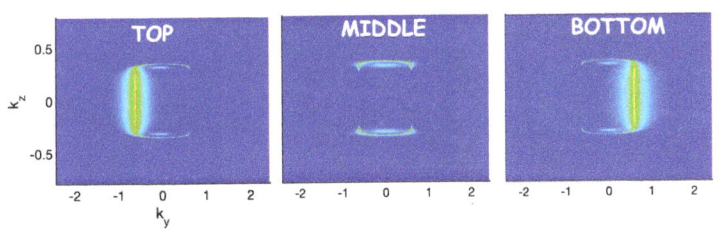

P21.3 **Surface states of Weyl semi-metals.**

Let us go back to the minimal model for a Weyl semi-metal $\boxed{\text{P9.4}}$, with time-reversal symmetry breaking interactions along the z-axis, and a finite size along the x-axis. Inverse transforming along x, by imposing $k_x \rightarrow i\partial/\partial x$, show that we get a familiar tridiagonal matrix representing a 1D wire in the x-direction, with $k_{y,z}$ dependent onsite and hopping matrices

$$\alpha_{k_y,k_z} = t[k_y\sigma_y + k_z^2\sigma_z], \quad \beta_{k_y,k_z} = t(\sigma_x - \sigma_z)/2i. \quad (21.29)$$

Calculate Green's function G and plot the layer resolved local density of states — top surface, bottom surface, and all bulk intermediate layers overlaid — for a Fermi energy near the Weyl points (I choose $E = 0.1$ eV, a bit away from the Weyl points, to give the results some 'size' for visualization, see above; I also enhance contrast by plotting log of the density of states) and show that the top and bottom surfaces have fractional Fermi arcs connecting the projections of the Weyl points onto the surfaces. Away from the Weyl point projections these surface states are pristine (bulk states are gapped at those $k_{y,z}$ values — imagine a vertical slice through the Dirac bands away from the Weyl points), while at the Weyl projections with zero gap the surface and bulk states hybridize, connecting the top and bottom x-surfaces straight through the Weyl points.

Plot the spin density of states $i\mathcal{T}r[\vec{\sigma}G - G^\dagger\vec{\sigma}]$ to show that electron spin is locked with momenta on the Weyl surface states. Which spin component?

Chapter 22

Non-Equilibrium Green's Functions (NEGF) for Current

22.1 From Equilibrium Charge to Non-Equilibrium Currents

To get the current, we evaluate the time-derivative of the correlation function and use the open boundary Schrödinger equation as well as defined quantities above to simplify. Note that the total current across all terminals must add up to zero, so we need to calculate the current at one terminal, and *use only the evolution equation involving that terminal to get the individual terminal current* (alternately, we can calculate the total current and then partition it into left terminal minus right terminal currents, but that requires an intuition about how to partition the current). We will now invoke definitions of in-scattering function Σ^{in}, electron correlation function G^n, broadening Γ and spectral function A from the previous chapter. Considering terminal α, and using $\text{Tr}(AB) = \text{Tr}(BA)$, we get

$$I_\alpha = \int dE q \frac{d}{dt} \text{Tr}\langle \psi \psi^\dagger \rangle$$

$$= \frac{q}{i\hbar} \int dE \text{Tr} \left\langle \left[\cancel{\mathcal{H}\psi} + \Sigma_\alpha \psi + S_\alpha \right] \psi^\dagger - \psi \left[\cancel{\psi^\dagger \mathcal{H}} + \psi^\dagger \Sigma_\alpha^\dagger + S_\alpha^\dagger \right] \right\rangle$$

$$= \frac{q}{i\hbar} \int dE \text{Tr} \left[\underbrace{(\Sigma_\alpha - \Sigma_\alpha^\dagger)}_{-i\Gamma_\alpha} \underbrace{\langle \psi \psi^\dagger \rangle}_{G^n} + \left\langle S_\alpha \underbrace{\psi^\dagger}_{S_\alpha^\dagger G^\dagger} - \underbrace{\psi}_{G S_\alpha} S_\alpha^\dagger \right\rangle \right]$$

$$= \frac{q}{i\hbar} \int dE \mathrm{Tr}\left(-iG^n\Gamma_\alpha + \underbrace{\langle S_\alpha S_\alpha^\dagger\rangle}_{\Sigma_\alpha^{in}} \underbrace{(G^\dagger - G)\rangle}_{iA}\right)$$

$$= \frac{q}{\hbar} \int dE \mathrm{Tr}(\Sigma_\alpha^{in} A - \Gamma_\alpha G^n) \quad (\mathrm{Meir - Wingreen}). \tag{22.1}$$

The first matrix in the bracket represents the rate of electron in-flow given by the influx from contact Σ_α^{in} (for instance, escape rate Γ_α times contact Fermi function f_α, Eq. (21.19)) times the density of states built into A. The second matrix $\Gamma_\alpha G^n$ represents the electron charge outflow, set by the escape rate Γ_α and the electron occupancy G^n inside the channel.

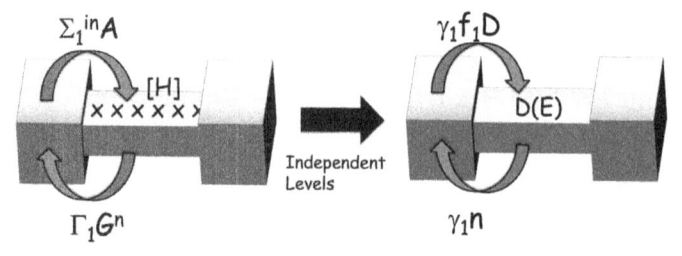

If we have a 'lumped circuit' treatment of the entire channel in terms of an overall density of states $D(E)$, for instance when the levels are independent (all matrices simultaneously diagonalizable), then the current simplifies to

$$I_\alpha = \frac{q}{h} \int dE \gamma_\alpha(f_\alpha D - n), \tag{22.2}$$

which makes sense, as the terminal current drops when the electron density in the channel reaches equilibrium with it, $n = f_\alpha D$.

22.2 Coherent NEGF $\Rightarrow$ Landauer

If we assume the self-energies are metallurgical contacts, and avoid any virtual contacts like Büttiker probes, then we can use $\Sigma_\alpha^{in} = \Gamma_\alpha f_\alpha$, $A = G(\Gamma_1 + \Gamma_2)G^\dagger$, $G^n = G\Sigma^{in}G^\dagger$, whereupon for a two-terminal device Eq. (22.1) simplifies to yield Landauer equation (Eq. (19.4))

$$I_1 = \frac{q}{\hbar} \int dE \mathrm{Tr}\left[\overbrace{\Gamma_1 f_1}^{\Sigma_1^{in}} \overbrace{G(\Gamma_1 + \Gamma_2)G^\dagger}^{A = G\Gamma G^\dagger} - \Gamma_1 \overbrace{G(\Gamma_1 f_1 + \Gamma_2 f_2)G^\dagger}^{G^n = G\Sigma^{in}G^\dagger}\right]$$

$$= \frac{q}{\hbar} \int dE M\mathcal{T}[f_1 - f_2] = \frac{1}{q} \int dE G(E)[f_1 - f_2] \quad (\mathrm{Landauer\ eqn.}),$$

$$\tag{22.3}$$

where

$$MT = \frac{G(E)}{q^2/h} = \text{Tr}[\Gamma_1 G \Gamma_2 G^\dagger] \quad \text{(Fisher–Lee formula)}. \qquad (22.4)$$

We used this equation in the previous chapter. We need the different Hamiltonian elements α, β, τ to calculate the self-energies $\Sigma_{1,2}$, which we often fit with experiments (e.g., empirical tight binding), or calculate 'first principles' as in density functional theory (DFT) or more advanced chemical techniques like configuration interaction (CI). Often though, we use model Hamiltonians P9.4, P22.1–P22.6.

22.3 Scattering Self-Energy and Current

The Büttiker probe approach earlier showed us how we can extend Landauer theory to capture incoherent scattering P22.7–P22.11, which cannot be simply woven into the Hamiltonian $\mathcal{H}$ with a perturbative potential, but needs an irreversible outflow of information, requiring thus a complex scattering self-energy Σ_ϕ with an imaginary (anti-Hermitian) part.

Using $G^n + G^p = A$ and $\Sigma_\alpha^{\text{in}} + \Sigma_\alpha^{\text{out}} = \Gamma_\alpha$ (Eq. (21.21)) we can rewrite the terminal current for a two-terminal device

$$I_\alpha = \frac{q}{\hbar} \int dE \, \text{Tr}(\Sigma_\alpha^{\text{in}} G^p - \Sigma_\alpha^{\text{out}} G^n). \qquad (22.5)$$

We can treat inelastic scattering as a similar 'terminal', with current

$$I_{sc} = \frac{q}{\hbar} \int dE \, \text{Tr}(\Sigma_{\text{sc}}^{\text{in}} G^p - \Sigma_{\text{sc}}^{\text{out}} G^n). \qquad (22.6)$$

For inelastic scattering, we can write down an expression for the scattering self-energies using the Born approximation. Electrons can enter from a higher energy $E + \hbar\omega$ by emitting a phonon with probability $N_\omega + 1$, or from a lower energy $E - \hbar\omega$ by phonon absorption with probability N_ω. The formal derivation is in NEMV (Section 26.2.3 explains underlying assumptions), but the result is a direct matricization of Eq. (20.14), in much the same way Fisher–Lee is a matrix generalization of Landauer.

$$\Sigma_{\text{sc}}^{\text{in}}(E, \omega) = \left[\mathcal{D}_0^n\right]\left[G^n(E + \hbar\omega)(N_\omega + 1) + G^n(E - \hbar\omega)N_\omega\right],$$

$$\Sigma_{\text{sc}}^{\text{out}}(E, \omega) = \left[\mathcal{D}_0^p\right]\left[G^p(E - \hbar\omega)(N_\omega + 1) + G^p(E + \hbar\omega)N_\omega\right]. \qquad (22.7)$$

Note that the calculation has to be done self-consistently, since Σ_{sc}^{in} depends on electron correlation G^n, while G^n depends on Σ^{in} through the Keldysh equation (Eq. (21.15)), where Σ_{sc}^{in} provides an additive component. In fact, it is easy to see that the self-consistency enforces the Büttiker probe condition that $I_{sc} = 0$. The energy arguments in the equation are different, but we can treat energy as a third axis of a matrix. From $\Sigma_{sc}^{in,out}(E)$ we can get Γ_{sc}, and then by Hilbert transform Σ_{sc} (Eq. (21.21)).

In fact, the terms $N_\omega + 1$, N_ω also come from the electron–phonon matrix terms including quantization, i.e., commutation relations. The interaction Hamiltonian depends on the impact of a displacement, i.e., a stretched spring, on the electronic energy U_{el}, namely, $\tau_{el-ph} = U_{el}(x_0+x) - U_{el}(x_0) \approx x[\partial U_{el}/\partial x]_{x_0}$ around the equilibrium atom position x_0, and is proportional to the electron density n and the displacement x. Consider an oscillator Hamiltonian ($\hat{x}$ in units of $\sqrt{\hbar/m\omega}$, $\hat{p}$ in units of $\sqrt{m\hbar\omega}$ and energy in units of $\hbar\omega$) $\boxed{\text{P5.3}}$, with commutator relation $[\hat{x},\hat{p}] = i$

$$H_{osc} = \hbar\omega\left(\frac{\hat{x}^2 + \hat{p}^2}{2}\right) = \hbar\omega \underbrace{\frac{(\hat{x} - i\hat{p})}{\sqrt{2}}}_{a^\dagger}\underbrace{\frac{(\hat{x} + i\hat{p})}{\sqrt{2}}}_{a} + \overbrace{\frac{\hbar\omega}{2}}^{\text{from } x-p \text{ commutator}}$$

$$= \underbrace{\hat{N}}_{\text{phonon number}} \hbar\omega + \overbrace{\hbar\omega/2}^{\text{0-pt energy}}, \quad \hat{N} = a^\dagger a, \tag{22.8}$$

where we have rotated our basis set in phase space $(\hat{x}, \hat{p}) \to (\hat{x} + i\hat{p}, \hat{x} - i\hat{p})$, while preserving commutation relation, $[a, a^\dagger] = 1$, in order to diagonalize the Hamiltonian. In fact, one can show that for a chain of coupled oscillators this rotation still diagonalizes the Hamiltonian and generates decoupled oscillator normal modes, which are called phonons. $(a, a^\dagger)$ are analogous to the angular momentum ladder operators $\hat{L}_\pm = \hat{L}_x \pm i\hat{L}_y$ $\boxed{\text{P6.2}}$, with $a^\dagger$ creating an extra phonon, $N \to N + 1$, a destroying it, $N + 1 \to N$

$$a^\dagger \Psi_N = \underbrace{\sqrt{N + 1}}_{C_{N+1}} \Psi_{N+1}, \quad a\Psi_N = \underbrace{\sqrt{N}}_{C_N} \Psi_{N-1}, \tag{22.9}$$

with coefficient C_N obtained since $\Psi_N^\dagger a$ is the Hermitian conjugate of $a^\dagger \Psi_N$.

$$(N + 1)|\Psi_N|^2 = \Psi_N^\dagger(a^\dagger a + 1)\Psi_N = \Psi_N^\dagger a a^\dagger \Psi_N$$

$$= C_{N+1}^2 |\Psi_{N+1}|^2 \Longrightarrow C_N = \sqrt{N} \tag{22.10}$$

Now, consider a phonon emission process between two states $(\epsilon_1, N) \rightarrow (\epsilon_2, N+1)$ (see Fig. in $\boxed{\text{P12.5}}$). In this many-body basis set,

$$
a^\dagger = \begin{array}{c} \\ N-1 \\ N \\ N+1 \end{array} \begin{array}{c} \cdots N-1 \quad N \quad N+1 \cdots \\ \left[\begin{array}{ccc} 0 & \sqrt{N} & 0 \\ 0 & 0 & \sqrt{N+1} \\ 0 & 0 & 0 \end{array} \right] \end{array}, \quad a = \begin{array}{c} N-1 \quad N \quad N+1 \\ \left[\begin{array}{ccc} 0 & 0 & 0 \\ \sqrt{N} & 0 & 0 \\ 0 & \sqrt{N+1} & 0 \end{array} \right] \end{array}
$$

$$
a^\dagger a = \begin{array}{c} \\ N \\ N+1 \end{array} \begin{array}{c} \cdots N \quad N+1 \cdots \\ \left[\begin{array}{cc} N & 0 \\ 0 & N+1 \end{array} \right] \end{array}, \quad \text{(number operator)}. \tag{22.11}
$$

In the $(\epsilon_1, N), (\epsilon_2, N+1)$ basis sub-set, the coupling matrix

$$
\tau_{\text{el}-\text{ph}}^{N+1} \sim x = K(a^\dagger + a) = \underbrace{K \left[\begin{array}{cc} 0 & \sqrt{N+1} \\ 0 & 0 \end{array} \right] e^{i\omega t}}_{\tau_{\text{el}-\text{ph}}^{\text{em},N+1}} + \underbrace{K \left[\begin{array}{cc} 0 & 0 \\ \sqrt{N} & 0 \end{array} \right] e^{-i\omega t}}_{\tau_{\text{el}-\text{ph}}^{\text{abs},N+1}},
$$

$$\tag{22.12}$$

where we have reinserted explicit time dependences. From these matrices, we can calculate the deformation potential. Invoking the familiar equation $\Sigma = \tau G \tau^\dagger$ (Eq. (21.4)), i.e., $\Sigma_{ij} = \sum_{kl} \tau_{ik} G_{kl} \tau_{lj}^\dagger = \sum_{kl} \tau_{ik} \tau_{jl}^* G_{kl} = \sum_{kl} [\mathcal{D}_0]_{ijkl}^n G_{kl}$, we can view the matrix $[\mathcal{D}_0]^n$ (the deformation potential) as the bilinear thermal ensemble average of the electron–phonon coupling

$$
[\mathcal{D}_0]_{ijkl}^n = \langle \tau_{ik} \tau_{jl}^* \rangle \implies [\mathcal{D}_0]^n = \langle \tau_{\text{el}-\text{ph}} \otimes \tau_{\text{el}-\text{ph}}^* \rangle, \tag{22.13}
$$

where we have written the 16 $[\mathcal{D}_0]^n$ elements $(i, j, k, l = 1, 2$ between the two electronic levels $\epsilon_{1,2})$ as a 4×4 matrix which in turn is a Kronecker product of two 2×2 τ matrices. Expanding into emission and absorption, and dropping cross terms that have random phases due to incoherence,

$$
[\mathcal{D}_0]^n = \langle \tau_{\text{el}-\text{ph}}^{\text{em}} \otimes \tau_{\text{el}-\text{ph}}^{\text{em}} \rangle + \langle \tau_{\text{el}-\text{ph}}^{\text{abs}} \otimes \tau_{\text{el}-\text{ph}}^{\text{abs}} \rangle. \tag{22.14}
$$

In the electron eigen-space, $\tau_{\text{el}-\text{ph}}$ is non-diagonal. However, in a real-space basis set such as a 1D tight binding wire, τ_{ij} is diagonal, $\tau_{ij} = U_i \delta_{ij}$, as it relates to the local x at a particular point. In the spatial coordinate basis set therefore, $[\mathcal{D}_0]_{ijkl}^n = \langle U_i U_j^* \rangle \delta_{ik} \delta_{jl} \times (N_\omega + 1/2 \pm 1/2)$, using $\langle a^\dagger a \rangle = N$, $\langle aa^\dagger \rangle = \langle a^\dagger a + [a, a^\dagger] \rangle = N+1$. We can then collapse $[\mathcal{D}_0]^n$ from a fourth to a second-rank tensor (four to two indices), so that $\Sigma_{ij}^{\text{in}} = [\mathcal{D}_0]_{ij}^n G_{ij}^n$, making it a product by product multiplication, e.g., $\mathcal{D}_0^n \otimes G^n$, in real space.

Furthermore, the nature of the elements of the second-rank $\mathcal{D}_0^n$ in real space are telling. Consider the following two forms for $[\mathcal{D}_0]$:

$$\mathcal{D}_0^{k,\phi} = D_0 \begin{bmatrix} 1 & 0 & 0 & 0 & 0 \\ 0 & 1 & 0 & 0 & 0 \\ \multicolumn{5}{c}{\dots\dots\dots\dots} \\ 0 & 0 & 0 & 0 & 1 \end{bmatrix}, \quad \mathcal{D}_0^{\phi} = D_0 \begin{bmatrix} 1 & 1 & 1 & 1 & 1 \\ 1 & 1 & 1 & 1 & 1 \\ \multicolumn{5}{c}{\dots\dots\dots\dots} \\ 1 & 1 & 1 & 1 & 1 \end{bmatrix}. \quad (22.15)$$

The first is localized (diagonal) in real space and spread out in Fourier space, connecting every k component equally with every other. It gives momentum scattering in addition to dephasing. The second, however, is a delta function in k-space, and thus has only dephasing without momentum scattering. The former will add a resistance, i.e., a slope to the voltage drop, while the latter will simply remove any quantum oscillations.

Finally, the physics gets more interesting when we have dephasing of electron spin $\vec{\sigma}$, for instance by a magnetic impurity $\vec{S}$, through a Hamiltonian $H_{\text{spin}} = J\vec{\sigma} \cdot \vec{S}$. The subtlety arises because the thermal average over the impurity spin density matrix ρ, $\langle O \rangle = \text{Trace}(\rho O)$, expressed in the electronic spin basis set, becomes

$$[\mathcal{D}_0]_{ijkl}^n = J^2 \langle (\vec{\sigma} \cdot \vec{S}) \otimes (\vec{\sigma} \cdot \vec{S})^* \rangle = J^2 \left\langle \begin{bmatrix} S_z & S_- \\ S_+ & -S_z \end{bmatrix}_{ik} \begin{bmatrix} S_z & S_+ \\ S_- & -S_z \end{bmatrix}_{jl} \right\rangle$$

$$= J^2 \langle M_{ik} M_{lj} \rangle, \quad (22.16)$$

while $[\mathcal{D}_0]^p$ involves its complex conjugate $J^2 \langle M_{lj} M_{ik} \rangle$. Note that terms like $[\sigma^z]_{ij} [\sigma^x]_{kl} \neq [\sigma^x]_{kl} [\sigma^z]_{ij}$, as the two σ operators do not commute. This means when we add the terms $\Sigma_{ij}^{\text{in}} = \sum_{kl} [\mathcal{D}_0]_{ijkl}^n \Gamma_{kl} f$ and $\Sigma_{ij}^{\text{out}} = \sum_{kl} [\mathcal{D}_0]_{ijkl}^p \Gamma_{kl} (1 - f)$ to extract the scattering rate/broadening (Eq. (21.21)), the Fermi function f does not cancel out due to this non-commutativity. The resulting uncompensated Fermi function in the broadening can create a resonance at the Fermi energy, which leads to a Kondo effect in metals with dilute magnetic impurities and quantum dots (NEMV Section 27.6). Such effects typically need inclusion of strong correlation effects P22.12 and 22.13 beyond a perturbative transport approach.

22.4   Numerical Modeling of Inelastic Transport

The following code shows how a scattering current flows between two levels at different energies P12.5 — which can be easily generalized to entire bands.

Matlab code for two-level scattering current

```
%% Commented parts not shown
%% Define parameterss hw, eps1, eps2, m, q, mu, hbar, kT, V, D0
%% Define H, sig1, sig2, gam1, gam2
%% Define energy grid, mu1, mu2, f1, f2, Nw
```
%% Define self-consistency limits scllim,sclum, $\pm \left[mod(\frac{dE}{N_\omega}) + 1 \right]$ from ends
```
%% construct tensors Dab, Dem
    Aabs = sqrt(Nw*D0); Aems = sqrt((Nw+1)*D0);
    Uabs1 = [0 Aabs;0 0]; Uabs2 = [0 0;Aabs 0];
    Uems1 = [0 Aems;0 0]; Uems2 = [0 0;Aems 0];
    Dab = kron(conj(Uabs1),Uabs1)+kron(conj(Uabs2),Uabs2);
    Dem = kron(conj(Uems1),Uems1)+kron(conj(Uems2),Uems2);
%%Initialize Gn,Gp,G,A,sigin1,sigin2,sigs,sigsr,sigsi,sigsin,sigsout
    Gn=zeros(2,2,NE); Gp=...
%% calculate equilibrium G, Gn, A, Gp (Eqs. 21.10, 21.19, 21.21)
    for k = 1:NE
        sigin1(:,:,k) = gam1*f1(k); sigin2(:,:,k) = gam2*f2(k);
        G(:,:,k) = inv((E(k)+zplus)*eye(2)-H-sig1-sig2);
        Gn(:,:,k) = G(:,:,k)*(sigin1(:,:,k)+sigin2(:,:,k))*G(:,:,k)';
        A(:,:,k) = 1i*(G(:,:,k)-G(:,:,k)');
        Gp(:,:,k) = A(:,:,k)-Gn(:,:,k);
    end
%% calculate G, Gn, A, Gp self consistently for all energies
count = 0; err = 1000;
while(err>errmax)
    err1 = 0; err2 = 0;
    for k = scllim:sculim
        % reshape (2,2,E) → (4,E), multiply D, reshape back (2,2,E)
        sigsinnew = reshape(Dem*reshape(Gn(:,:,k+hwn),4,[]),2,2)+...
        Dab*reshape(Gn(:,:,k-hwn),4,[]),2,2);  % Eq. 22.7
        sigsoutnew = reshape(Dab*reshape(Gp(:,:,k+hwn),4,[]),2,2)+...
        Dem*reshape(Gp(:,:,k-hwn),4,[]),2,2);  % Eq. 22.7
        err1 = err1+sum(sum(abs(sigsout(:,:,k)-sigsoutnew)))/...
```

```
            sum(sum(abs(sigsout(:,:,k)+sigsoutnew)));
         err2 = err2+sum(sum(abs(sigsin(:,:,k)-sigsinnew)))/...
            sum(sum(abs(sigsin(:,:,k)+sigsinnew)));
         sigsout(:,:,k) =(sigsout(:,:,k)+sigsoutnew)/2;
         sigsin(:,:,k) = (sigsin(:,:,k)+sigsinnew)/2;
         sigsi(:,:,k) = (-1i/2)*(sigsin(:,:,k)+sigsout(:,:,k));%  Eq. 21.21
         G(:,:,k) = inv((E(k)+zplus)*eye(2)-H-sig1-sig2-sigs(:,:,k));
         Gn(:,:,k)=G(:,:,k)*(sigin1(:,:,k)+sigin2(:,:,k)+sigsin(:,:,k))*G(:,:,k)';
         A(:,:,k) = 1i*(G(:,:,k)-G(:,:,k)');
         Gp(:,:,k) = A(:,:,k)-Gn(:,:,k);
      end
      %Hilbert tranform for real part of sigs (Eq. 21.26)
      for ii = 1:2
         for jj = 1:2
            sigsr(ii,jj,:) = -imag(hilbert(imag(sigsi(ii,jj,:))));
         end
      end
      sigs = sigsr+sigsi;
      err = err1 + err2;
   end

%calculate I (Eq. 22.1), similarly I2, Is and plot..
for k = 1:NE
   I1(k) = (q/h)*real(trace(sigin1(:,:,k)*A(:,:,k))-trace(gam1*Gn(:,:,k)));
end
```

Homework

P22.1 **1D wire physics.** Let us first set up the current through a 1D chain with $N = 101$ atoms. We use the Hamiltonian Eq. (8.3) ($\epsilon_0 = 2$ eV, $t_0 = 1$ eV) with an E–k given by Eq. (8.5) and self-energies Eq. (5.11). Choose an energy grid of 51 points from -2 to 5 eV. Use Eqs. (21.10), (21.7), (21.11), (22.4) to calculate Green's function G, broadening $\Gamma_{1,2}$, density of states D and transmission $\mathcal{T}$. If you do things right, you should get an answer that should look 'obvious'!

Repeat now for a few variations:

(a) Put weak contact couplings by scaling down $\Sigma_{1,2}$ by a factor of 10. Explain what you see.

(b) Put a defect by making one diagonal term in $[H]$ different, say $2t_0 + U$ with $U = 1$ eV, like a dopant atom. You should see a dopant band split-off from the regular band, as expected.

(c) Put a barrier by putting $U = 1$ eV over say 10 consecutive atoms. We should now see resonances and tunneling. Show by varying the barrier width upward from 10 that below barrier transmission (outside the band-width) drops exponentially (Problem P20.4).

(d) Put two barriers separated by 10 atoms, and you should again see resonances. Also plot the energy-position dependent current density $J_i = (i/\hbar)([G^n]_{i,i\pm1}[H]_{i\pm1,i} - [H]_{i,i\pm1}[G^n]_{i\pm1,i})$.

(e) Finally, put an atom parallel to the chain, like a T-junction. Treat the two connected atoms as a channel with a 2×2 Hamiltonian, write down the self-energies (which only couple to one atom), and find the transmission. Show that as we gate the onsite atom energies, the transmission changes shape from maxima to minima, describing the so-called *Fano resonance*.

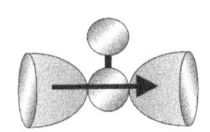

P22.2 **Quantum-classical transition for a barrier.** Recall that electrons around slowly varying potentials act classically (Section 4.1). We demonstrated this analytically with a smooth step function, Eq. (4.4). Work out the NEGF-based quantum transmission for a smooth barrier potential $U(x) = \left[\tanh\{(l/2 - x)/2a\} + \tanh\{(l/2 + x)/2a\}\right]/2\tanh(l/2a)$ and show how the transmission evolves from quantum to classical as the smoothness a/L increases. Results for $a/L \to 0$ (blue) and $a/L = 1/5$ (black) are shown compared to the classical expectation (red). To keep the flat top of the same width, we chose $l = 2.3L$. Note how the transmission looks classical for a smooth barrier even when it's wider.

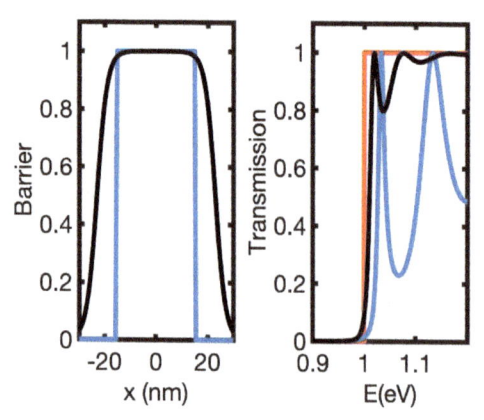

P22.3 **Multi-dimensional wires.** Let us calculate transmission for the wire geometry in Fig. 8.5. You can treat one two atom unit cell as the device, and find the DOS (recall the Batman plot in **P11.2**!) and transmission. Find its current.

Let us now look at a dimer geometry, with hoppings $t_1 = 2$ eV and $t_2 = 1$ eV. Vary the onsite energies from left to right smoothly over a depletion width to mimic a silicon PN junction (assume $N_D = N_A = 10^{16}$cm^{-3} to work out the depletion width and built-in potential). Calculate the transmission and energy–position-dependent current density $J_i = i/\hbar\left([G^n]_{i,i\pm1}[H]_{i\pm1,i} - [H]_{i,i\pm1}[G^n]_{i\pm1,i}\right)$.

P22.4 **Tunnel transistor.**

A tunnel field effect transistor (TFET) has a p-i-n junction, where a central gate drives the intrinsic 'i' region below the 'p' region, till the valence band of the p region adjoins the conduction band of the i-region, and electrons can tunnel between the two. This happens abruptly as the bandedges cross each other, so that the subthreshold swing (steepness of the gate transfer I–V_G curve) is less than the $(k_B T/q)\ln 10 \approx 60$ mV/decade limit thermodynamically imposed (in reality, it is compromised by higher-order effects).

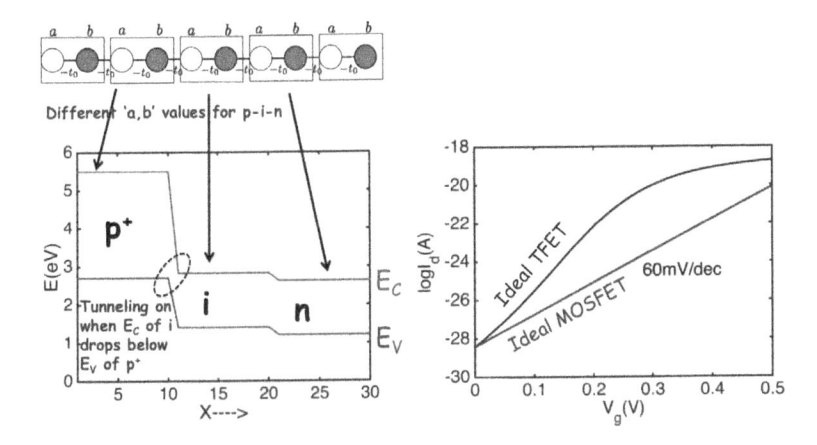

We will build a 'toy' model for a TFET. To create valence and conduction bands, let us create a 1D p-type source of 20 dimers with onsite and hopping terms (all in eV) $a = 2.7, b = 5.5, t_0 = 1$, an i-type channel of 20 dimers with $a = \sqrt{2}, b = 2 \times \sqrt{2}, t_0 = 1$, and finally an n-type drain of 20 dimers with $a = \sqrt{2} - 0.2, b = 2 \times \sqrt{2} - 0.2, t_0 = 1$. Show that we now have a p-i-n junction. Assume finally a drain voltage $V_D = 0.3$ eV, with no drain potential in the i-channel (perfect gate control). How much gate bias V_{th} do we need to turn on tunneling? Use NEGF to get the I–V above.

P22.5 **Majorana Fermions: a quantum 'horcrux' (courtesy Prof. Bhaskaran Muralidharan).** There is an active field called topological quantum computing, which proposes to use as quantum bits two spatially separated, entangled states, so that each is protected by non-locality. One such qubit pair consists of two Majorana fermions generated at the ends of a quantum wire with p-wave superconductivity (in practice, this can be generated using a quantum wire with strong spin–orbit coupling that splits the bands laterally P10.6 , imposing an axial magnetic field to separate the crossing bands, and adding an s-wave superconductor like Al or Nb that creates a coherent superposition of an electron and a hole). In addition to the conventional hopping term, we now have a delocalized pairing term that looks like $\Delta\Psi_i^\dagger\Psi_{i+1}^\dagger$. We can, however treat the electron and hole as our basis set (the Nambu representation), keeping in mind that electron creation $\Psi_{el}^\dagger$ implies hole annihilation Ψ_{hole}), which then looks like two cross-coupled wires (Fig. 8.5) — one for the electron, and one for the hole.

The onsite and hopping terms, ignoring spins, then look like

$$\alpha = \begin{pmatrix} \epsilon & 0 \\ 0 & -\epsilon \end{pmatrix}, \quad \beta = \begin{pmatrix} -t & \Delta \\ -\Delta & t \end{pmatrix}, \tag{22.17}$$

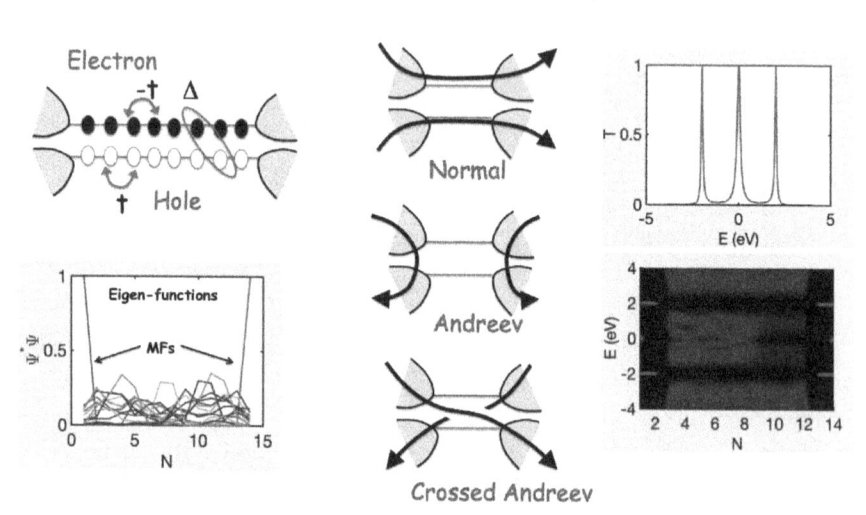

where the sign flip comes because of exchange antisymmetry of electrons and holes, $\Psi^\dagger \Psi = -\Psi \Psi^\dagger$. Now, assume there are separate leads to the electron and hole channels with coupling $\gamma = 0.2$ eV, and calculate the (a) normal electron conductance (electron enters from left lead and escapes out the right lead), (b) Andreev conductance (electron enters from left, hole exits at left) and (c) crossed Andreev (electron enters from left, hole exits at right). Show in particular that the NEGF evaluated Andreev transmission has a peak at zero bias of magnitude unity, and that this peak is robust even if there is site disorder along the chain (bottom right). Also plot the wavefunctions to show the two localized end states (bottom left). We assumed 14 atoms, $\Delta = t = 1$ eV, $\epsilon = 0$ and $\eta = 1e - 9$.

> **P22.6** **Spin transfer and spin–orbit torque.** Just like charge current, we can define a spin current at spatial location i, coupled to nearest neighbor sites $i \pm 1$, as

$$\vec{I}_i^S = i\frac{\vec{\sigma}}{2}\Big[G^n_{i,i\pm1}H_{i\pm1,i} - H_{i,i\pm1}G^n_{i\pm1,i}\Big]. \tag{22.18}$$

In presence of a spin-dependent Hamiltonian $H = \xi\vec{\sigma}\cdot\vec{B}$, we can get a divergence of the spin current density. Assuming $\vec{I}_S$ gets completely absorbed by a soft ferromagnet, we can get the torque by just the incoming spin current from the non-magnet to the ferromagnet.

(a) Calculate and plot the spin torque assuming $\vec{B} = \vec{M}$, the magnetization of a fixed magnet, and compare with Eq. (10.13) for spin transfer torque (STT).

(b) Calculate and plot the spin torque assuming $\vec{B} = \vec{k}$, representing for instance spin–orbit torque from a heavy metal adjoining the soft magnet for SOT.

(c) Numerically solve the 2-d (θ, ϕ) polar plot for the magnetization in the LLG equation for a magnet with anisotropy and STT. For SOT, model a metal with a 2D cross-section on a tight binding grid, to allow electrons injected at one side into the metal to veer to the top and bottom of the metal with opposite spins, generating a spin Hall voltage, and allowing one species of spin to preferentially enter the magnet and torque it.

P22.7 **Tunnel magnetoresistance and spin-flip scattering.** A spin valve consists of a channel sandwiched between two ferromagnetic contacts, one hard one soft (easily flippable), so that an external magnetic field can orient them mutually parallel or antiparallel. Let us redo the 1D chain first. To account for spin, we double the size of the Hamiltonian, where each entry is expanded to a 2×2 matrix for the 2 spin states. In other words, instead of our original basis of points $(1, 2, 3, \ldots, N)$, we now have the basis set $(1 \uparrow, 1 \downarrow, 2 \uparrow, 2 \downarrow, \ldots, N \uparrow, N \downarrow)$. We assume no spin effects in the channel material, so that up spins at site i only couple with up spins at sites $i \pm 1$, and same for down spins. This means your off diagonal terms $-t_0$ should couple the $i \uparrow$ ONLY with the $i \pm 1, \uparrow$ terms, which aren't nearest neighbor basis sets any more because we list every up next to its down. Set up the new 101×101 Hamiltonian.

Now to set up the self-energies. In presence of spins, Σ_1 will have a non-zero 2×2 block at its top left diagonal corner, and Σ_2 will have a non-zero 2×2 block at its bottom right diagonal corner. Assuming an angle θ between the two spins and polarization P, we can write these two blocks as

$$-t_0 e^{ika} \begin{bmatrix} 1+P & 0 \\ 0 & 1-P \end{bmatrix} \text{ and } -t_0 e^{ika} \begin{bmatrix} 1+P\cos\theta & 0 \\ 0 & 1-P\cos\theta \end{bmatrix}, \quad (22.19)$$

respectively, since the two contacts are oppositely polarized.

Instead of an energy loop, set $E = 2t_0$ (mid-band), and loop over a polarization P from 0 to 1, and angle θ from 0 to π. Plot the transmission as a function of P for antiparallel contacts ($\theta = \pi$). Also write down $\mathcal{T}$ analytically and compare.

Now, let us add a spin flip term $A\sigma_x$, where σ_x is the Pauli spin matrix along x, to each diagonal 2×2 block in H. Replot your $\mathcal{T} - P$ plot for three different A values between 0 and 0.5. Explain what is happening here. Show that we get the same result from the following circuits for parallel and antiparallel combinations, where the polarization $P = (R_{\text{maj}} - r_{\text{min}})/(R_{\text{maj}} + r_{\text{min}})$ involves the majority and minority spin resistances, and R_{sc} is the impurity spin resistance. Show that $G_P = (G_{\text{maj}} + g_{\text{min}})/2$, and $G_{AP}/G_P = 1 - P^2(G_{\text{max}} + g_{\text{min}})/(G_{\text{max}} + g_{\text{min}} + 2G_{\text{sc}})$. Also note that if we can maintain a polarization difference between impurity down and up spins, $F_u > F_d$, like a spin battery, then there is an entropy-driven current $I_{\text{sc}} \propto F_u - F_d$ (Eq. (20.10)) between the branches until the spins get depolarized by thermal forces. There is a wonderful *gedanken* experiment by Supriyo Datta that explains this physics very elegantly.

P22.8 **Charge and Spin voltages.**

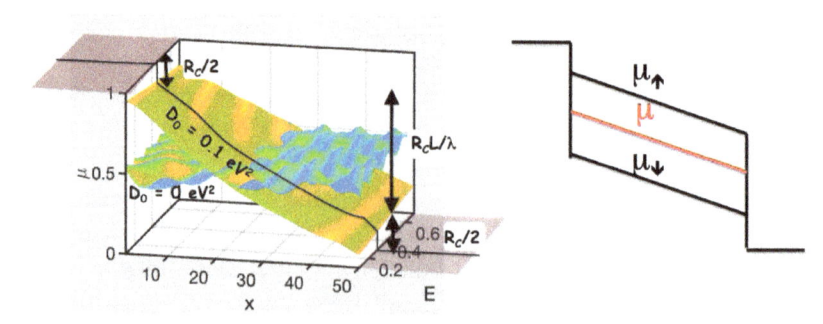

Repeat the 1D wire ($t = 1$ eV, 101 atoms) along the x axis, under 1 V bias, with elastic dephasing $\Sigma_{\text{sc}} = \mathcal{D}_0 \otimes G$, $\Sigma_{\text{sc}}^{\text{in}} = \mathcal{D}_0 \otimes G^n$ self-consistently, assuming momentum and dephasing scattering (Eq. (22.15),

$D_0 = 0$, 0.05, 0.1 eV^2 times I). From the result, plot the diagonal entries of G^n/A, which represent f. Assume zero temperature, where $f_1 = 1$, $f_2 = 0$, and plot f vs x, E, interpreting $\delta f = \delta \mu$. Show that we move from ballistic to diffusive, similar to $\boxed{\text{P20.4}}$, as we increase scattering (top left).

Now, solve for the spin potentials across FM/NM, FM/SC/FM and NM/TI/NM pairs (FM: Ferromagnet, SC: Semiconductor, NM: Non-magnetic metal) as in $\boxed{\text{P19.4}}$ by calculating $\uparrow$ and $\downarrow$ entries of diag(G^n)/diag(A) as $\mu_{\uparrow,\downarrow}$ and plotting. For FMs, we add different onsite energies for the up and down spins. For TI, we add spin–orbit coupling $i\lambda\sigma_x\partial/\partial x$ in 1D as a Kronecker product.

$\boxed{\text{P22.9}}$ **Spin dephasing and the Kondo effect.** Start with Eq. (21.16),
$$\Sigma_{ij}^{\text{in}} = \sum_{ABkl} \tau_{iA;kB} \underbrace{\langle \psi_{kB}\psi_{lB}^{\dagger} \rangle}_{\rho G_{kl}^n} \tau_{jA;lB}^{\dagger} = \sum_{kl} \mathcal{D}_{ijkl}^n G_{kl}^n, \text{ where } A, B \text{ are}$$
the internal impurity spin states. Simplify $\mathcal{D}^n$, using the Hamiltonian in Eq. (22.16), and assuming the impurity spin density matrix is diagonal, $\rho = \begin{bmatrix} F_u & 0 \\ 0 & F_d \end{bmatrix}$. Write down a similar equation for $\Sigma^{\text{out}} = \tau^{\dagger}\langle\psi_I^{\dagger}\psi_I\rangle\tau$ and show that their sum, the broadening, retains a Fermi function f, which leads to the Kondo effect (NEMV, Section 27.6).

$\boxed{\text{P22.10}}$ **Incoherent scattering simplified.** Assume a 2-state system coupled separately to left and right contacts labeled α, β, and a scattering deformation potential with only non-zero components
$$\mathcal{D}_{ii,jj}^n = M^2 \begin{pmatrix} 1 & P_{\alpha\beta} \\ P_{\beta\alpha} & 1 \end{pmatrix} \text{ and } \mathcal{D}_{ii,jj}^p = M^2 \begin{pmatrix} 1 & P_{\beta\alpha} \\ P_{\alpha\beta} & 1 \end{pmatrix}.$$

Using $\Gamma_{\alpha,\beta} = \Sigma_{\alpha,\beta}^{\text{in}} + \Sigma_{\alpha,\beta}^{\text{out}}$, show that the scattering current $I_{\alpha,\text{sc}} = -I_{\beta,\text{sc}} = (q/h)\left(\Sigma_{\alpha}^{\text{in}}A_\alpha - \Gamma_\alpha G_\alpha^n\right)$ can be written as $(qM^2/h)A_\alpha A_\beta \Big[P_{\alpha,\beta}f_\beta(1-f_\alpha) - P_{\beta\alpha}f_\alpha(1-f_\beta)\Big]$, where $f_{\alpha,\beta} = G_{\alpha,\beta}^n/A_{\alpha,\beta}$. This formally derives Eq. (20.10).

$\mathcal{D}$ can be derived from an off-diagonal coupling of the form $U = M\vec{\sigma}\cdot\vec{S}$, using $\mathcal{D}_{ij,kl} = \langle U_{ik}U_{lj}^*\rangle$, where $\langle\ldots\rangle$ involves a trace over the density matrix of the form $\boxed{\text{P22.9}}$ for the impurity spin $\vec{S}$. Derive $\mathcal{D}_{ii,jj}$ above.

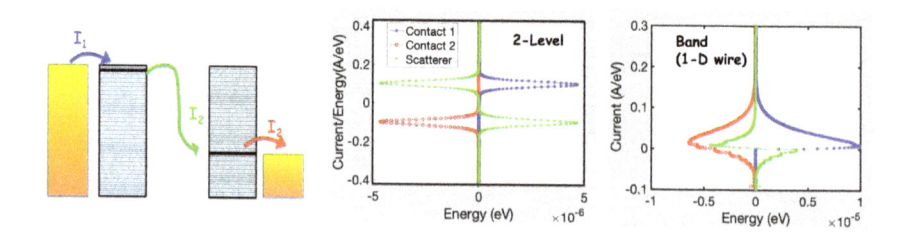

P22.11 **NEGF with electron–phonon scattering (courtesy Dr. Samiran Ganguly).** Consider two energy levels $\epsilon = \pm 1$ eV, separated by a phonon energy and electron-phonon deformation potential $D_0 = 10^{-4}$ eV. Under a large enough applied bias, electrons can enter the first level, emit a phonon and escape from contact 2 $\boxed{\text{P12.5}}$. Using the code toward the end of this chapter, show the energy dependence of the currents (bottom center). Next, apply to a 1D wire, and show how the two-level current expands into bands where the electrons enter at high energy and exit at lower (bottom right). Show that we have conservation of current $I_1 + I_2 + I_s = 0$ and energy currents $I_{E1} + I_{E2} + I_{Es} = 0$, where the latter have an added $(E - \mu_1)$, $(\mu_2 - E)$ and $(\mu_1 - \mu_2)$ factor inside their energy integrals, respectively $\boxed{\text{P19.1}}$. The net current is $(I_1 - I_2)/2$.

Now, apply this to a single energy level ϵ_0 and show that near its peak the conductance dI/dV produces phonon emission and absorption sidebands (NEMV, Section 24.5 — you see signatures above right). Also apply far from resonance, and show that we get inelastic electron tunneling spectra, where the second-derivative d^2I/dV^2 shows peaks (NEMV, Section 24.4).

P22.12 **NEGF with strong interaction: Coulomb Blockade.** When we fill an energy level ϵ_0 with an up electron, Coulomb repulsion will push the down electron to a higher energy $\epsilon_0 + U_0$, creating a pronounced plateau in a conductance vs voltage plot. Capturing non-equilibrium correlations (NEMV, Chapters 25–27) requires working in the eigensector (Fock space) of a many-body Hamiltonian, and setting up rate equations among the various particle number sectors. However, approximations can be made using equation of motion techniques. One approximation for an electron occupying an energy level ϵ_0 is to add a Coulomb cost

$$U_\uparrow = U_0 \Delta n_\downarrow, \quad U_\downarrow = U_0 \Delta n_\uparrow, \tag{22.20}$$

which has to be calculated self-consistently with $\Delta n_{\uparrow\downarrow}$, including contact self-energies as well. When $n_{\bar{\sigma}} = 1$, we pay an added energy cost U_0.

Consider $\epsilon_0 = 1$ eV, $U_0 = 5$ eV, $kT = 0.0005$ eV, $\eta = 10^{-4}$, assuming slight initial bias in up vs down spin occupancy for computational convergence. Show that the charge density vs drain bias N–V shows a jump when the up spin state is filled, then a Coulomb Blockade plateau as the down state is pushed away, and then another jump when the latter is again filled. Show a similar plateau for I–V. Plot for V from 0 to 8 V. For the broadenings, first assume equilibrium with injecting contact ($\gamma_2 = 0.1$ eV, $\gamma_1 = 0.01$ eV), then with removing contact ($\gamma_2 = 0.01$ eV, $\gamma_1 = 0.1$ eV) and finally strongly non-equiibrium ($\gamma_1 = \gamma_2 = 0.05$ eV).

The plateau heights, however, need the full machinery of Fock space.

P22.13 **Trap-channel interactions.** Let us see how a trap can block the current through a channel and create a drop in current. This is very relevant for present day devices. The trap is essentially a non-conducting level that communicates very weakly with the contacts (small γ_s), so that once a charge is injected onto it, it stays there for a very long time. Filling the trap, however, pushes the conducting level (channel, with larger γ_s) out of the conduction window through Coulomb repulsion, thus blocking the channel. Our aim in this problem is to create a 'toy' numerical model for this trap-channel interaction.

Consider a conducting channel with a single energy level centered around ϵ_0, with equal couplings $\gamma_1 = \gamma_2$ to the contacts. The channel density of states is described by a Lorentzian $\boxed{\text{P20.3}}$ around $E = \epsilon_0$. The Fermi energy of the channel lies below the level, so that the channel is initially empty. Let us raise the drain electrochemical potential $\mu_2 = E_F + qV$ by putting a negative voltage relative to the source, held at $\mu_1 = E_F$. At some point, the channel is filled by the drain, and starts conducting.

(a) Set up NEGF to compute the I–V for a voltage range between zero and 0.6 V. Use the following parameters: $E_F = 0, \epsilon_0 = 0.3$ eV, $\gamma_1 = \gamma_2 = 0.002$ eV, $kT = 0.025$ eV. Choose 101 voltage points in the voltage range between 0 and 0.9 V. For the energy integration needed to compute the current, choose 5501 energy grid points between -4 to $+4$ eV.

(b) Let us now include a trap, another Lorentzian with different parameters ($\epsilon_1 = 0.5$ eV, and weak couplings $\gamma_{1t} = \gamma_{2t} = 0.0002$ eV to the contacts). Once the drain voltage is high enough to fill the trap, you will get additional current from the trap.

(c) Now, let's add the Coulomb repulsion from the trap onto the channel. In the numerical code, add a Coulomb potential $U_0 n_t$ to the channel energy ϵ_0, where $U_0 = 5$ eV is the single-electron charging energy due to the trap, and n_t is the trap charge. Compute n_t first and then use the corresponding potential as an added term to ϵ_0 to compute the shifted channel density of states, and then calculate current through both. Your I–V should show what is called *Negative Differential Resistance*. Explain.

Let us evaluate the subthreshold swing, the inverse slope of the I–V_G for $V_G < V_{\text{th}}$. Fix $V_D = 0.3$ V, vary V_G from 0 to V_{th}, and plot the log (I_D) with V_G and also the 60 mV/dec line. Assume $E_F = 2.6$ eV, $k_B T = 0.025$ eV. Integrate E in the valence band range of the source. Use a dense energy grid of 201 points. Use a metallic contact self-energy $\Sigma_{1,2} = [-i, 0; 0, -i]$. In order to eliminate coherent oscillations in 1D, add a scattering factor $i\eta$ in to Green's function, $G = [(E+i\eta)I - H - \Sigma_1 - \Sigma_2]^{-1}$, with $\eta = 5 \times 10^{-2}$ eV.

Chapter 23

Time-Dependent Effects

23.1 Photoconductivity

The interaction of electrons with photons lies at the very foundation of quantum mechanics. It was the discrete photoinduced electronic transitions in an atomic spectrum $\hbar\omega = E_2 - E_1$ that lead to the realization of 'quantization' as a basic property of nature. It is these discrete electronic transitions that contribute to the bright colors of chlorophyll, indigo, haemoglobin and alkaloids. Photoemission measurements (and inverse PE) are routinely used to map out energy bands of solids. Solar cells and photovoltaic devices are used to harness energy from the photons to convert to electrical voltage and do useful work. Much of the efficiency in coupling photons with electronic bands relies on two factors — the energy eigenvalues of the electronic system (e.g., the bandgap) to achieve resonance with the photon energy, and the energy eigenvectors of the electronic system (the relative shapes of the conduction and valence band wavefunctions, specifically their electric dipole matrix element) that allows us to conserve the overall chiral symmetries of the photons and electrons in order to make the transition possible.

The study of photoconductivity lies in a broad general field of AC response, where an electron is interrogated by an external AC field $\vec{\mathcal{E}} = \hat{n}\mathcal{E}_0 \cos\omega t$, the frequency ω ranging from RF and IR to visible. The analysis is, on the one hand, complex, because we need to take into account the 3D polarization of the photons, their coupling with the electrons, and their thermalization with an external bath. However, there are two simplifying features that make this analysis easier. At high frequency, the corresponding short wavelengths means that for modest sized devices the contacts are not sampled by the electron waves. Instead of metallurgical contact self-energies, we can focus then on the intrinsic photon properties and the dissipative effects of the bath. Secondly, we often look at 'small signal' AC fields, for instance in RF devices, and this allows to linearize our equations and simplify them further.

The Hamiltonian for an electron in an AC field can be written by including the electromagnetic momentum that gives us the Lorenz magnetic force $\vec{F} = q\vec{v} \times \vec{B}$ upon spatial differentiation (NEMV, Chapter 21)

$$\mathcal{H} = \frac{(\vec{p} - q\vec{A})^2}{2m} = \mathcal{H}_0 - \underbrace{\frac{q\vec{A} \cdot \vec{p}}{m}}_{\mathcal{H}_{\text{el-photon}}} + O(A^2), \qquad (23.1)$$

with some caveats (we invoke a 'gauge degree of freedom' that makes $\vec{\nabla} \cdot \vec{A} = 0$). How would this Hamiltonian create an optical transition? To estimate that, we can go back to (what else?) Fermi's Golden Rule (Eq. (14.2)). The delta functions $\delta(E_k^C - E_{k'}^V - \hbar\omega)$ enforce energy conservation — for instance for silicon that has indirect bandgap, it will need a photon to connect between the two different k points where E_C and E_V reside. We know however that this is not allowed in silicon because of the widely different velocities of electrons and photons — on the scale of the silicon Brillouin zone, optical transitions must look vertical, meaning they cannot bridge the E_C and E_V points. The constraint is ensured by the matrix element that enforces symmetries, including momentum conservation. A typical matrix element between the conduction and valence band Bloch states reads $\langle \vec{k} | \mathcal{H} | \vec{k}' \rangle = \int d^3\vec{r} \int d^3\vec{r}' u_{kC}^*(\vec{r}) e^{-i\vec{k} \cdot \vec{r}} \mathcal{H}_{\text{el-photon}} u_{k'V}(\vec{r}') e^{i\vec{k}' \cdot \vec{r}'}$ $\boxed{\text{P23.1}}$. Translating to center of mass and difference coordinates, the exponents can be seen to have the form $e^{i(\vec{k}' - \vec{k}) \cdot (\vec{r} - \vec{r}')}$ integrated over $d^3(\vec{r} - \vec{r}')$. This term is zero for $\vec{k}' \neq \vec{k}$, rendering the delta function term zero and eliminating the probability of photoabsorption in bulk silicon. The electron–photon matrix element itself integrates over $\vec{p} = m\vec{r}/T$, where $T = \hbar/\Delta E$ is the transition time between the two energy states. Thus, the transition is driven by the dipole matrix element $\propto q\langle \vec{r} \rangle$.

23.2 Time-Dependent NEGF

The NEGF equations can be modified fairly easily to include time-dependent processes. The general matrix structure of the equations stays intact, except time appears as an added matrix index (Eq. (23.5)). We start with the open boundary time-dependent Schrödinger equation for the αth contact

$$\left[i\hbar\frac{\partial}{\partial t} - \mathcal{H}(t)\right]\psi(t) - \int dt_1 \Sigma_\alpha(t,t_1)\psi(t_1) = S_\alpha(t), \qquad (23.2)$$

whose solution is

$$\psi(t) = \int dt_1 G(t,t_1) S_\alpha(t_1),$$

where $\left[i\hbar\frac{\partial}{\partial t} - \mathcal{H}(t)\right] G(t,t') - \int dt_1 \Sigma_\alpha(t,t_1) G(t_1,t') = \delta(t-t').$ (23.3)

This gives us the current for the αth terminal as

$$I_\alpha(t,t) = q\left[\frac{\partial}{\partial t} + \frac{\partial}{\partial t'}\right]\text{Tr}\langle\psi(t)\psi^\dagger(t')\rangle\Big|_{t\,=\,t'}$$

$$= q\text{Tr}\langle\frac{\partial\psi(t)}{\partial t}\psi^\dagger(t') + \psi(t)\frac{\partial\psi^\dagger(t')}{\partial t'}\rangle\Big|_{t\,=\,t'}$$

$$= \frac{q}{i\hbar}\text{Tr}\langle\left[\mathcal{H}(t)\psi(t) + \int dt_1 \Sigma_\alpha(t,t_1)\psi(t_1) + S_\alpha(t)\right]\psi^\dagger(t)$$

$$-\psi(t)\left[\psi^\dagger(t)\mathcal{H}(t) + \int dt_1 \psi^\dagger(t_1)\Sigma_\alpha^\dagger(t,t_1) + S_\alpha^\dagger(t)\right]\rangle$$

$$= \frac{q}{i\hbar}\text{Tr}\left[\int dt_1 \Sigma_\alpha(t,t_1)\underbrace{\langle\psi(t_1)\psi^\dagger(t)\rangle}_{G^n(t_1,t)} - \int dt_1 \underbrace{\langle\psi(t)\psi^\dagger(t_1)\rangle}_{G^n(t,t_1)}\Sigma_\alpha^\dagger(t,t_1)\right.$$

$$\left. + \int dt_1 \underbrace{\langle S_\alpha(t)S_\alpha^\dagger(t_1)\rangle}_{\Sigma_\alpha^{in}(t,t_1)} G^\dagger(t,t_1) - \int dt_1 G(t,t_1)\underbrace{\langle S_\alpha(t_1)S_\alpha^\dagger(t)\rangle}_{\Sigma_\alpha^{in}(t_1,t)}\right],$$

$$(23.4)$$

so that

$$I_\alpha(t) = \overbrace{\frac{q}{i\hbar}\int dt_1 Tr\left[\Sigma_\alpha^{in}(t,t_1)G^\dagger(t_1,t) - \text{h.c.}\right]}^{I_\alpha^{in}(t)}$$

$$\underbrace{-\frac{q}{i\hbar}\int dt_1 Tr\left[G^n(t,t_1)\Sigma_\alpha^\dagger(t_1,t) - \text{h.c.}\right]}_{I_\alpha^{out}(t)}$$

$$I(t) = \left[I_1(t) - I_2(t)\right]/2 \quad \text{(for two leads)}. \qquad (23.5)$$

h.c. is Hermitian conjugate. As expected, getting rid of the time index at steady-state reverts us back to the regular NEGF equations (Eq. (22.1)).

23.3 Fourier Transform

Converting to energy arguments, we use the symbol

$$O(\omega) = \int dt O(t) e^{i\omega t},$$

$$O(E_1, E_2) = \int dt_1 \int dt_2 O(t_1, t_2) e^{-i[E_1 t_1 - E_2 t_2]/\hbar},$$

$$O(t_1, t_2) = \int \frac{dE_1}{2\pi} \int \frac{dE_2}{2\pi} O(E_1, E_2) e^{i[E_1 t_1 - E_2 t_2]/\hbar}. \tag{23.6}$$

This gives us

$$\begin{aligned} I(\omega) = \frac{q}{i\hbar} Tr \int dt e^{i\omega t} \int dt' \int \frac{dE_1}{2\pi} \int \frac{dE_2}{2\pi} \int \frac{dE_3}{2\pi} \int \frac{dE_4}{2\pi} \\ \times e^{i(E_1 t - E_2 t')/\hbar + i(E_3 t' - E_4 t)/\hbar} \\ \times \left[\Sigma_\alpha(E_1, E_2) G^n(E_3, E_4) - G(E_1, E_2) \Sigma_\alpha^{in}(E_3, E_4) + \cdots \right]. \end{aligned} \tag{23.7}$$

The two time integrals over t and t' give us $2\pi\delta([E_1 - E_4]/\hbar + \omega)$ and $2\pi\delta([E_3 - E_2]/\hbar)$, so the integral simplifies to

$$\begin{aligned} I_\alpha(\omega) &= \frac{q}{i\hbar} \int \frac{dE_3}{2\pi} \int \frac{dE_4}{2\pi} [\Sigma_\alpha(E_4 - \hbar\omega, E_3) G^n(E_3, E_4) + \cdots] \\ &= \frac{q}{\hbar} \int_{-\infty}^{\infty} \frac{dE}{2\pi} \int_{-\infty}^{\infty} \frac{dE_1}{2\pi} \mathrm{Tr}[\{G(E + \hbar\omega, E_1) \Sigma_\alpha^{in}(E_1, E) \\ &\quad - \Sigma_\alpha^{in}(E_1, E) G^\dagger(E, E_1 - \hbar\omega)\} \\ &\quad + \{G^n(E_1, E) \Sigma_\alpha^\dagger(E, E_1 - \hbar\omega) - \Sigma_\alpha(E + \hbar\omega, E_1) G^n(E_1, E)\}] \\ &= \frac{q}{\hbar} \iint_{-\infty}^{\infty} \frac{dE \, dE_1}{2\pi \, 2\pi} \mathrm{Tr}[\{G(E + \hbar\omega, E_1) \Sigma_\alpha^{in}(E_1, E) \\ &\quad - \Sigma_\alpha^{in}(E + \hbar\omega, E_1) G^\dagger(E_1, E)\} + \{G^n(E + \hbar\omega, E_1) \Sigma_\alpha^\dagger(E_1, E) \\ &\quad - \Sigma_\alpha(E + \hbar\omega, E_1) G^n(E_1, E)\}]; \end{aligned} \tag{23.8}$$

where in the last line we swapped the variables E and E_1 in terms 2 and 3, and then made the transformation $E \to E + \hbar\omega$.

23.4 AC Field and Linear Response

We are interested in steady-state properties of the electrons under a continuously operatiing AC field. Green's functions $G(E_1, E_2)$ can be rewritten as an integral over time (Eq. (23.6)) and then converted to average $T = (t_1 + t_2)/2$ and relative $\tau = t_1 - t_2$ coordinates. If we ignore transients, meaning that $G(T, \tau)$ is independent of the time origin when the AC turns on and thus of T, then the dT integral enforces equality of energy arguments $E_1 = E_2$, and the AC current expression (Eq. (23.8)) can be simplified so that the G and Σ terms have the same argument, $G(E + \hbar\omega) = G(E + \hbar\omega, E + \hbar\omega)$ and so on. Let us also ignore the energy-dependence of Σ assuming wide band contacts, so that the algebra simplifies.

$$
I_\alpha(\omega) = \frac{q}{\hbar} \int \frac{dE}{2\pi} \text{Tr}[i \left\{ G(E + \hbar\omega) - G^\dagger(E) \right\} \Sigma_\alpha^{in}(E + \hbar\omega, E)
$$
$$
- G^n(E + \hbar\omega, E)\Gamma_\alpha]. \tag{23.9}
$$

Let us now look at specifics.

Oscillating channel under optical input. Consider first the case where the contacts are unaffected by a high frequency short wavelength AC field, as the channel alone is irradiated quasi-optically as in a solar cell. In this case, the contacts satisfy

$$
\Sigma_\alpha^{in}(E, E') = \Gamma_\alpha(E)f_\alpha(E)\delta(E - E'),
$$
$$
\Sigma_\alpha(E, E') = [-i\Gamma_\alpha(E)/2 + H(\Gamma_\alpha(E))]\delta(E - E'), \tag{23.10}
$$

H being the Hilbert transform. The entire time-dependence sits in tho Hamiltonian, which we can evaluate using the Crank–Nicolson technique, as we did for a two-state system earlier $\boxed{\text{P10.2}}$, albeit now in linear response.

Oscillating contacts under RF input. Here, we assume a low frequency (e.g., RF) AC signal is electronically fed to contact β, varying its electrochemical potentials sinusoidally. This means in-scattering Σ_β^{in}s get affected by the AC signal, coupling with resonant sidebands energetically. Recall $\Sigma_\beta^{in}(t, t') = \langle S_\beta(t)S_\beta^\dagger(t')\rangle$. In an AC field, we expect these contact wavefunctions to be modified trivially by unitary evolution (Eq. (5.9)),

$S_\beta(t) = S_\beta(0) \exp\left[-iqV_{ac} \int dt' \sin \omega t'/\hbar\right]$. We then get

$$\Sigma_\beta^{in}(E_1, E_2) = \int dt_1 \int dt_2 e^{-iqV_{ac}\cos\omega t_1/\hbar\omega}$$

$$\times \quad \underbrace{\Sigma_\beta^{in}}_{\langle S_\beta(0)S_\beta^\dagger(0)\rangle} \quad e^{iqV_{ac}\cos\omega t_2/\hbar\omega}$$

$$\times e^{-i(E_1 t_1 - E_2 t_2)/\hbar}$$

$$= \sum_{m,n} J_m\left(\frac{qV_{ac}}{\hbar\omega}\right) J_n\left(\frac{qV_{ac}}{\hbar\omega}\right) \int \frac{dE}{2\pi} \int \frac{dE'}{2\pi} \Gamma_\beta f_\beta(E)\delta(E - E')$$

$$\times \underbrace{\int dt_1 e^{i(E - E_1 - n\hbar\omega)t_1/\hbar}}_{2\pi\hbar\delta(E - E_1 - n\hbar\omega)}$$

$$\times \underbrace{\int dt_2 e^{i(m\hbar\omega + E_2 - E')t_2/\hbar}}_{2\pi\hbar\delta(E' - E_2 - m\hbar\omega)}, \tag{23.11}$$

where we have used the generating equation for Bessel functions in terms of the cosine of a cosine $\boxed{\text{P23.2}}$, $e^{iA\cos\omega t} = \sum_{n=-\infty}^{\infty} J_n(A)e^{in\omega t}$, to show the *Tien–Gordon formula*, the emergence of sidebands at $\pm n\hbar\omega$ with weight $J_n(qV_{ac}/\hbar\omega)$ (NEMV, Section 24.7).

For small field amplitudes $\Theta = qV_{ac}/\hbar\omega \ll 1$, the only Bessel functions that survive are $J_0(\Theta) \approx 1$ and $J_{\pm 1}(\Theta) \approx \pm\Theta$. Executing the energy integrals, we get

$$\Sigma_\alpha^{in}(E_1, E_2) = \pm\delta_{\alpha\beta}\Theta\Gamma_\beta\left[f_\alpha(E_1) - f_\alpha(E_2)\right]\delta(E_1 - E_2 \mp \hbar\omega)$$

$$+ \Gamma_\alpha f_\alpha(E_1)\delta(E_1 - E_2). \tag{23.12}$$

Substituting we finally get for an RF input into contact β

$$I_\alpha(\omega) \approx \frac{q^2}{2h} \int dE\,\text{Tr}[i(G(E + \hbar\omega) - G^\dagger(E))\Gamma_\alpha\delta_{\alpha\beta}$$

$$- \Gamma_\alpha G(E + \hbar\omega)\Gamma_\beta G^\dagger(E)]$$

$$\times V_{ac}\left[\frac{f_\beta(E + \hbar\omega) - f_\beta(E)}{\hbar\omega}\right]. \tag{23.13}$$

For two leads, we get $I_1 + I_2 = i\omega Q$, where the AC charge $Q(\omega) = q \int dE \mathrm{Tr}\left[G^n(E + \hbar\omega, E)\right]$.

For a sharp resonant level with contact 1 oscillating at ω the equation suggests that the AC current at the other terminal $I_2 = (q^2 V_{\mathrm{ac}}/\hbar)\Gamma_1\Gamma_2/(\Gamma_1 + \Gamma_2 - i\hbar\omega)$, evaluated at the resonant energy $\boxed{\text{P23.3}}$.

23.5 Quantifying Current Noise

The Fermi–Dirac distribution is the average occupancy of a fluctuating electron number in an open system connected to contacts. There is thus a fluctuation quantified by a standard deviation. The autocorrelation function

$$C(T) = \frac{\int_0^T dt' \langle I(t')I(t' + T)\rangle}{T} \tag{23.14}$$

is a convolution (average of a sliding product). By a theorem in mathematics (Weiner–Khinchine theorem), easily verified by substitution, we can show that this quantity is the Fourier transform of the power spectrum $|I^2(\omega)|/2$ for large times, so that

$$S(\omega) = \lim_{T\to\infty} \frac{2}{T}\langle|I^2(\omega)|\rangle. \tag{23.15}$$

In other words, Fourier transform of a convolution is the product of Fourier transforms. Note, the units are $\mathrm{Amp}^2 - \mathrm{Hz}$, so that we are basically measuring the current fluctuation per unit band-width. Let us now try to estimate this current fluctuation using Landauer theory.

23.6 Poissonian Shot Noise

Let us first consider a channel that allows electrons to trickle in at very low doses. In general, the charge distribution under nonequilibrium thermal conditions is fairly complex, involving a set of Bessel functions and exponentials (see NEMV). For independent stochastic tunneling events at low temperature, the charge fluctuations follow a Poisson process in N, the number of tunneling events within a time t. Let us assume a mean free time τ between two tunneling events, considered independent, so that in a small time dt the probability of an event is dt/τ. Let $P_N(t)$ be the probability of N tunneling events within time t. To get N events in time $t + dt$, we must either have N events already in time t and none in the remaining time dt/τ, or $N - 1$ events in time t and the Nth event in time dt/τ (we assume dt is small enough that a double event is unlikely). Thus

$$P_N(t + dt) = P_N(t)\left(1 - \frac{dt}{\tau}\right) + P_{N-1}(t)\frac{dt}{\tau}, \tag{23.16}$$

where the independence of tunneling events asserts through the product of probabilities, such as $P_N \times (1 - dt/\tau)$. By dividing through by dt and Taylor expanding $P(t + dt)$, we get a recursive differential equation

$$\frac{dP_N}{dt} = -\frac{P_N - P_{N-1}}{\tau}. \tag{23.17}$$

By recursion, it is straightforward to show that the solution is a Poisson process $\boxed{\text{P23.4}}$

$$P_N(t) = \frac{\left(\frac{t}{\tau}\right)^N e^{-t/\tau}}{N!}. \tag{23.18}$$

From knowledge of Poisson statistics, it is easy to show that the average and variance are both equal,

$$\sigma_N^2 = \langle N^2 \rangle - \langle N \rangle^2 = \langle N \rangle = \frac{t}{\tau}. \tag{23.19}$$

We can thus write the average current and spectral function at low temperature

$$I_{\text{LT}} = \frac{q\langle N \rangle}{t} = \frac{q}{\tau},$$

$$S_{\text{LT}} = \frac{2q^2\sigma_N^2}{t} = \frac{2q^2\langle N \rangle}{t} = 2qI. \tag{23.20}$$

The last equation is a defining characteristic of shot noise.

The equations make sense — consider plus-minus square pulses of current of height I and width τ, each carrying charge $q = \pm I\tau$. Convolution, i.e., sliding product of two square pulses would be a triangle of width 2τ, height I^2, with area q^2/τ for positive pulses, and same for negative, giving $C = 2q^2/\tau = 2qI$.

Based on our knowledge of Landauer theory, we can postulate that the occasional current flickers have an amplitude at equilibrium

$$I = \frac{q}{h}\int dE \mathcal{T}(E)M(E)f_0(E)[1 - f_0(E)]. \tag{23.21}$$

Recall that in Landauer theory, *conductance is transmission*,

$$G = \frac{q^2}{h} \int dE \, T(E) M(E) \left(-\frac{\partial f}{\partial E} \right). \tag{23.22}$$

Using the relation $C = 2qI$, and $-\partial f/\partial E = f(1-f)/k_B T$, we get the fluctuation–dissipation result, *conductance is correlation*

$$G = C/2k_B T. \tag{23.23}$$

23.7 Generalize to Thermal Noise

Note that a Poissonian process implies that the variance matches the mean occupancy which equals f_0, the Fermi–Dirac distribution, at equilibrium. This is only true at low temperature. As the temperature rises, the fluctuations change accordingly. In reality, N can only take on values of zero or unity, meaning $N^2 = N$ from charge quantization. This means $\langle N^2 \rangle = \langle N \rangle = f$, distinct from a Poisson process, and

$$\sigma_N^2 = \langle N^2 \rangle - \langle N \rangle^2 = \langle N \rangle - \langle N \rangle^2 = f_0(1 - f_0). \tag{23.24}$$

The exclusion term $1 - f_0$ was ignorable when f_0 was small. This term will also make an appearance in the shot noise term, $S \approx 2qI(1 - f_0)$.

Let us estimate the standard deviation under non-equilibrium conditions. The only times we get a charge is when $\Delta N = 1$, with probability $f_1(1 - f_2)T$ (electron in left contact, empty state in right), and when $\Delta N = -1$, with probability $f_2(1 - f_1)T$. When $\Delta N = 0$, i.e., electrons moving from both sides, the net current contribution is zero. The standard deviation is

$$\langle \Delta N^2 \rangle = (+1)^2 \times f_1(1 - f_2)T + (-1)^2 \times f_2(1 - f_1)T,$$

$$\langle \Delta N \rangle^2 = [(+1)f_1(1 - f_2)T + (-1)f_2(1 - f_1)T]^2,$$

$$\langle \Delta N^2 \rangle - \langle \Delta N \rangle^2 = T(f_1 + f_2 - 2f_1 f_2) - T^2(f_1 - f_2)^2$$

$$= T\Big[f_1(1 - f_1) + f_2(1 - f_2)\Big] + T(1 - T)(f_1 - f_2)^2. \tag{23.25}$$

Plugging in the usual terms in Landauer theory — we then get the Blanter-Büttiker formula

$$\boxed{ S(\omega = 0) = \frac{q^2}{h} \int dE M \{ T[f_1(1 - f_1) + f_2(1 - f_2)] + T(1 - T)(f_1 - f_2)^2 \} }.$$

$$\tag{23.26}$$

Assuming constant transmissions independent of energy, this equation can be written for multiple single-moded channels ($M = 2$ for spins) as

$$S(0) = \frac{4q^2}{h} \left[2k_B T \sum_\alpha \mathcal{T}_\alpha^2 + qV \coth\left(\frac{qV}{2k_B T}\right) \sum_\alpha \mathcal{T}_\alpha \left(1 - \mathcal{T}_\alpha\right) \right]. \quad (23.27)$$

It is easy to show then that

$$S(0) = \begin{cases} 4k_B T\left(\dfrac{2q^2 \mathcal{T}}{h}\right) = 4k_B T G & (qV \ll k_B T, \quad \text{thermal Johnson noise}) \\ 2qI\left(1 - \mathcal{T}\right) & (qV \gg k_B T, \quad \text{shot noise}). \end{cases}$$

$$(23.28)$$

The Fano factor $S(0)/2qI$ is a measure of the signal-to-noise ratio. In the high bias shot noise limit, this is simply the reflection coefficient $1 - \mathcal{T}$.

Note that there is a way to matrix generalize the Blanter–Buttiker equations in presence of inelastic scattering, analogous to how Meir–Wingreen generalizes Landauer equation. The process, however, is a bit involved (Crepieux *et al.*, *PRL*, 120, 107702 (2018)). The distribution function, also evolves from the Poissonian process, as phonons are generated and recombined. The room temperatured distribution is given by the so-called Huang–Rhys model, discussed in detail in NEMV, Section 24.5.

23.8   Noise with Gain: Excess Noise Parameter

One of the major concerns in photodetection is to ensure that we amplify the signal more than the noise. Let us assume we reach a current gain of M. We would then naively expect the power spectrum to increase by M^2 for uncorrelated noise, i.e., $S = 2qI\langle M \rangle^2$, i.e., the noise increases according to the average gain squared. Correlation, however, can alter the noise, making $\langle M^2 \rangle \neq \langle M \rangle^2$. As we saw in quantum physics, adding two signals of equal intensity I_0 and extracting their probability

$$f(t) = f_1 e^{i\omega_1 t} + f_2 e^{i\omega_2 t}$$
$$\Longrightarrow I = \langle |f(t)|^2 \rangle = \underbrace{|f_1|^2}_{I_0} + \underbrace{|f_2|^2}_{I_0} + 2\text{Re}\langle f_1 f_2^* e^{i(\omega_1 - \omega_2)t} \rangle, \quad (23.29)$$

where $\langle \ldots \rangle$ implies a time integral over a probability distribution $P(t)$ for noise signals with stochastic frequencies $\omega_{1,2}$. If the signals are completely uncorrelated, meaning the difference frequency $\omega_1 - \omega_2$ is a random variable that averages the time integral to zero, we get $I = 2I_0$. If however, they are correlated, for instance, $\omega_1 = \omega_2$, then the time integral gives unity, and we get $I = 4I_0$, a factor of two higher. We can thus write

$$S = 2qI\langle M \rangle^2 F(M) \tag{23.30}$$

where the excess noise factor $F = \langle M^2 \rangle / \langle M \rangle^2$ is a measure of the signal-to-noise ratio, with the averages done the usual way, $\langle M^2 \rangle = \sum_m m^2 P(m)$, etc. In fact, we can write $F(M) = 1 + \sigma_M^2 / \langle M \rangle^2$. The range of F can be obtained by setting $F(M) - 1 = (M-1)/M$. This result is easy to interpret. If we have a gain $M = 3$ for instance, we have two excess particles, and a fraction $2/3$ of the total particle number contributes to the noise.

Avalanche processes however create a chain reaction $\boxed{\text{P23.5}}$, where a multiplied electron current also creates a cascaded hole current. In the absence of the latter, for instance, if holes are rendered immobile, we simply get a factor of two enhancement for excess noise at high gain. However, things get complicated because electron avalanche usually comes from impact ionization, the opposite of Auger processes, where energized electrons in the conduction band lose their energy by pulling one out of the valence band, leaving a hole in its place. This hole in turn starts to drift because of the in-built electric field, creating its own avalanche process. Since this second stream comes from a dual noise process, the probability of this dual noise stream increases quadratically with $M - 1$. This physics is captured by the *Mc Intyre formula* $\boxed{\text{P23.6}}$

$$F(M) - 1 = \underbrace{\left(\frac{M-1}{M} \right)}_{\text{unipolar}} + \underbrace{\frac{\beta}{\alpha} \left[\frac{(M-1)^2}{M} \right]}_{\text{bipolar}}, \tag{23.31}$$

where $k = \beta/\alpha$ is the ratio of hole to electron ionization coefficients

$$\boxed{\text{Homework}}$$

P23.1 One of the striking properties of graphene is its high optical transmittivity — only about 2% of incident light is actually absorbed by monolayer graphene (twice that for bilayer). The electron–photon Hamiltonian looks a bit different for graphene, which has a linear rather than a parabolic band.

$$\mathcal{H} = v_F \vec{\sigma}.(\vec{p} - q\vec{A}/c) \Rightarrow \mathcal{H}_{\text{el-photon}} = -q v_F \vec{\sigma} \cdot \vec{A}/c \approx -\frac{q v_F A_0 \sigma_x}{2c} e^{-i\omega t} \tag{23.32}$$

σ_x connects the two orthogonal pseudospin states between conduction and valence band,

(a) Show that the dipole matrix element $\psi_C^\dagger \sigma_x \psi_V = i \sin\theta$.

(b) Work out the matrix element $M_{CV} = \psi_C^\dagger \mathcal{H}_{\text{el-photon}} \psi_V$ and then the scattering time using FGR

$$\frac{1}{\tau} = \frac{2\pi}{\hbar} \sum_f |M_{CV}|^2 \delta(E_c - E_v - \hbar\omega). \tag{23.33}$$

Do the integrals assuming a sheet of graphene of area S, and find the power absorbed per unit area $P_a = \hbar\omega/S\tau$.

(c) We can calculate the incident power as

$$P_I = c\epsilon_0 \mathcal{E}_0^2/2. \tag{23.34}$$

The electric field and the vector potential are related by

$$\vec{\mathcal{E}} = -\frac{1}{c}\frac{\partial \vec{A}}{\partial t} \implies \mathcal{E}_0 = A_0 \omega/c. \tag{23.35}$$

Substituting, we can get the ratio of absorbed to incident power density

$$A = \frac{P_a}{P_I}. \tag{23.36}$$

Show that $A = \alpha\pi \approx 0.023$, where $\alpha = q^2/4\pi\epsilon_0 \hbar c \approx 1/137$ is the fine structure constant.

The transmission $\mathcal{T} = 1 - \alpha\pi$, explaining the ultrahigh, frequency-independent optical transmittivity of graphene. If we go to finite doping, we insert a frequency dependence $\mathcal{T}(E_F) = (1 - \alpha\pi)\Theta(\hbar\omega - 2E_F)$, which means that photons with energies lower than the Fermi-energy gap are Pauli blocked and the absorption drops. At finite temperature, we get a set of Fermi functions replacing the Θ functions, smearing out the sharp Pauli edges $\mathcal{T}(E_F) = (1 - \alpha\pi)[\tanh{(\hbar\omega + 2E_F)}/k_B T + \tanh{(\hbar\omega - 2E_F)}/k_B T]$.

(d) For bilayer graphene, show that $\mathcal{T} \approx 1 - 2\alpha\pi$.

P23.2 **Dynamic localization in AC fields.** Let us recap how in a DC field a ballistic electron in a superlattice with period d shows Bloch oscillation $\boxed{\text{P8.8}}$. Solve for a 1D tight binding chain with a DC field $\mathcal{E}_0$, and show that the band breaks into a set of equally spaced steps (a Wannier–Stark ladder) with separation $\hbar\omega_B = q\mathcal{E}_0 d$, as follows. This is analogous to Landau levels in a magnetic field $\boxed{\text{P11.8}}$.

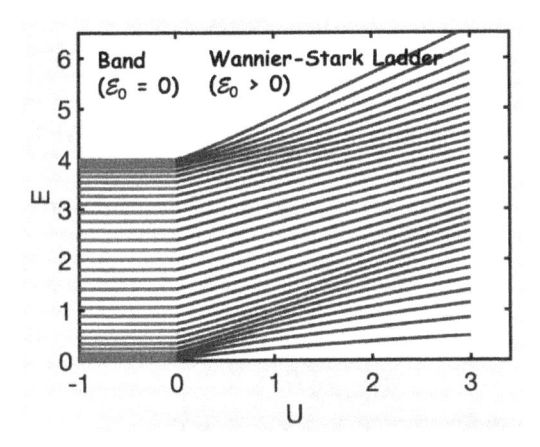

In an AC field, the added switching of field direction can synchronize with the turnaround of the oscillating electron to make it freeze classically. Write down the semiclassical equations of motion

$$\hbar dk/dt = q\mathcal{E}_0 \cos\omega t, \quad v = \partial E/\hbar dk \tag{23.37}$$

and show that the electron current density $J = qnv$ involves a Bessel function arising from the cosine of a cosine, $J_0(\Theta)$, $\Theta = q\mathcal{E}_0 d/\hbar\omega$ (ratio of Bloch oscillation frequency $q\mathcal{E}_0 d/\hbar$ and driving frequency ω), so that when Θ reaches one of the roots of J_0, we force the electron to *dynamically localize*. Show that the bandwidth Δ gets renormalized by the Bessel function $J_0(\Theta)$ (Eq. (23.11)), so that at the Bessel roots the frequency-renormalized quasi-bandwidth collapses and the electron freezes.

P23.3 Verify the maths of the AC NEGF current and apply it to (a) a single energy level, and (b) two energy levels, to see how the transitions are controlled by the AC resonant field.

P23.4 Show that the solution to the recursive equation $dP_N/dt = -[P_N - P_{N-1}]/\tau$ for integer N is a Poisson process. You can do it backward — simply by substitution. Also try it forward. You do it by observing that $P_{-1} = 0$ (why?) and then work out $P_0(t)$, with the initial condition $P_0(0) = 1$ (why?). Then use the recursive argument — assume the solution and show that given the Poisson solution for P_{N-1}, P_N will also have a Poisson solution.

P23.5 **Avalanche threshold.** Let's start with a semiconductor (a 1D dimer) with contacts at the end, like we did for a tunnel-FET — except the i-region is short compared to the depletion width so the voltage drop is linear. Add a linear voltage drop along it from the drain. Finally, add a scattering self-energy $\Sigma_S(E)$ very similar to phonons, except from Coulomb attraction between electrons and holes. This self-energy allows a high kinetic energy electron to drop down and transfer its energy to an electron in the valence band, and multiply the electron current (and hole current).

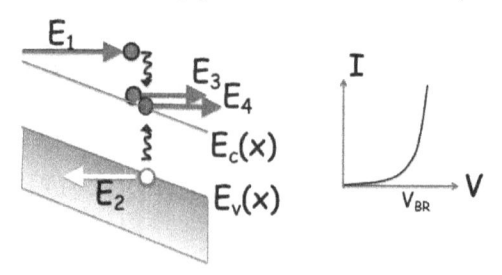

Note that impact ionization (inverse Auger) processes involve scattering terms like $\sim n^2 p$ or $\sim p^2 n$ $\boxed{\text{P15.3}}$. A careful look at the above picture gives terms like $G^n(E_1)G^n(E_2)G^p(E_3)G^p(E_4)$, where the valence band electron has near unit occupancy, so that this term can be written as $\Sigma^{in}(E)G^p(E_4)$, with $\Sigma^{in}(E) \approx MG^n(E_1)G^n(E_2)G^p(E_3)$, analogous to $c_n n(np - n_i^2)$. The energies are connected by conservation, and the Auger matrix element depends on field-dependent electron–hole overlap. Use self-consistent Born with the above process to show breakdown at reverse bias, i.e., an exponential jump in current at a voltage threshold. Show that it changes in a predictable way with variation in ratio of electron to hole effective mass. Try it also for a 4-level dot, and make sure there is no current leakage.

P23.6 **McIntyre distribution in avalanche photodetectors (APD).** Let us revisit gain in the APD. We chose $n - 1$ ionization sites $x_1, x_2, \ldots, x_{n-1}$ for an injected electron, leading to a gain of n. After each ionization at site x_i say, the electron and hole have a chance of reaching the electrodes with non-ionizing probabiliity $P_{ni}(x_i) = \exp\left[- \int_{x_i}^{W} \alpha(x)dx \right] \times \exp\left[- \int_0^{x_i} \beta(x)dx \right]$, where α and $\beta = k\alpha$, $k < 1$ are the electron and hole ionization coefficients.

Let us also account for s out of the $n-1$ ionizations that create secondary electrons from impact ionizing holes ($0 < s < n-2$). Let us say these subset of points are $(x_j, x_{j+1}, \ldots, x_{j+s-1})$ in no particular order. The ionizing holes at those sites each give a factor $\beta = k\alpha$, while the primary electrons at the remaining sites each give a factor α. The combinatorial product then is

$$C(n, s) = \beta(x_j)\beta(x_{j+1})\ldots\beta(x_{j+s-1}) \times \alpha(x_{n-j})\alpha(x_{n-j-1})\ldots\alpha(x_{n-j-s})$$
$$\times \underbrace{[j(j+1)\ldots(j+s-1)] \times [(n-j)(n-j-1)\ldots(n-j-s)]}_{(n-1)!}.$$

$$(23.38)$$

(a) Show that the C product, summed over all s, can be written as $L(n) = \sum_{s=0}^{n-2} C(n, s) = K(n)\Pi_{i=1}^{n-1}\alpha(x_i)$, where

$$K(n) = \frac{(1 - k)^{n-1}\Gamma\left(\dfrac{n}{1 - k}\right)}{[1 + k(n - 1)]\Gamma\left(\dfrac{1 + k(n - 1)}{1 \quad k}\right)} = (n - 1)! \quad \text{if } k = 0.$$

$$(23.39)$$

Hint: Pair the first square bracket above with the k^s term from the β product, and the second with the α, then write out the product as terms like $a + (n - a)k$ factors, and then use binomial expansion.

(b) From the two products above, show that the probability distribution

$$P(n) = \int_0^{W} dx_1 \int_{x_1}^{W} \ldots \int_{x_{n-1}}^{W} dx_1 \ldots dx_{n-1} P_{ni}(x_1) \ldots P_{ni}(x_{n-1})L(n)$$

$$= \frac{\Gamma\left(\dfrac{n}{1-k}+1\right)}{n!\,\Gamma\left(\dfrac{nk}{1-k}+2\right)}\left[\frac{1+k(M-1)}{M}\right]^{\frac{nk}{1-k}+1}$$

$$\times \left[\frac{(1-k)(M-1)}{M}\right]^{n-1}, \tag{23.40}$$

where the average

$$\langle n \rangle = M = \frac{\alpha - \beta}{\alpha e^{-(\alpha - \beta)W} - \beta}. \tag{23.41}$$

(c) From here, show that the excess noise

$$F(M) = \frac{\langle n^2 \rangle}{M^2} = kM + (1-k)\left(2 - \frac{1}{M}\right) \tag{23.42}$$

which agrees with Eq. (23.31). Plot the McIntyre distribution $P(n)$ for $M = 5, 50, 300$ and $k = 0.1, 0.2, 0.3$. Also plot $F(M)$ vs M for $k = 0.1, 0.5, 0.9$.

References

This book is a short and simplified version of more detailed material covered in A. Ghosh *Nanoelectronics — a Molecular View*, World Scientific Series on Nanoscience and Nanotechnology, 2016. I frequently refer to that reference as NEMV throughout this book. As a result, NEMV serves as a go-to reference for much of the material covered in the chapters, especially formalism, derivations and some of the more detailed physics. In addition, I would recommend the following reading materials:

Chapter 1

For an overview of digital switching and its energy cost, some classic references:

(1) R. E. Landauer, 'Irreversibility and heat generation in the computing process,' *IBM J. Res. Develop.*, 40, 183–191 (1961).
(2) C. H. Bennett, 'The thermodynamics of computation — a review', *Int. J. Theoret. Phys.*, 21, 905–940 (1982).
(3) C. H. Bennett, 'Logical reversibility of computation,' *IBM J. Res. Develop.*, 17(6), 525–532 (1973).
(4) R. P. Feynman, *Feynman Lectures on Computing*, 1st edition, Boulder: Westview Press, 2000.

For an overview of the energy of nanomagnetic switching, see
(5) Y. Xie, J. Ma, S. Ganguly and A. W. Ghosh, 'From materials to systems: a multiscale analysis of nanomagnetic switching', *J. Comp. Electron.*, 16, 1201 (2017).

Chapters 2–7

For an introduction to first quantization, I will recommend:

(1) C. Cohen-Tannoudji, B. Diu and F. Laloe, *Quantum Mechanics, Vol. 1, Vol. 2*, Wiley-Interscience, 1977.
(2) J. J. Sakurai and J. J. Napolitano, *Modern Quantum Mechanics*, 2nd edition, London: Pearson Education Limited, 2014.
(3) D. J. Griffiths, *Introduction to Quantum Mechanics*, 2nd edition, London: Pearson Education Limited, 2014.
(4) A. Szabo and N. S. Ostlund, *Modern Quantum Chemistry: Introduction to Advanced Electronic Structure Theory*, New York: Dover Publications, 1996.

Chapters 8–11

For solid state theory and introduction to magnetism:

(1) N. W. Ashcroft and N. D. Mermin, *Solid State Physics*, 1st edition, Cengage Learning, 1976.
(2) C. Kittel, *Introduction to Solid State Physics*, 8th edition, Wiley, 2004.
(3) C. Kittel, *Quantum Theory of Solids*, 2nd edition, Wiley, 1987.

Chapters 9–18

For classical solid state devices:

(1) R. F. Pierret, *Advanced Semiconductor Fundamentals*, 2nd edition, London: Pearson, 1992.
(2) R. F. Pierret, *Semiconductor Device Fundamentals*, 2nd edition, Boston: Addison Wesley, 1996.
(3) S. M. Sze and K. K. Ng, *Physics of Semiconductor Devices*, 3rd edition, Wiley-Interscience, 2007.
(4) B. Streetman and S. Banerjee, *Solid State Electronic Devices*, 6th edition, London: Pearson, 2000.
(5) J. Singh, *Semiconductor Devices: Basic Principles*, Wiley, 2000.
(6) Y. Taur and T. H. Ning, *Fundamentals of Modern VLSI Devices*, 2nd edition, Cambridge: Cambridge University Press, 2009.
(7) B. Van Zeghbroeck, *Principles of Semiconductor Devices*, The Oxford Series in Electrical and Computer Engineering, Prentice Hall, May, 2010.

For mobility data in $\boxed{\text{P16.3}}$, we use

(8) N. D. Arora, J. R. Hause and D. J. Roulston, 'Electron and hole mobilities in silicon as a function of concentration and temperature', *IEEE Trans. Electron Dev.*, ED-29, 292 (1982).

Chapters 19–23

For select references on quantum transport:

(1) A. Ghosh, NEMV — 'Nanoelectronics - a Molecular View', Singapore: World Scientific, 2016.
(2) S. Datta, *Electronic Transport in Mesoscopic Systems*, Cambridge Studies in Semiconductor Physics and Microelectronic Engineering, Cambridge: Cambridge University Press, 1997.
(3) S. Datta, *Lessons From Nanoelectronics: A New Perspective On Transport (Lessons from Nanoscience: A Lecture Notes Series)*, Singapore: World Scientific, 2012.
(4) S. Datta, *Quantum Transport: Atom to Transistor*, 2nd edition, Cambridge: Cambridge University Press, 2005.
(5) D. K. Ferry, S. M. Goodnick and J. Bird, *Transport in Nanostructures*, 2nd edition, Cambridge: Cambridge University Press, 2009.
(6) J. Rammer and H. Smith, 'Quantum field-theoretical models in transport theory of metals', *Rev. Mod. Phys.*, 58, 323 (1986).
(7) G. Stefanucci and R. van Leeuwen, *Nonequilibrium Many-Body Theory of Quantum Systems: A Modern Introduction*, 1st edition, Cambridge: Cambridge University Press, 2013.
(8) H. Haug and A.-P. Jauho, *Quantum Kinetics in Transport and Optics of Semiconductors*, Springer Series in Solid-State-Sciences, 1996.
(9) Y. Meir and N. S. Wingreen, 'Landauer formula for the current through an interacting electron region', *Phys. Rev. Lett.*, 68, 2612 (1992).
(10) Y. Meir, N. S. Wingreen and P. A. Lee, 'Transport through a strongly interacting electron system: theory of periodic conductance oscillations', *Phys. Rev. Lett.*, 66, 3048 (1991).
(11) D. S. Fisher and P. A. Lee, 'Relation between conductivity and transmission matrix', *Phys. Rev. B*, 23, 6951 (1981).
(12) A. Caldeira and A. J. Leggett, 'Influence of dissipation on quantum tunneling in macroscopic systems, *Phys. Rev. Lett.*, 46, 211 (1981).
(13) R. P. Feynman and F. L. Vernon, 'The theory of a general quantum system interacting with a linear dissipative system', *Ann. Phys. (N. Y.)*, 24, 118 (1963).

(14) R. Golizadeh-Mojarad and S. Datta, 'Nonequilibrium Green's function based models for dephasing in quantum transport', *Phys. Rev. B*, 75, 081301(R) (2007).

For spin dynamics including potentials, and for 2D materials:

(1) S. Hong, V. Diep, S. Datta, and Y. P. Chen, 'Modeling potentiometric measurements in topological insulators including parallel channels', *Phys. Rev. B*, 86, 085131 (2012).

(2) G. Schmidt, D. Ferrand, L. W. Molenkamp, A. T. Filip and B. J. van Wees, 'Fundamental obstacle for electrical spin injection from a ferromagnetic metal into a diffusive semiconductor', *Phys. Rev. B*, 62, R4790 (2000).

(3) A. Ghosh, 'Spin control with a topological semimetal', *Physics*, 13, 38 (2020).

(4) A. H. Castro Neto, F. Guinea, N. M. R. Peres, K. S. Novoselov and A. K. Geim, 'The electronic properties of graphene', *Rev. Mod. Phys.*, 81, 109 (2009).

(5) S. Datta and B. Das, 'Electronic analog of the electro-optic modulator', *Appl. Phys. Lett.*, 56, 665 (1990).

(6) Y. Xie, K. Munira, I. Rungger, M. T. Stemanova, S. Sanvito and A. W. Ghosh, 'Spin transfer torque: a multiscale picture', in S. Bandyopadhyay and J. Atulasimha (eds.), *Nanomagnetic and Spintronic Devices for Energy Efficient Memory and Computing*, Wiley, 2016.

(7) R. Sajjad and A. W. Ghosh, 'Manipulating chiral transmission with gate geometry: switching with graphene with transmission gaps', *ACS Nano* 7, 9808 (2013).

(8) S. Chen, Z. Han, M. M. Elahi, K. M. M. Habib, L. Wang, B. Wen, Y. Gao, T. Taniguchi, K. Watanabe, J. Hone, A. W. Ghosh and C. R. Dean, 'Electron optics with ballistic graphene junctions', *Science* 353, 1522 (2016).

(9) K. Wang, M. M. Elahi, K. M. M. Habib, T. Taniguchi, K. Watanabe, A. W. Ghosh, G-H. Lee and P. Kim, 'Graphene transistor based on tunable Dirac–Fermion-optics', *Proc. Nat. Acad. Sci.*, 116, 6575 (2019).

(10) Y. Tan, M. M. Elahi, H.-Y. Tsao, K. M. Masum Habib, N. Scott Barker and A. W. Ghosh, 'Graphene Klein tunnel transistors for high speed analog RF applications', *Nat. Sci. Rep.*, 7, 9714 (2017).

(11) B. Yan and C. Felser, 'Topological materials: Weyl semimetals', *Annu. Rev. Condens. Matter Phys.*, 8, 337 (2017).

(12) B. Q. Lv *et al.*, 'Observation of Fermi-arc spin texture in TaAs', *Phys. Rev. Lett.*, 115, 217601 (2015).

For energy conversion including thermoelectrics and solar cells:

(1) A. F. Ioffe, *Semiconductor Thermoelements, and Thermoelectric Cooling*, 1st edition, London: Infosearch Limited, 1957.
(2) M. Wolf, R. Hinterding and A. Feldhoff, 'High power factor vs. high zT — a review of thermoelectric materials for high-temperature application', *Entropy*, 21, 1058 (2019).
(3) W. Shockley and H. J. Queisser, 'Detailed balance limit of efficiency of pn junction solar cells', *J. Appl. Phys.*, 32, 510 (1961).
(4) L. Siddiqui, A. W. Ghosh, and S. Datta, 'Phonon runaway in carbon nanotube quantum dots', *Phys. Rev. B* 76, 085433 (2007).

For optoelectronics and AC response:

(1) A. S. Huntington, *InGaAs Avalanche Photodiodes for Ranging and Lidar*, Elsevier, 2020.
(2) R. J. McIntyre, 'Multiplication Noise in Uniform Avalanche Diodes', *IEEE Trans. Electr. Dev.*, 164, ED-13 (1966).
(3) R. J. McIntyre, 'A new look at impact ionization — part I: a theory of gain, noise, breakdown probability, and frequency response', *IEEE Trans. Electr. Dev.*, 46, 1623 (1999).
(4) Y. M. Blanter and M. Buttiker, 'Shot noise in mesoscopic conductors', *Phys. Rep.* 336, 1 (2000).
(5) M. Anantram and S. Datta, 'Effect of phase breaking on the ac response of mesoscopic systems', *Phys. Rev. B*, 51, 7632 (1995).
(6) B. J. Keay, S. Zeuner, S. J. Allen, Jr., K. D. Maranowski, A. C. Gossard, U. Bhattacharya, and M. J. W. Rodwell, 'Dynamic localization, absolute negative conductance, and stimulated, multiphoton emission in sequential resonant tunneling semiconductor superlattices', *Phys. Rev. Lett.*, 75, 4102 (1995).
(7) A. W. Ghosh, A. V. Kuznetsov, and J. W. Wilkins, 'Reflection of THz radiation by a superlattice', *Phys. Rev. Lett.*, 79, 3494 (1997).

For emerging CMOS devices:

(1) J. Burghartz (ed.), *Electron Devices: An Overview by the Technical Area Committee of the IEEE Electron Devices Society*, John Wiley and Sons Ltd, 2013.

(2) A. W. Ghosh, 'Transmission engineering as a route to subthermal switching', Special issue on low subthreshold switching, *IEEE JEDS*, 3, 135 (2015).

(3) T. M. Conte, E. Track and E. DeBenedictis, 'Rebooting computing: new strategies for technology scaling', *Computer*, 48, 10 (2015).

(4) M. M. Sabry Aly *et al.*, 'Energy-efficient abundant-data computing: the N3XT 1,000x', *Computer*, 48, 12 (2015).

Index

Lessons from Nanoscience: A Lecture Note Series

Print ISSN: 2301-3354
Online ISSN: 2301-3362

Series Editors: Mark Lundstrom and Supriyo Datta
(Purdue University, USA)

"Lessons from Nanoscience" aims to present new viewpoints that help understand, integrate, and apply recent developments in nanoscience while also using them to re-think old and familiar subjects. Some of these viewpoints may not yet be in final form, but we hope this series will provide a forum for them to evolve and develop into the textbooks of tomorrow that train and guide our students and young researchers as they turn nanoscience into nanotechnology. To help communicate across disciplines, the series aims to be accessible to anyone with a bachelor's degree in science or engineering.

More information on the series as well as additional resources for each volume can be found at: http://nanohub.org/topics/LessonsfromNanoscience

Published:

More information on this series can also be found at
https://www.worldscientific.com/series/lnlns